NanoScience and Technology

Series Editors

Phaedon Avouris
Bharat Bhushan
Dieter Bimberg
Klaus von Klitzing
Hiroyuki Sakaki
Roland Wiesendanger

For further volumes:
http://www.springer.com/series/3705

The series NanoScience and Technology is focused on the fascinating nano-world, mesoscopic physics, analysis with atomic resolution, nano and quantum-effect devices, nanomechanics and atomic-scale processes. All the basic aspects and technology-oriented developments in this emerging discipline are covered by comprehensive and timely books. The series constitutes a survey of the relevant special topics, which are presented by leading experts in the field. These books will appeal to researchers, engineers, and advanced students.

Axel Lorke · Markus Winterer
Roland Schmechel · Christof Schulz
Editors

Nanoparticles from the Gas Phase

Formation, Structure, Properties

Editors

Axel Lorke
Department of Physics and CENIDE
University of Duisburg-Essen
Lotharstraße 1
47057 Duisburg
Germany

Markus Winterer
Nanoparticle Process Technology,
  and CENIDE
University of Duisburg-Essen
Lotharstraße 1
47057 Duisburg
Germany

Roland Schmechel
Faculty of Engineering, and CENIDE
University of Duisburg-Essen
Bismarckstraße 81
47057 Duisburg
Germany

Christof Schulz
Institute for Combustion and Gasdynamics
  (IVG), and CENIDE
University of Duisburg-Essen
Lotharstraße 1
47057 Duisburg
Germany

ISSN 1434-4904
ISBN 978-3-642-28545-5   ISBN 978-3-642-28546-2   (eBook)
DOI 10.1007/978-3-642-28546-2
Springer Heidelberg New York Dordrecht London

Library of Congress Control Number: 2012940460

© Springer-Verlag Berlin Heidelberg 2012
This work is subject to copyright. All rights are reserved by the Publisher, whether the whole or part of the material is concerned, specifically the rights of translation, reprinting, reuse of illustrations, recitation, broadcasting, reproduction on microfilms or in any other physical way, and transmission or information storage and retrieval, electronic adaptation, computer software, or by similar or dissimilar methodology now known or hereafter developed. Exempted from this legal reservation are brief excerpts in connection with reviews or scholarly analysis or material supplied specifically for the purpose of being entered and executed on a computer system, for exclusive use by the purchaser of the work. Duplication of this publication or parts thereof is permitted only under the provisions of the Copyright Law of the Publisher's location, in its current version, and permission for use must always be obtained from Springer. Permissions for use may be obtained through RightsLink at the Copyright Clearance Center. Violations are liable to prosecution under the respective Copyright Law.
The use of general descriptive names, registered names, trademarks, service marks, etc. in this publication does not imply, even in the absence of a specific statement, that such names are exempt from the relevant protective laws and regulations and therefore free for general use.
While the advice and information in this book are believed to be true and accurate at the date of publication, neither the authors nor the editors nor the publisher can accept any legal responsibility for any errors or omissions that may be made. The publisher makes no warranty, express or implied, with respect to the material contained herein.

Printed on acid-free paper

Springer is part of Springer Science+Business Media (www.springer.com)

Foreword

During the last decades no area in natural sciences and engineering has developed more rapidly than nanostructured materials. The reason is that in the size regime between a few atoms and the bulk limit, many properties vary in a dramatic way such as melting temperature, band gap, or magnetic and optical properties. This rich variety of size effects by nanoparticles has inspired many practical applications and is considered to be one of the driving forces behind future technical solutions. The importance of the nanometer size regime was already mentioned by W. Ostwald in his famous book "Die Welt der vernachlässigten Dimensionen" published in 1915. But only at the end of the twentieth century, a systematic preparation of nanoparticles in colloidal and gaseous systems and their characterization by a variety of physical and chemical tools started. New theoretical concepts exploring the measured size effects were accompanying the starting nano-age. The developing nanoscience is a highly interdisciplinary field that sheds light on many fundamental issues but must also show its practical relevance by inspiring new or improved processes and products.

Nanoparticles from the gas phase have their special properties because the formation and transport space is in this case restricted. Typical birth processes are nucleation from supersaturated vapor and/or reactive formation due to a chain of gaseous decomposition and synthesis reactions. Laser or plasma ablation from solid surfaces and subsequent nucleation and surface growth is a kind of rearrangement of material in the form of nanoparticles. A reactive formation of nanostructures normally proceeds in a fluid flow containing the gaseous or liquid dissolved precursors. The partly endothermic reactions must be initiated by increasing the temperature by various energy sources. Typical devices used in the science community are hot wall reactors, plasma and laser reactors, or different types of flames. They all open the possibility of staged reaction processes for coated or mixed particles.

In industry, gas-phase combustion synthesis of carbonaceous and inorganic particles is used routinely today to make a variety of commodities like various kinds of soot or the oxides SiO_2, TiO_2, Al_2O_3, etc. amounting to millions of tons annually. They are industrially used in particular for tires and as pigments, opacities, catalysts, flowing aids, for optical fibers, and telecommunication. The flame reactor is today the workhorse of this technology. It is cost-effective and offers advantages over other material synthesis processes, e.g., wet-phase chemistry. The characteristic of gas-phase-made particles is self-purification with the final powder product as a consequence of normally high synthesis temperature.

The detailed knowledge of the gas-phase synthesis of nanoparticles is still limited. The kinetics of e.g. precursor reactions is not understood in all details. Radicals, intermediates, and product molecules are formed, which polymerise or nucleate to first clusters. Their thermal stability determines in many cases the further particles evolution process. The clusters can grow by gas kinetic collisions between each other or by the addition of monomers to the cluster surface. The coalescence of cluster–cluster ensembles is normally very fast resulting in compact nearly spherical structures, which could be called "particle". The further particle dynamics is determined by surface growth and by the interdependence of coagulation and coalescence. Depending on their typical time scales, either "soft" or "hard" fractal aggregates are formed.

To get insights into the structure and properties of gas phase synthesized particles, modern diagnostics have been developed in the last decades. In situ size measurements during the birth period of the particles are confronted with the problem of sizing structures form nearly molecular dimensions to about 50 nm and more. Sampling techniques with online characterization of nanoparticles, which have been developed for some time, are calibrated only with spheres. But synthesized nanoparticles are often occurring as agglomerates, which need a detailed description of their structure. As the smallest particles have a very high mobility and reactivity, also ex situ probing on surfaces, e.g. TEM grids, must be performed very carefully. A representative particle sampling, which does not falsify or change the particle properties during the collection itself, is necessary. The further analysis of deposited particles depends on the available instrumentation or better on the structural or morphological information which is of interest.

Applications of nanoparticles collected from the gas phase are still under debate. Obviously, flame-made powders produced in industrial scales are needed and the respective technologies will be improved. On the other hand, gas-phase synthesis will contribute to the development of new materials and products, which previously have not been made by other techniques. In situ doped particles, mixed metals, and mixed oxides will find technical and medical applications. Unusual combinations of materials in segregated, dissolved, or core–shell structures wait for practical usage. The question of bringing such functional particles into respective matrixes or on surfaces will be a scientific and technical challenge for the future.

This book summarizes scientific results, obtained from the "Sonderforschungsbereich 'Nanopartikel aus der Gasphase' ", which was over many years financially supported by the Deutsche Forschungsgemeinschaft. It was initiated in 1998 by scientists from mechanical and electrical engineering and from physics.

Foreword

The different experiences of the participants in reactive flow phenomena, aerosol technology, semiconducting, magnetic, and optoelectronic materials was successfully bundled and was the interdisciplinary basis of the resulting research program.

Duisburg, Germany

Paul Roth
Heinz Fissan
Eberhard Wassermann

Contents

Part I   Formation

1   Synthesis of Tailored Nanoparticles in Flames:
Chemical Kinetics, In Situ Diagnostics, Numerical
Simulation, and Process Development   3
 1.1   Introduction   3
  1.1.1   State of the Art in Flame-Based Synthesis
    of Nanoparticles   5
 1.2   Principles of Particle Formation and Growth
   in the Gas Phase   6
 1.3   Shock-Tube Studies of Precursor Reactions   8
  1.3.1   Formation of Iron Clusters From $Fe(CO)_5$ 9
  1.3.2   Reaction of Gallium Atoms
    with Nitrogen and Oxygen   11
 1.4   Flame Reactor Studies of Particle Formation and Growth   13
  1.4.1   Molecular-Beam Sampling   15
  1.4.2   Thermophoretic Sampling with Ex Situ Analysis   16
 1.5   Laser-Based Diagnostics in Particle Synthesis Reactors   16
  1.5.1   Diagnostics for Temperature   17
  1.5.2   Diagnostics for Species Concentration   19
  1.5.3   Optical Diagnostics for Particle Sizes   22
 1.6   Chemistry Modeling and Simulation of Reactive Flows   23
  1.6.1   Modeling and Validation of the Fe
    Chemistry in Flames   25
  1.6.2   CFD Simulation of Reactive Flows
    in Nanoparticle Synthesis   30
 1.7   Examples of Flame-Synthesized Particles   32
  1.7.1   Semi-Conducting Single Oxide Particles   34
  1.7.2   Mixed Oxide and Composite Nanoparticles   37
 1.8   Conclusions   42
 References   42

2 Chemical Vapor Synthesis of Nanocrystalline Oxides 49
   2.1   Introduction 49
   2.2   Influence of Pulsed Precursor Delivery on Particle
         Size and Size Distribution 50
         2.2.1   Experimental Methodology 51
         2.2.2   Results and Discussion 53
   2.3   Influence of the Time-Temperature Profiles
         on Powder Characteristics 59
         2.3.1   Experimental Methodology 60
         2.3.2   Results and Discussion 61
   2.4   Control of Composition by Laser Flash Evaporation:
         Crystal and Local Structure of Cobalt Doped Zinc
         Oxide Nanoparticles 64
         2.4.1   Experimental Methodology 65
         2.4.2   Results and Discussion 67
   2.5   Summary and Conclusions 74
   References 75

3 Nucleation, Structure and Magnetism of Transition Metal
Clusters from First Principles 77
   3.1   Introduction 77
   3.2   Magnetism and Structural Transformations
         in Nanoparticles 79
   3.3   Structure and Magnetism in Binary Nanoparticles 85
   3.4   MAE of Clusters 89
         3.4.1   MAE for Perfect Clusters 91
         3.4.2   MAE for Relaxed Clusters 93
   3.5   Summary and Outlook 94
   References 95

4 Synthesis and Film Formation of Monodisperse
Nanoparticles and Nanoparticle Pairs 99
   4.1   Introduction 99
   4.2   Synthesizing Monodisperse Nanoparticles by Means
         of Spark Discharge 103
         4.2.1   Spark Discharge 103
         4.2.2   Synthesis of Monodisperse Au Nanoparticles 105
         4.2.3   Synthesis of Ge Nanoparticles 108
   4.3   Formation of Alloy and Pair Nanoparticles 110
         4.3.1   Motivation and Synthesis Approach 110
         4.3.2   Synthesis of Au-Ge Pair Nanoparticles 111
         4.3.3   Synthesis of AuGe Alloyed Nanoparticles
                by Co-Sparking 114
   4.4   Film Formation by Electrostatic Means 116
   References 119

Contents   xi

Part II Structure and Dynamics

**5 Diffusion Enhancement in FePt Nanoparticles
for L1$_0$ Stability** 123
   5.1   Introduction 123
   5.2   Gas-Phase Preparation of FePt Nanoparticles 125
   5.3   Diffusion Enhancement in FePt 129
      5.3.1   Oxygen Mediated Destabilization of Twinned
             Structures 129
      5.3.2   L1$_0$ Stabilization by Interstitial
             Nitriding–Denitriding 132
      5.3.3   L1$_0$ Stabilization by the Mobility Enhancement
             of a Substitutional Element 135
   5.4   Conclusion. 137
   References 137

**6 Simulation of Cluster Sintering, Dipolar Chain Formation,
and Ferroelectric Nanoparticulate Systems** 139
   6.1   General Introduction 139
      6.1.1   Methods 140
   6.2   Molecular-Dynamics Simulations of the Dipolar-Induced
         Formation of Magnetic Nanochains and Nanorings 142
   6.3   Molecular-Dynamics Simulations of Metal Cluster
         Agglomeration and Sintering 146
      6.3.1   Sintering of Nickel Nanoparticles. 146
      6.3.2   Agglomeration of Icosahedral Iron Nanoparticles 150
   6.4   Density Functional Simulations of Dielectric Nanoparticles:
         Agglomeration and Ferroelectric Trends 152
   6.5   Summary. 157
   References 158

7 Nanopowder Sintering 161
   7.1   Introduction 161
   7.2   Particle Coalescence 163
      7.2.1   Phenomenological Theory 164
      7.2.2   Atomistic Modeling 167
      7.2.3   Coalescence of Agglomerates. 168
      7.2.4   Coalescence of Two Particles of Different Size 169
   7.3   The Effect of Grain Boundaries 171
      7.3.1   Rigid Body Dynamics Combined with KMC 173
      7.3.2   Competition Between Reorientation and Neck
             Growth for Two Particles 174
      7.3.3   Reorientation Effects in Porous Agglomerates 178

7.4	Conclusion and Outlook	180
	Appendix: Activation Energies Used in the KMC-RBD Hybrid Model	181
	References	182

8 Material and Doping Contrast in III/V Nanowires Probed by Kelvin Probe Force Microscopy ... 185

8.1	Introduction	185
8.2	Instrumental Setup	187
8.3	Material and Doping Contrast in Single GaAs Based Nanowires	190
	8.3.1 Material Transitions in Single GaAs Based Nanowires	190
	8.3.2 KPFM on Single p-Doped GaAs Nanowires	194
	8.3.3 Localization of Doping Transitions in Single p-Doped GaAs Nanowires	197
8.4	GaAs p–n Junction Nanowire Devices	200
8.5	Conclusion	204
	References	204

Part III Properties and Applications

9 Optical Properties of Silicon Nanoparticles ... 209

9.1	Introduction	209
9.2	Vibrational Properties	212
9.3	Recombination Dynamics	214
9.4	Electroluminescence	224
	References	228

10 Electrical Transport in Semiconductor Nanoparticle Arrays: Conductivity, Sensing and Modeling ... 231

10.1	Introduction	231
10.2	Principles of (Nano) Particle-Based Conduction Processes	233
10.3	Electrical Measurement of (Nano) Particle Arrays	235
	10.3.1 Impedance Spectroscopy	235
	10.3.2 Conductivity Measurements During Powder Compaction	236
10.4	Formation of Nanoparticle Arrays	237
	10.4.1 Compaction of Nanoparticle Powders	237
	10.4.2 Printing of Nanoparticle Thin Films	237
	10.4.3 Molecular Beam-Assisted Deposition	239
10.5	Modeling of Electrical Transport in (Nano) Particulate Networks	241

Contents   xiii

10.5.1   How the Macroscopic Impedance Depends
          on Sample Geometry 243
10.5.2   A Simple Model for Current-Assisted
          Powder Compaction 246
10.6   Examples of Nanoparticle Array Conductivity
       and Sensitivity 250
       10.6.1   Tin Dioxide 250
       10.6.2   Tungsten Oxide 254
       10.6.3   Zinc Oxide 259
       10.6.4   Electrical Properties of Nanoscale
                 Powders During Compaction 267
10.7   Conclusions 269
References 270

**11   Intrinsic Magnetism and Collective Magnetic Properties
       of Size-Selected Nanoparticles** 273
11.1   Introduction 273
11.2   Structural Characterisation 276
       11.2.1   Fe/Fe-Oxide Nanocubes 276
       11.2.2   FePt Nanoparticles 277
11.3   Element-Specific, Site-Selective Magnetism 281
       11.3.1   Influence of Local Composition and Crystal
                 Symmetry on the Magnetic Moments 281
       11.3.2   Magnetic Response of Fe on Different Lattice
                 Sites in Fe/FeO_x Nanocubes 285
11.4   Spectro-Microscopy of Individual Nanoparticles 287
       11.4.2   Magnetic Hysteresis and Spectroscopy
                 of Monomers, Dimers, Trimers
                 and Many Particle Configurations 288
11.5   Magnetisation Dynamics of Nanoparticle Ensembles 294
11.6   Summary 297
References 298

**12   Optical Spectroscopy on Magnetically Doped
       Semiconductor Nanoparticles** 303
12.1   Introduction 303
12.2   Magnetically Doped ZnO Nanoparticles 304
       12.2.1   ZnO Nanoparticles Doped with Chromium 305
       12.2.2   ZnO Nanoparticles Doped with Cobalt 311
12.3   Magnetically Doped CdSe Nanoparticles 313
       12.3.1   Characterization of Mn-Doped
                 CdSe Nanoparticles 314
       12.3.2   Exciton Magnetic Polaron Formation
                 in Mn-Doped CdSe Nanoparticles 317

12.4	Conclusion		323
References			324

13 Gas Sensors Based on Well-Defined Nanostructured Thin Films ... 329

13.1	Introduction		329
13.2	Experimental Details		330
13.3	Results and Discussion		334
	13.3.1	Differential Mobility Analyser (DMA) Measurements	334
	13.3.2	Sintering Behaviour of Generated SnO_x Nanoparticles	334
	13.3.3	Synthesis of Monodispersed SnO_x, Pd and Ag Nanoparticles	335
	13.3.4	Low Pressure Impaction (LPI)	336
	13.3.5	Gas Sensor Preparation	338
	13.3.6	Sensing Results on SnO_x: M Mixed Nanoparticle Layers	341
	13.3.7	Pure Pd-Nanoparticle Layers for Concentration Specific H_2 Sensing at Room Temperature	352
References			354

14 III/V Nanowires for Electronic and Optoelectronic Applications ... 357

14.1	Introduction		357
14.2	Growth of III/V Nanowires		358
	14.2.1	Vapour–Liquid–Solid Growth	359
	14.2.2	InAs	361
	14.2.3	GaAs	362
14.3	InAs Nanowire MISFET		370
	14.3.1	Nanowire FET Design	370
	14.3.2	Device Performance	372
14.4	Properties of GaAs Nanowire p–n Junction		375
	14.4.1	Electrical Properties	375
	14.4.2	Optoelectronic Properties	376
14.5	Conclusion		382
References			382

15 Metal Oxide Thin-Film Transistors from Nanoparticles and Solutions ... 387

15.1	Introduction	387
15.2	Operation Principle of Thin-Film Transistors	389
15.3	TFTs with Semiconducting Metal Oxides	393

		15.3.1	General Remarks	393
		15.3.2	TFTs with Nanoparticles from a Carrier Gas Stream	394
		15.3.3	TFTs with Nanoparticles from Dispersion	398
		15.3.4	TFTs from Solutions (Liquid Precursors)	401
	15.4		Summary	405
	15.5		Conclusion	406
	References			407

Index 411

Contributors

Dr. Mehmet Acet Faculty of Physics and CENIDE, University of Duisburg-Essen, Lotharstraße 1, 47057 Duisburg, Germany, e-mail: mehmet.acet@uni-due.de

Dr. Carolin Antoniak Faculty of Physics and CENIDE, University of Duisburg-Essen, Lotharstraße 1, 47057 Duisburg, Germany, e-mail: carolin.antoniak@uni-due.de

Prof. Dr. Gerd Bacher Faculty of Engineering and CENIDE, Electronic Materials and Nanodevices and CENIDE, University of Duisburg-Essen, Bismarckstraße 81, 47057 Duisburg, Germany, e-mail: gerd.bacher@uni-due.de

Prof. Dr. Peter Entel Faculty of Physics and CENIDE, University of Duisburg-Essen, Lotharstraße 1, 47057 Duisburg, Germany, e-mail: entel@thp.uni-duisburg.de

Anna Grünebohm Faculty of Physics and CENIDE, University of Duisburg-Essen, Lotharstraße 1, 47057 Duisburg, Germany, e-mail: anna@thp.uni-due.de

Prof. Dr. Frank Einar Kruis Faculty of Engineering and CENIDE, University of Duisburg-Essen, Bismarckstraße 81, 47057 Duisburg, Germany, e-mail: einar.kruis@uni-due.de

Prof. Dr. Axel Lorke Faculty of Physics and CENIDE, University of Duisburg-Essen, Lotharstraße 1, 47057 Duisburg, Germany, e-mail: axel.lorke@uni-due.de

Prof. Dr. Cedrik Meier Physics Department and CeOPP and CENIDE, University of Paderborn, Warburger. Str. 100, 33098 Paderborn, Germany, e-mail: cedrik.meier@uni-paderborn.de

Dr. Wolfgang Mertin Faculty of Engineering and CENIDE, Electronic Materials and Nanodevices and CENIDE, University of Duisburg-Essen, Bismarckstraße 81, 47057 Duisburg, Germany, e-mail: wolfgang.mertin@uni-due.de

Dr. Ralf Meyer Laurentian University, 935 Ramsey Lake Road, Sudbury, ON P3E 5Y5, Canada, e-mail: rmeyer@cs.laurentian.ca

Dr. Werner Prost Faculty of Engineering, Solid-State Electronics Department and CENIDE, University of Duisburg-Essen, Lotharstraße 55, 47057 Duisburg, Germany, e-mail: werner.prost@uni-due.de

Prof. Dr. Roland Schmechel Faculty of Engineering and CENIDE, University of Duisburg-Essen, Bismarckstraße 81, 47057 Duisburg, Germany, e-mail: roland.schmechel@uni-due.de

Prof. Dr. Christof Schulz Faculty of Engineering, Institute for Combustion and Gasdynamics (IVG) and CENIDE, University of Duisburg-Essen, Lotharstraße 1, 47057 Duisburg, Germany, e-mail: christof.schulz@uni-due.de

Dr. Marina Spasova Faculty of Physics, University of Duisburg-Essen, Lotharstraße 1, 47057 Duisburg, Germany, e-mail: marina.spasova@uni-due.de

Dr. Hartmut Wiggers Faculty of Engineering, Institute for Combustion and Gasdynamics (IVG) and CENIDE, University of Duisburg-Essen, Lotharstraße 1, 47057 Duisburg, Germany, e-mail: hartmut.wiggers@uni-due.de

Prof. Dr. Markus Winterer Faculty of Engineering, Nanoparticle Process Technology and CENIDE, University of Duisburg-Essen, Lotharstraße 1, 47057 Duisburg, Germany, e-mail: markus.winterer@uni-due.de

Prof. Dr. Dietrich E. Wolf Faculty of Physics and CENIDE, University of Duisburg-Essen, Lotharstraße 1, 47057 Duisburg, Germany, e-mail: dietrich.wolf@uni-due.de

Part I
Formation

Chapter 1
Synthesis of Tailored Nanoparticles in Flames: Chemical Kinetics, In Situ Diagnostics, Numerical Simulation, and Process Development

Hartmut Wiggers, Mustapha Fikri, Irenaeus Wlokas, Paul Roth and Christof Schulz

Abstract Flame synthesis of nanoparticles provides access to a wide variety of metal oxide nanoparticles. Detailed understanding of the underlying fundamental processes is a prerequisite for the synthesis of specific materials with well-defined properties. Multiple steps from gas-phase chemistry, inception of first particles and particle growth are thus investigated in detail to provide the information required for setting up chemistry and particle dynamics models that allow simulating particle synthesis apparatus. Experiments are carried out in shock wave and flow reactors with in situ optical diagnostics, such as absorption, laser-induced fluorescence, and laser-induced incandescence, with in-line sampling via mass spectrometry as well as with thermophoretic sampling for ex situ microscopic analysis and electronic characterization. Focus is on tuning particle size as well as crystallinity and stoichiometry, with a specific focus on sub-stoichiometric materials with tunable composition.

1.1 Introduction

Understanding and improving flame-based synthesis of nanoparticles uses many strategies that have been introduced by "classical" combustion science and technology. Nearly every flame produces particles, which are sometimes quite visible and obvious in a sooting flame, but sometimes nearly invisible. By adding specific gaseous precursors, we have learned to consider flames not only as a reactive flow with internal energy transfer, but also as a reactor for synthesizing desired particulate materials with well-defined properties.

H. Wiggers · M. Fikri · I. Wlokas · P. Roth · C. Schulz (✉)
IVG, Institute for Combustion and Gasdynamics and CENIDE,
Center for Nanointegration, University of Duisburg-Essen,
47057 Duisburg, Germany
e-mail: christof.schulz@uni-due.de

A. Lorke et al. (eds.), *Nanoparticles from the Gas Phase*, NanoScience and Technology,
DOI: 10.1007/978-3-642-28546-2_1, © Springer-Verlag Berlin Heidelberg 2012

Similar like in combustion processes a series of different processes is closely linked. A comprehensive understanding and—if possible—a theoretical description of the entire system is required to direct the process towards the desired products, to optimize its efficiency and to potentially scale it from lab scale that is ideal to understanding the basics to production scale that is of industrial interest and makes new materials available for numerous applications, e.g. in technologies for energy conversion, storage, and efficiency.

Within this article, we describe the investigation of the process on different stages. The first reaction steps are relevant in the homogeneous gas phase where the volatile meta-organic precursors decompose and form the initial clusters. These high-temperature reactions tend to form a complicated network of parallel and subsequent reactions that can furthermore interfere with the combustion reaction in the flame itself. Studying these ultrafast reactions requires specific apparatus. Shock-tubes allow to study these reaction systems with microsecond time resolution, In this article examples for iron and gallium precursors will be presented in Sect. 1.3 as examples that then lead to the formation of chemical kinetics mechanisms.

Section 1.3.2 focuses on the particle formation in one-dimensional low-pressure flames. These flames are equipped with molecular-beam sampling that allows to extract samples at various positions after the initiation of the reaction in the flame front. The rapid expansion of the gases into the vacuum freezes all reactions and allows to study the particle size distribution and allows to deposit size-selected particles for further characterization. Thermophoretic sampling from the reactor provides material for various ex situ analysis methods as well as for the further investigation of properties in various application fields.

Besides the above mentioned sampling measurements with subsequent in-line and ex situ analysis, optical in situ measurement techniques have been proven extremely helpful in combustion science and have been transferred also to particle synthesis flames. They allow to measure temperature and species concentration within the process with spatial resolution (Sect. 1.5).

Information from the various experimental approaches is then combined to develop, support, and validate models that—based on computational fluid dynamics simulations—allow to describe the entire process with the final aim to develop predictive power for materials properties as well as for the process and reactor design (Sect. 1.6). The ultimate goal of the entire activity, therefore, is a complete understanding of the nanoparticle formation in flames that allows to determine process parameters for the synthesis of highly-specific tailored materials and allows to scale-up particle synthesis from the lab to production scale.

Within this paper a restriction is made with respect to the material of the particles. The synthesis of carbonaceous particles is not considered, although they also represent a desirable material, industrially produced in big quantities. We focus on the synthesis of oxidic inorganic particles. Examples for various materials synthesized in the low-pressure flame reactor are given in Sect. 1.6.2.

1.1.1 *State of the Art in Flame-Based Synthesis of Nanoparticles*

Flame synthesis is used routinely today to make a variety of commodities like SiO_2, TiO_2, Al_2O_3, etc. amounting to millions of tons annually. They are used industrially as pigments, opacities, catalysts, flowing aids, for optical fibers and telecommunication. In some cases flame synthesis has already superceded production routes by wet-phase chemistry. Evonik has e.g. developed a H_2/O_2 flame process for synthesizing titania (Degussa P25), which is used in the expanding area of photocatalysis, as well as cosmetics applications. They have demonstrated the ability to control the particle morphology, whilst achieving high production yields and large production rates. The high temperature flame reactor can be designed for a wide range of operating conditions. The process is self-purifying with respect to the final powder product. The characteristics of flame-made particles are controlled by the following: the mixing of the reactants and precursor, the overall composition, and the time-temperature behavior, including rapid quenching of the gas/particle flow. The required powders should be of high purity with a well controlled size distribution and morphology, which depend on the particular application.

Beside the large scale industrial flame synthesis reactors, combustion scientists mostly from academia have studied particle synthesis in nearly all types of flames, including burner-stabilized premixed flat flames, stagnation point premixed flames, coflow flames, counter flow flames, and multi-diffusion flames. Also well-stirred reactors and non-stationary flames in closed vessels have been used to synthesize particles. Self-sustaining flames of e.g. hypergolic type, as well as normal flames doped with various particle precursors, have been used. The early studies were focused on the development of new technologies and had to demonstrate control over the process. The characterization of the particulate product with respect to size, structure, and morphology was quite limited. It was gradually further developed in parallel with the gas-phase diagnostics of flame species and has profited much from ideas coming from aerosol science. Also new devices for characterizing materials properties (TEM, XRD, AFM, and others) have contributed to the understanding and fine-tuning of particle synthesis.

With the interest in nanostructured materials, flame-synthesized particles—as also those produced by other routes—were more and more focused on the investigation of size effects of nanostructured materials. Particles with a very narrow size distribution and well controlled phase composition and morphology called "functional nanoparticles" became desirable products. The reason for the size effects exhibited by nanoparticles compared to the bulk, is their large surface to volume ratio. For a particle of about 4 nm, half of the molecules forming the nanostructure are at the surface with consequences for the lattice structure. This causes dramatic changes in the physical and chemical properties compared to the bulk material and changes e.g. the melting temperature, the mechanical properties, the band gap for semiconducting particles, the magnetic or the optical properties, as well as the catalytic behavior.

Useful studies on flame or combustion synthesized particles are contained in comprehensive review papers of Pratsinis [1], Wooldridge [2], Roth [3], Pratsinis and coworkers [4, 5], and Rosner [6].

1.2 Principles of Particle Formation and Growth in the Gas Phase

Nanoparticles from gas-phase processes are mostly synthesized in laboratory flames by adding a precursor dopant, in a gaseous or liquid state, to the unburned gas. Such precursors are often metal halides, metal-organic or organometallic compounds. They can also be dissolved in water or in liquid hydrocarbons and sprayed into a flame [4], which usually leads to evaporation and subsequent decomposition of the respective precursor material. The combustion reaction mainly provides the temperature necessary for the decomposition of the tracer and in many cases the kinetics of the combustion reactions are only loosely coupled to the precursor's decomposition and the nucleation of particles. The energy of the exothermic combustion reactions is used to increase the temperature of the fluid flow, thus driving the chemical reactions of the precursor gas. Nuclei and clusters are formed, which further grow to nanoparticles by surface growth and/or coagulation and coalescence.

Such synthesis of particles in a fluid flow can also be established by using energy sources other than combustion processes to start or sustain the precursor reactions. Particularly advantageous for the synthesis of non-oxidic or metallic particles are hot-wall reactors, plasma reactors, or laser reactors. In these cases, the energy for increasing the temperature and initiating the reaction is transferred by convection and radiation from a hot wall or is directly coupled into the flow by microwave or laser energy. These reactors also open the possibility of staged reaction processes for coated or mixed particles, e.g. [7–10].

A typical sequence of basic steps illustrating particle formation in a flowing gas is shown in Fig. 1.1. It is based on an early representation by Ulrich et al. [11, 12] who studied very carefully the SiO_2 particle formation in flames. The precursor is injected as a gas or a spray into the flow, which is rapidly heated up by either external or internal energy transfer, e.g. by heat of combustion. A sprayed precursor rapidly evaporates and starts to decompose like a primary gas-phase compound. Consumption of the gaseous precursors can proceed either by gas-phase or by surface reactions or both. A complete description of the decomposition kinetics and the subsequent oxidation/hydrolysis reactions is rarely obtained.

The main interaction between decomposition kinetics and combustion kinetics is expected to originate from highly reactive and/or quenching processes, e.g., via radical reactions. Radicals, intermediates and product molecules are formed, which polymerize or nucleate to the first clusters, whose thermal stability and evolution of a critical size determines in many cases the further particle-forming processes. The clusters can grow either at the gas kinetic collision rate with sticking coefficients

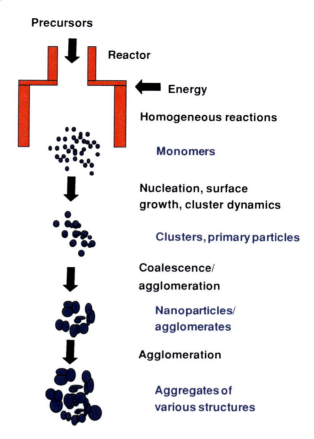

Fig. 1.1 Principle of the reactive particle formation sequence in a high temperature fluid flow, adapted from [3]

often assumed to be equal to one, or by heterogeneous growth processes such as the addition of monomers or oligomers to the cluster's surface. The coalescence of ensembles consisting of clusters and/or small nuclei is normally very fast resulting in compact and dense, usually spherical structures, which could be called particles. The typical time scale of the gas-phase chemical reactions including the cluster processes is very short compared to that for the subsequent particle growth.

When typical particle diameters become larger than several nanometers, the further development of a particle is determined by surface growth and by the interdependence of collision, coagulation, and coalescence. The importance of sintering was demonstrated e.g. by Helble and Sarofim [13] and by Matsoukas and Friedlander [14]. Brownian coagulation starts to form fractal structures, which merge again into spheres by rapid coalescence. As coalescence rates show a strong dependence on particle size and temperature, Brownian coagulation finally wins the race in the cooler parts of the flow due to the fact that the characteristic times for coalescence and sintering dramatically increase. As a result, fractal agglomerates are formed. They are called "soft agglomerates" or simply "agglomerates" if the primary particles are only interconnected by van-der-Waals forces resulting in small inter-particle point contacts while "hard agglomerates", so called "aggregates", are formed, when partial sintering takes place ending up in fractal structures with interconnecting sinter necks. Materials properties, residence time, temperature of the flowing gas as well as the time-temperature profile and the characteristic time scales for collision, coalescence and sintering are the key properties which determine the morphology and crystallinity of the agglomerates.

Besides the well known fluid mechanical forces determining the convective and diffusive mobility of the particles, thermophoretic forces (due to large temperature gradients) also influence a particle's residence time. Photoionization, thermionization or the addition of ionic species or charges can influence the charge of primary particles leading to Coulomb attraction or repulsion. This can affect both the size of the primary particles and the structure and size of aggregates, especially in external electric fields, see e.g. [15, 16]. Magnetic properties of primary particles also can affect the final structure of agglomerates [17], leading to the formation of long chain-like aggregates that consist of more than 50 primary particles as a result of self-organization of magnetic nanoparticles.

1.3 Shock-Tube Studies of Precursor Reactions

Over the last decades shock-tubes have been applied to several different research fields. Though the main thrust was focused aerodynamic and gas dynamic problems, the kinetics of high-temperature gas-phase reactions have been studied, mostly related to combustion processes. The shock tube as a high-temperature wave reactor enables the investigation of reaction rate coefficients under diffusion free conditions because it provides a nearly one-dimensional flow, with practically instantaneous heating of the reactants.

Shock-tube kinetics experiments that aim at investigating elementary reactions in complex reaction mechanisms typically apply high dilution of the reactants by inert gases (usually argon). At the same time, high sensitivity diagnostics techniques, such as Atomic Resonance Absorption Spectroscopy, ARAS, are employed to monitor species concentrations as a function of time after the initiation of the reaction through the shockwave. The low reactant concentration prevents the influence of secondary reactions and therefore facilitates the investigation of elementary reactions without or with limited interference. Additionally, in diluted systems the thermal effects of the reaction do not significantly alter the temperature during the measurement time. Radical species and atoms can be generated in a shock tube trough the thermal decomposition of suitable precursors. If it is ensured that radical formation is faster than the reaction under investigation, elementary reactions that involve radical and atom species can be studied in a straightforward way.

Though shock tube research of homogenous systems has been making a steady progress, heterogeneous studies in shock tube are still at an early stage. The difficul-ties are the degree of the increased complexity of the reaction mechanisms and the

1 Synthesis of Tailored Nanoparticles in Flames

choice of diagnostics. A pre-condition of experimentation requires that the particles are homogeneously dispersed in the carrier gas. Also, lack in the understanding of interference between the complexities of two phase flow and kinetics are additional hindrance to overcome. Roth and co-workers have made significant contributions in heterogeneous shock tube kinetics in dispersed systems [18, 19].

The shock tube as wave reactor provides also an excellent environment for the study of nucleation and growth of particles from the vapor phase at high temperatures. Apart from providing nearly instantaneous and uniform heating of the reactant, it allows for rapid quenching of products leading to particle condensation and growth. The effect of varying the initial temperature, pressure, and mixture composition on the size and yield of the particles produced, can be conveniently studied in shock tubes. Frenklach [20] studied silicon particle nucleation and growth using light extinction measurement behind reflected shock waves at 900–2000 K. Herzler et al. [21] have investigated the formation of both TiN molecule and particles in $TiCl_4/NH_3/H_2$ systems behind reflected shock waves. The relative TiN particle concentrations were determined by ring dye laser light extinction. For brevity, we will discuss in the following only a few exemplary synthesis routes: nucleation of iron clusters and bimolecular reactions of precursor initiated reactions.

1.3.1 Formation of Iron Clusters From Fe(CO)$_5$

The magnetic and electronic properties of iron open a wide field of practical applications for iron nanoparticles. Gas-phase synthesis allows to produce of very small particles of uniform composition. For a better understanding of the formation process, kinetics data are necessary. The gas-phase synthesis of iron particles is strongly controlled by the decomposition of the precursor and the kinetics of iron cluster formation.

Roth et al. studied the condensation of iron atoms in a shock tube in mixtures of iron pentacarbonyl (IPC) highly diluted in argon [22]. The formation and consumption of Fe atoms behind incident shock waves was followed by ARAS at 271.9 nm using a hollow cathode lamp as light source. Also, molecular resonance absorption spectroscopy (MRAS) at 151 nm (from a microwave-excited plasma lamp) was applied to measure the side product CO during the reaction. For the interpretation of the signals, a simplified reaction mechanism was proposed to describe the measured iron concentration profiles at $T > 730$ K. A typical example illustrating the temperature dependency of the Fe-atom and CO-molecule resonance absorption in a 100 ppm Fe(CO)$_5$/Argon mixture is given in Fig. 1.2. All Fe and CO absorption profiles show a fast increase due to the thermal decomposition of IPC to form Fe atoms and CO molecules. In the case of the highest temperature of 1,110 K, the Fe absorption shows after a few microseconds a constant absorption level, indicating no further reaction within 1 ms. A decrease in temperature leads to a decreasing Fe-atom absorption with an inverse temperature dependence which was attributed to the cluster formation and the thermal stability of critical clusters. Although, the

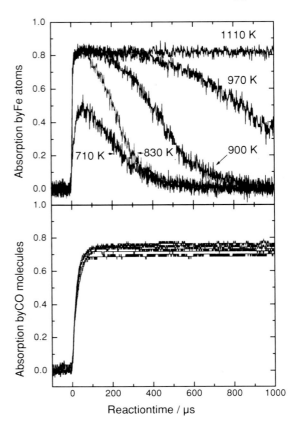

Fig. 1.2 Fe- and CO-absorption profiles for a 100 ppm Fe(CO)$_5$/Ar mixture at different post-shock temperatures, adapted from [22]

thermal decomposition of IPC to Fe and CO is known to be very fast, an influence of intermediates cannot be excluded. It is accepted in the literature that the further growth of clusters to form particles mainly proceeds by coagulation with rates nearly equal to the collision frequency [23]. It is expected that those coagulation processes only weakly affect the disappearance rate of Fe atoms. A simplified reaction mechanism has been proposed, which contains the above mentioned subsystems: Fe(CO)$_5$ decomposition, formation, and dissociation of small clusters and coagulation of clusters. The most sensitive reactions during the nucleation are the recombination of Fe atoms and the reverse reaction:

$$Fe + Fe + M \rightarrow Fe_2 + M \tag{R1}$$

$$Fe_2 + M \rightarrow Fe + Fe + M \tag{R2}$$

Woiki et al. [24] showed that the IPC decomposition proceeds via CO abstraction. At temperatures above 1,000 K, the time for the total decomposition takes a

1 Synthesis of Tailored Nanoparticles in Flames

few microseconds and increases significantly at the lower temperature end of this study. At low temperature the decomposition of IPC is not fast enough compared to the consumption process. Therefore, a detailed knowledge about the kinetics of the IPC decomposition is necessary. As the major Fe-cluster-growth mechanism, Fe addition to iron clusters is considered. The formation and removal of clusters is represented by the reaction mechanism described in [22] which is initiated by (R1) and (R2). The rate coefficients of the exothermic formation reaction (R2) was determined by fitting the decay of an experiment at conditions where the reverse reaction is negligible. The value obtained is $k_2 = 1.0 \times 10^{19} \, \text{cm}^6 \, \text{mol}^{-2} \, \text{s}^{-1}$, which is in quite good agreement with the theoretically determined value of Bauer and Frurip ($\sim 1.7 \times 10^{19} \, \text{cm}^6 \, \text{mol}^{-2} \, \text{s}^{-1}$) [25]. The rate coefficients for the other reactions of single clusters with Fe atoms were estimated to be of the same magnitude like (R1) and were defined as $k = 1.0 \times 10^{19} \, \text{cm}^6 \, \text{mol}^{-2} \, \text{s}^{-1}$ and $5.0 \times 10^{14} \, \text{cm}^3 \, \text{mol}^{-1} \, \text{s}^{-1}$ for termolecular and for bimolecular reactions, respectively. The value $n = 5$ at which the transition from termolecular to bimolecular kinetics is assumed to occur is based on theoretical considerations of Jensen [26].

1.3.2 Reaction of Gallium Atoms with Nitrogen and Oxygen

The reaction of trimethylgallium ($Ga(CH_3)_3$) and NH_3 has emerged as the leading candidate for the synthesis of GaN for commercial applications and is mainly used in metal-organic chemical vapor deposition and nanoparticle generation. The mechanism of this reaction can be described by two competing reaction routes: the thermal decomposition of $Ga(CH_3)_3$ at high temperatures and the adduct formation between $Ga(CH_3)_3$ and NH_3 at relatively low temperatures. However, no information has been known about the reaction of Gallium atoms with ammonia. Fikri et al. studied the kinetics of the bimolecular reaction of Gallium atoms with ammonia in shock-heated mixtures [27]. For this study, knowledge about the thermal decomposition of $Ga(CH_3)_3$ in the gas phase is required [28]. The high-temperature decomposition of $Ga(CH_3)_3$ occurs via three sequential first-order reactions (R3–R5) leading first to $Ga(CH_3)_2$ and then to $Ga(CH_3)$. Similarly, the last decomposition step generates a Ga atom and a methyl radical.

$$Ga(CH_3)_3 \rightarrow Ga(CH_3)_2 + CH_3 \qquad (R3)$$

$$Ga(CH_3)_2 \rightarrow Ga(CH_3) + CH_3 \qquad (R4)$$

$$Ga(CH_3) \rightarrow Ga + CH_3 \qquad (R5)$$

For the simulation of the Ga-atom formation additional measurements on the decomposition of $Ga(CH_3)_3$ in Ar in the absence of NH_3 were necessary and have been evaluated [28]. Experiments were conducted between 1,100 and 1,560 K at pressures of approximately 0.4, 1.6, and 4 bar. Because the reaction steps (R3) and (R4) are

Fig. 1.3 Measured and simulated Ga-atom concentration profiles with and without NH$_3$. The mixtures contain 5 ppm Ga(CH$_3$)$_3$ in Ar and 6 ppm Ga(CH$_3$)$_3$ + 2, 300 ppm NH$_3$ in Ar, respectively, adapted from [28]

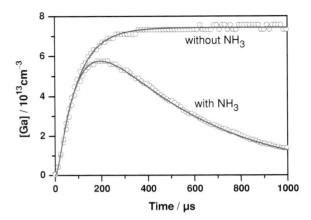

fast, the Ga-atom formation in the NH$_3$-free system is approximately proportional to $1 - \exp(-k_5 t)$. Figure 1.3 shows the comparison of typical concentration profiles with and without the presence of ammonia.

Based on the assumption of the successive Ga-C bond dissociation (R3–R5) and the presence of excess NH$_3$, a pseudo-first-order evaluation can be applied for the measured Ga-atom concentration profiles. Therefore, the posterior Ga-atom concentration profiles were transferred into first-order plots by considering the calibration equation. The rate coefficients k_6, e.g., Ga + NH$_3$ → products (R6), are directly related to the measured pseudo-first-order rate coefficient. A pressure dependence of the reaction was not noticeable. A least-squares fit yielded the following expression over the temperature range of 1380–1870 K:

$$k_6(T) = 10^{14.1 \pm 0.4} \exp(-11900 \pm 700 \, \text{K}/T) \, \text{cm}^3 \, \text{mol}^{-1} \, \text{s}^{-1}$$

where the error limits are at the 1σ standard deviation level and are statistical only. A comparison of the Arrhenius expression for the reaction of Ga atoms with ammonia with that obtained for Ga atoms with oxygen for temperatures below 1,600 K [27] shows a similarity in activation energies and pre-exponential factors. In this previous study a collisionally-stabilized adduct was assumed to explain the pressure dependence of the reaction of Ga + O$_2$. Similarly, both pre-exponential factors are in the order of 10^{14} cm^3 mol^{-1} s^{-1}. This confirms the possibility of such a high frequency factor in this type of reaction system. The apparent activation energies are for both reactions approximately 100 kJ mol^{-1}. Thermochemical considerations and the relatively high activation energy would imply a collision stabilization of the association complex. Because the observed rate coefficients are pressure independent we conclude that the reaction is near the high-pressure regime. Complementary quantum chemical calculations have been performed to gain insight into the energetics and the possible distribution. The first step is the addition of Ga to NH$_3$ which leads

1 Synthesis of Tailored Nanoparticles in Flames 13

to the well known $GaNH_3$ stable adduct which lies $29.3\,kJ\,mol^{-1}$ compared to the reactants [27].

In the same manner the reaction of $Ga+O_2$ was studied with mixtures containing 13–23 ppm $Ga(CH_3)_3$ and 200–2895 ppm O_2 behind reflected shock waves at temperatures between 1,290 and 2,320 K and pressures of approximately 1.5 and 3.7 bar by applying atomic resonance absorption spectrometry for time-resolved measurements of Ga atoms at 403.299 nm [29]. Two reaction channels were found for the reaction of Ga atoms with O_2, the recombination reaction

$$Ga + O_2 + M \rightarrow GaO_2 + M \qquad (R6)$$

$$Ga + O_2 \rightarrow GaO + O \qquad (R7)$$

The trimethylgallium decomposition experiments showed that above 1600 K the decomposition is complete after a few μs. Together with the high oxygen excess and the absence of other Ga-consuming reactions this allows a first-order evaluation of the Ga-atom concentration profiles. Therefore, all measured Ga-atom absorption profiles at $T > 1600$ K were transferred into first-order plots by applying the calibration relation. For the reactions of methyl$+O_2$ the GRI 3.0 [30] mechanism was used. In most cases these reactions are of minor importance for the Ga-atom concentration because the reduction of the O_2 concentration is negligible due to the high O_2 excess.

The evaluated rate coefficients of the Ga-consuming reaction are shown in the Arrhenius diagram of Fig. 1.4. It can be seen that two Ga-atom-consuming reactions occur, a reaction with a negative temperature and a total pressure dependence. The appearance of two reaction channels for the reaction of a metal atom with O_2 was only observed by Giesen et al. [31] for the reaction of $Fe+O_2$. The temperature region of the two channels was similar to $Ga+O_2$ but for the recombination reaction a positive activation energy of 9 kJ was observed. Considering an extrapolation of the observed experimental values at the upper and lower temperature end of the measurement, where only one channel is dominant, a separation of both channels was possible.

$$Ga + O_2 + M \rightarrow GaO_2 + M \qquad (R8)$$

$$k_9 = 10^{12.95\pm0.41} \times \exp(12400 \pm 1367\,K/T)\,cm^6\,mol^{-2}\,s^{-1}$$

and

$$Ga + O_2 \rightarrow GaO + O \qquad (R9)$$

$$k_{10} = 10^{14.57\pm0.36} \exp(12460 \pm 1535\,K/T)\,cm^3\,mol^{-1}\,s^{-1}$$

1.4 Flame Reactor Studies of Particle Formation and Growth

The investigation of particle formation and growth in flames requires a simple reactor with well-defined boundary conditions and a homogeneous flow field. Therefore, a

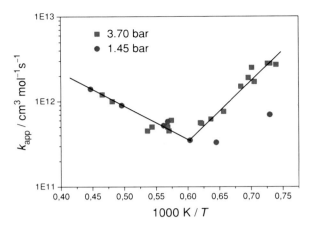

Fig. 1.4 Arrhenius diagram of the rate coefficient of the reaction $Ga + O_2 \rightarrow$ products. *Blue circles* experiments at about 1.45 bar. *Red squares* experiments at about 3.70 bar

premixed flame reactor where the flame gases emerge from a porous, flat burner head was developed to enable an almost one-dimensional reaction system with the distance from the burner ("height above burner", HAB, measured in cm) as the relevant coordinate [32]. This setup requires a homogeneous mixture of gaseous species fuel (usually hydrogen), oxidizer (oxygen), and vaporized precursor. For modifying the flame temperature, inert gases (typically argon) can be added. All gases are mixed within a water-cooled mixing chamber directly mounted in front of the burner head. This system allows for the investigation of several precursor materials as long as they provide the required vapor pressure. For technical reasons liquid precursors for spray pyrolysis cannot be used with this setup.

This construction provides homogeneous gas mixtures and it ensures that within the center area of the flame (about one centimeter in diameter) no gradients in temperature and flow conditions occur perpendicular to the flow direction. The distance between the burner head and a sampling area downstream can be adjusted by moving the burner head. Sophisticated sampling and measuring techniques such as laser spectroscopy, gas analysis, thermophoretic and molecular beam sampling are mounted in such a way that they can deliver detailed information about gas-phase composition and temperature, particle size, and particle morphology. A significantly increased spatial resolution along the flow coordinate can be obtained at low pressures leading to an elongated flame zone compared to atmospheric pressures. Therefore, the experimental setup was optimized to operate at around 30 mbar. This pressure range allows for the formation of sufficient amount of particles and it offers the possibility to also investigate precursor materials with a very low vapor pressure down to about 1 mbar. The different measurement techniques that were applied to the premixed flame reactor are explained in detail within the next chapters.

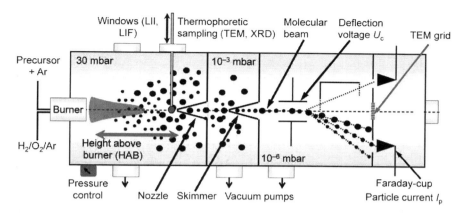

Fig. 1.5 Flame synthesis apparatus with molecular beam sampling and particle mass spectrometer (PMS)

1.4.1 Molecular-Beam Sampling

In particle synthesis, information about particle size and morphology is required, which can best be obtained from collected particles. However, this requires careful and representative particle sampling, which does not falsify or change the particles' properties during the collection procedure itself. The technically most fastidious method is molecular beam sampling in combination with a particle mass spectrometer (PMS). It is a variant of the classical mass spectrometer for gaseous flame species, including radicals, see e.g. [33, 34]. The setup is schematically shown in Fig. 1.5, see also [32]. Nanoparticles in a flame are partly charged either by the particle synthesis process itself or by an appropriate source. A sample of the aerosol is supersonically expanded through an electrically-grounded nozzle into a first vacuum chamber. The supersonic free jet formed by the expanding flow contains both, particles and gaseous species. The flow conditions are such that the gas temperature decreases extremely rapidly, thus, freezing any physical or chemical processes. The center of the free jet is extracted by a skimmer and moves as a particle-laden molecular beam into a high-vacuum chamber.

The molecular beam is directed through a capacitor with an adjustable electric field where the charged particles are deflected from the beam according to their mass, velocity and charge, thus forming a fan-shaped particle beam. As within one experiment the particle velocity of all particles is nearly the same [35] and particle charge is mostly unity for the relevant particle size range (see Fuchs' theory [36]), the crucial criteria for the deflection of the particles within the charged capacitor is their mass. The classification of the particles due to their mass downstream the deflection capacitor is performed by introducing a grounded plate acting as collimator into the fan-shaped beam carrying two symmetrical slits and one central slit. By varying the voltage of the capacitor, the fan of charged particles is scanned over the outer slits allowing particles of different masses to pass through. The PMS offers three possibilities for further particle processing:

1. Collection of charges carried by the particles by two Faraday cups. This enables on-line determination of the particle mass distribution without further calibration.
2. Particle collection on TEM grids either behind the central slit (neutral particles or representative particle ensemble in case of $U_C = 0$) or behind the two other slits (charged particles) with further ex situ size and structural imaging by, e.g., high resolution electron microscopy (HR-TEM).
3. Utilization of the PMS as a mass filter by applying a constant deflection voltage and deposition of size-selected particles on any substrate placed behind the eccentric slits.

All possibilities have been successfully applied to spatially-resolved nanoparticle analysis in low pressure flames and in other devices, see [8, 37–40].

1.4.2 Thermophoretic Sampling with Ex Situ Analysis

A simple way for collecting particles out of a flowing fluid is thermophoretic sampling, introduced by George et al. [41] and further improved by Dobbins and co-workers [42] for soot particles. A cold surface, e.g. a TEM grid, placed parallel to the flow direction causes a strong temperature gradient, along which the particles move towards the surface, driven by thermophoretic forces which in the Knudsen regime are independent of particle size. It must be kept in mind, however, that the inserted surface to some extent influences the flow. The subsequent ex situ analysis depends on the available instrumentation and can be made, e.g., by atomic force microscopy (AFM), by TEM or even by optical diagnostics. It is advantageous to "shoot" the cold surface through the flame, thereby avoiding any restructuring of the collected particles by the influence of the hot flame gases. Typically, the residence time of the TEM grids within the hot zone is a few ten milliseconds and below, which is not sufficient to heat TEM grid and grid holder significantly. A comparison between particles deposited from the molecular beam and thermophoretically sampled particles from the same flame show surprisingly good agreement [43].

1.5 Laser-Based Diagnostics in Particle Synthesis Reactors

The properties of gas-phase-synthesized particles strongly depend on the reaction conditions like flame temperature, fuel/oxygen equivalence ratio, pressure, and precursor concentration. The final properties of the particles depend on the temporal history of the above mentioned parameters. Therefore, if the generation of particles with well-defined conditions is desired, well-controlled reaction conditions are

1 Synthesis of Tailored Nanoparticles in Flames

essential. In situ measurements of concentration and temperature distribution, thus, give important input for modeling the synthesis process and for apparatus design. Laser-based techniques enable non-intrusive in situ measurements of the conditions during nanoparticle formation. Many of the techniques applied here to the in situ diagnostics in particle synthesis are adapted from combustion-related diagnostics where quantitative laser-based imaging measurements and infrared absorption strategies find widespread application (e.g. [44]).

In the following paragraphs we focus on three aspects, temperature imaging via multi-line nitric oxide laser-induced fluorescence (LIF) thermometry [45], spatially-resolved iron atom measurements with LIF in iron oxide nanoparticle synthesis in low-pressure flames [46], and the determination of particle-sizes via laser-induced incandescence (LII) [47]. The results of the Fe-atom and temperature measurements will be further evaluated in Sect. 1.6.1.

1.5.1 Diagnostics for Temperature

Laser-based imaging techniques have the capability to provide multi-dimensional temperature information without perturbing the investigated system in contrast to sampling techniques or thermocouple measurements. Two-dimensional temperature imaging can be obtained by two-line [48, 49] and multi-line [50] LIF as well as Rayleigh scattering [51] and filtered Rayleigh thermometry [52]. Vibrational and rotational Raman techniques [53] are often limited to line measurements and CARS [54] yields point measurements only. Infrared absorption techniques [55] can provide line-integrated information and therefore do not resolve inhomogeneous temperature distributions.

Multi-line LIF thermometry is based on the temperature-dependent population of rotational and vibrational energy levels of the nitric oxide (NO) molecule. In contrast to conventional two-color LIF thermometry for gas-temperature imaging [48, 49], the multi-line technique yields absolute temperatures without calibration [56] and can be applied even in systems with strong scattering and fluorescence background as well as in the presence of pressure broadening [50, 57]. It has been used in stoichiometric and sooting flames [50] as well as in spray flames at atmospheric pressure [58, 59] and high-pressure flames up to 6 MPa [57]. The technique is based on the measurement of LIF-excitation spectra of NO, which is added to the fresh gases as a fluorescent tracer. The laser beam is formed to a light sheet, which illuminates a plane in the area of interest. The pulsed laser is tuned over a part of the NO A-X(0,0) absorption band at $\sim$225 nm while individual images are taken with an ICCD camera for each excitation wavelength. The camera is equipped with filters to suppress the detection of elastically scattered light and interference from other laser-excited species. From the resulting stack of pictures (each with the laser tuned to the next wavelength) LIF excitation spectra can be extracted for each pixel. Figure 1.6 shows two example spectra for 750 and 1,400 K. From the strong temperature dependence of the spectra, the local temperature is derived. Simulated spectra are then fitted to

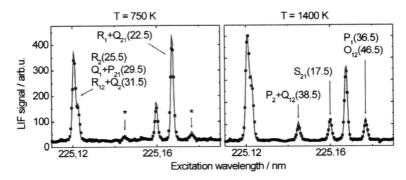

Fig. 1.6 Experimental (*symbols*) and the fitted simulated (*lines*) excitation spectra for two different temperatures. The labels show the respective rotational transition in the NO A-X(0,0) band

the experimental data with absolute temperature, broad-band background as baseline and signal intensity as free parameters using LIFSim [60].

In the present experiment we applied multi-line NO-LIF thermometry for the first time to a low-pressure nanoparticle flame reactor. 200 ppm NO were added to the fresh gases since there is no natural NO production in a $H_2/O_2/Ar$ flame. This small amount avoids disturbance of the lean flame and ensures negligible laser attenuation by NO inside the reactor.

A tunable, narrowband KrF excimer laser (248 nm, $\Delta\nu \sim 0.3\,\text{cm}^{-1}$, Lambda Physik EMG 150 TMSC) is frequency-shifted to 225 nm in a Raman cell filled with 5 bar hydrogen, enabling NO excitation in the A-X(0,0) band. The laser beam is expanded in the horizontal direction and compressed in the vertical direction with two cylindrical lenses $f = 1,000$ and 300 mm to form a light sheet of $50 \times 10\,\text{mm}^2$. The LIF signal is recorded with an intensified CCD camera (LaVision). Elastically-scattered light is suppressed by three long pass filters (230 nm, LayerTec). This ensures a high signal-to-noise ratio of the excitation spectra even with low NO concentrations of 200 ppm and low laser fluence of 4 kW/cm^2. The general setup, as well as the data evaluation procedure, is similar to the one in [50]. The observed field is located in between the (moveable) burner and the sampling nozzle of the PMS (cf. Fig. 1.7). Measurements at different distances relative to the burner head are carried out through shifting the burner relative to the laser light sheet within the combustion chamber. Therefore, the measurements are in fact carried out under slightly different conditions While this is taken into account when comparing the results to the simulations, the assembled images shown later show some discontinuities of the measured results due to this effect.

The flat-flame reactor is described in detail elsewhere [61]. The 36-mm diameter sintered metal-plate burner head is centered in a 300 mm long, horizontally mounted, metal tube with a diameter of 100 mm. Optical access is possible through fused silica windows (50 mm diameter) from three sides. The burner head can be moved horizontally with respect to a reference position, where a skimmer nozzle extends

1 Synthesis of Tailored Nanoparticles in Flames

Fig. 1.7 Horizontal cut through the flat-flame reactor. Optical access is given through fused silica windows. The light sheet illuminates a 10 mm wide section in the combustion chamber. Measurements with different burner positions are used to assess the entire temperature field

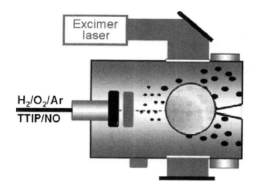

into the reactor to enable online particle size measurements. Hydrogen is used as fuel, oxygen as oxidizer and the mixture is diluted with argon. Typical flow rates are 600 sccm H_2, 800 sccm O_2, 500 sccm Ar, and 200 ppm of the respective precursor. The fuel/air ratio of this lean mixture is $\phi = 0.375$. The flow velocity inside the reactor reaches a few m/s.

The gas-phase temperature field was measured in the $H_2/O_2/Fe(CO)_5$ flat flame of the reactor by moving the burner nozzle relative to the window position and with signal detection at the same position for five different burner distances. Figure 1.8a shows the resulting temperature distribution in a 175×50 mm^2 horizontal area of the flame inside the reactor for typical operating conditions at a pressure of 3 kPa. The spatial resolution in the horizontal plane is 1×1 mm^2. The burner head is situated on the left side where the incoming cold fresh gas is seen in the temperature distribution.

Because NO is added as a tracer, the effect of different NO concentrations on the temperature measurements was investigated. The NO concentration was varied between 100 and 1,000 ppm while all other parameters were kept constant. No effect on the temperature results could be detected.

1.5.2 Diagnostics for Species Concentration

Using a similar arrangement of tunable pulsed UV laser and camera, many atomic and molecular species can be imaged in the flame. While OH imaging is a standard diagnostics in combustion research (e.g., [44]), the additional detection of atoms and molecular intermediates that occur in nanoparticle synthesis of metal oxides provides valuable input for the understanding of the reaction system. However, these materials systems are unusual for laser-based diagnostics and thus, many details of the measurement strategies had to be specifically developed. In the following we present the detection of iron atoms during the synthesis of iron oxide nanoparticles as one example.

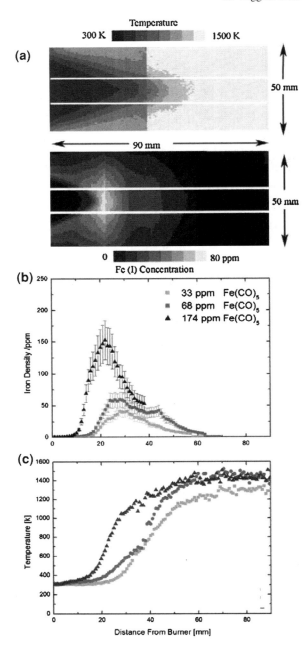

Fig. 1.8 **a** Temperature and neutral iron (Fe(I)) concentration distribution with 70 ppm of Fe(CO)$_5$ in the fresh gases for three positions of the burner within the reactor chamber. **b** Iron-atom density averaged in the central 10 mm along the y-axis (marked in a) for three Fe(CO)$_5$ concentrations in the fresh gases; **c** Temperature profiles for the same region marked in **a**. Discontinuities in the images and profiles result from the combination of measurements with different burner positions

1 Synthesis of Tailored Nanoparticles in Flames

During the flame synthesis of Fe_2O_3, iron atoms exist in different oxidation from 0 (neutral, in spectroscopy usually called Fe(I)) to 3. The fluorescence signal of iron has been used in medical and astronomical applications [62]. Several spectroscopic data bases for iron atoms exist [63]. Metal organic compounds are known to strongly interact with flame chemistry. Ironpentacarbonyl ($Fe(CO)_5$), that is frequently used as a precursor for iron oxide nanoparticles, is known to be a flame inhibitor [64]. These effects motivated several studies of iron atom concentration in the past where measurements were carried out using absorption spectroscopy [65] and LIF [66]. The application to nanoparticle synthesis, however, is a topic of recent development [46]. Therefore, it is important to study its influence on the temperature field as a function of precursor concentration. Additionally, temperature information (as provided from the strategy presented in the previous section) is required to correct Fe-atom LIF measurements for temperature-dependent variations of the ground-state population.

For the measurement of the spatial distribution of Fe(I) the laser system was tuned to absorption lines of iron atoms and the same detection system was used as described above. For measurements of emission spectra, an imaging spectrograph (ARC, $f = 155\,mm$, $f_\# = 4300$ lines/mm grating) was used. Two excitation wavelengths of iron were chosen within the fundamental (248 nm) and the Raman-shifted (225 nm) tuning range of the KrF excimer laser. At $\sim$225 nm ($44415\,cm^{-1}$) a weak transition from the ground state ($3p^63d^64S^2$) to an excited state ($3p^63d^6(a^3F)4s4p(^3P^0)$) is used [63]. Due to its weak transition probability, the laser beam is not significantly attenuated inside the reactor for the present experimental conditions. Additionally, emission upon 248 nm excitation is observed. This is due to inter-system crossing to the $3p^63d^6(^5D)4s4p(^1P^0)$ system. Despite the fact that the transition probability for the emission process is higher in the latter case by a factor of 200, the observed signal is weaker because of the underlying spin-forbidden inter system crossing (ISC). Additional weak emissions from lower lying states populated by further ISC can be observed around 300 nm (III). The transition at 248 nm ($40257\,cm^{-1}$) also originates from the ground state, but reaches a different excited state ($3p^63d^6(^5D)4S4P(^1P^0)$). This transition is the strongest of all iron transitions in this spectral window. Hence, at the relevant Fe concentration levels the laser beam is nearly completely absorbed within the first 10–20 mm inside the reactor.

For correction for temperature effects on the Fe-LIF diagnostics the temperature distribution was used obtained from the method described in the previous section. The 225 and 248 nm transitions used in this investigation originate from the same electronic ground state. Its temperature-dependent population was evaluated to allow for quantitative measurements also in systems with inhomogeneous temperature distribution. The partition function of iron atoms was calculated, using the five states with the lowest energy from the NIST database [63]. Higher states have negligible population at the temperatures of interest. With this partition function, the fractional population of the ground state at temperature T was calculated.

The calibration of the Fe-LIF signal intensities I_λ for excitation at wavelength λ with respect to concentration is based on the measured attenuation of the laser radiation at 248 nm during passage through the flame gases [67]. Because the Fe-atom concentration and temperature vary in the observed region simultaneously, imaging measurements with 248 nm excitation were related to measurements at 225 nm excitation where the laser attenuation was negligible. The relative signal thus depends on the local variation in laser intensity of the 248 nm beam. With an iterative procedure, local concentrations can be determined from the local attenuation. For further details of this procedure, see [46]. From repeated measurements a standard error of 20 % was evaluated for this method.

Figure 1.8 shows measured two-dimensional temperature and Fe(I)-concentration distributions for a flame with 68 ppm $Fe(CO)_5$. One-dimensional profiles along the flow coordinate were taken by averaging the temperature and concentration data for each y position within the central 10 mm of the burner chamber. Figure 1.8b shows the resulting Fe(I)-concentration profiles for three different $Fe(CO)_5$ concentrations. The different zones of the flame were investigated by translating the burner to different positions within the chamber which slightly changes the flame conditions. Therefore, the graphs show steps at the limits of the individual measurement sections. Temperature profiles for the same conditions are shown in Fig. 1.8c. It can be observed that the formation of iron atoms starts earlier (i.e., closer to the burner surface) when higher initial concentrations of $Fe(CO)_5$ are added to the fresh gases. Higher $Fe(CO)_5$ initial concentrations are also associated with a faster rise of the temperature and higher end gas temperatures, i.e., the flame front moves closer to the burner surface.

For all $Fe(CO)_5$ concentrations investigated, the iron atom formation starts at temperatures around 380–420 K. The results of these measurements will be further discussed in comparison the model results in Sect. 1.6.1.

1.5.3 Optical Diagnostics for Particle Sizes

Diagnostic techniques for sizing particles are confronted with the problem of coping with structures ranging from nearly molecular dimensions to approximately 100 nm. Additional to the molecular-beam techniques described above in Sect. 1.4, there is a need for optical diagnostics providing in situ measurements of particle sizes. These could be applied to a wide variation of particle synthesis processes.

The interaction of light with particles in the form of static or dynamic light scattering has been used, see, e.g. [68]. Good spatial resolution can be obtained, but according to Mie theory, the scattering decreases rapidly with decreasing particle size and can reach values comparable to those of gas-phase species. From sooting flames, light extinction, especially laser light extinction, is known to be a useful detection technique for small particles, see e.g. [69]. For high particle concentrations, good signal quality has been obtained in systems with nanoparticles. However, these line-of-sight techniques require signal deconvolution and an assumption of specific symmetries of the problem. A quantitative interpretation also depends on the size- and material-dependent value of the refractive index. A spectrally resolved light extinction technique has been developed by D'Anna et al. [70] and successfully applied to sooting flames.

In the last years, an additional laser-based strategy for measuring nanoparticles has been developed called LII, see e.g. [71, 72]. In its simplest form, the particles are heated up by a laser through light absorption and their laser-induced radiation is a measure of the number density of the particles. The time-resolved version (TiRe-LII) is in principle able to deliver particle size information, see e.g. [18, 47, 72–76]. In this case, the particles are heated up by a nanosecond laser pulse. The emission of light from the particles during their subsequent cooling is recorded. The temporal behavior of this signal contains information on the particle sizes, as the cooling rate of small particles is faster than that of big ones. The interpretation of the measured data to obtain particle sizes is complex and is based on assumptions e.g. for heat and mass transfer properties. This particle sizing technique has been significantly improved in recent years and its application is not restricted to soot. Variants of this technique, like two-color LII, are in use. A series of international workshops on LII [77] has demonstrated the problems and perspectives of this diagnostic technique for particles [72, 78].

The sizing of Fe_2O_3 nanoparticles synthesized in a low pressure $H_2/O_2/Ar$ flame was demonstrated by applying simultaneously TiRe-LII and PMS. From the I/U-spectrum of the PMS (I = flux of charged particles arriving at the Faraday cup, U = deflection voltage) a mean particle size of $d_p = 7.1\,nm$ can be determined. The TiRe-LII raw signal shows a specific feature: As the particles are very small, the whole process of particle heat-up by the pulsed laser and the subsequent particle cooling takes only a few ns. A certain time overlap of both process steps is also very obvious, which makes the evaluation of such signals more complicated. A first attempt to determine the mean particle size from the TiRe-LII signal based on a modified theory results in a value which is very close to the PMS particle size. The simultaneous application of both these particle sizing techniques is expected to be quite helpful in the further development and improvement of the TiRe-LII diagnostics.

1.6 Chemistry Modeling and Simulation of Reactive Flows

The formation, growth and transport of nanoparticles in a flame can be formally described with the conservation laws of continuum mechanics and the population balance equations. Here, particle sizes and numbers allows a treatment of the gas-borne particles as a continuous, dispersed phase. Furthermore, the low precursor concentration allows to neglect the influence of the dispersed phase on the main flow. These assumptions lead to a decoupled calculation of the chemically reacting flow and the passively transported dispersed phase.

The flames presented here have low Reynolds and Mach numbers. As precursor typically metal-organic species are used which are known to affect the combustion kinetics already at very low concentrations [79]. Therefore, a detailed calculation of the transport process and of the finite rate chemistry is required. In this section a detailed and a reduced mechanism are presented that describe the effects caused by

IPC ($Fe(CO)_5$) addition on lean H_2/O_2 flames. Subsequently, simulation results and the calculated nanoparticle formation and growth dynamics are presented.

The particle dynamics are described by the population balance equations, which are transport equations for probability distribution functions describing the particle properties. In component notation the population balance equation for a distribution function $n(t, \boldsymbol{x}, \varphi)$ reads as

$$\frac{\partial n(t, \boldsymbol{x}, \varphi)}{\partial t} + \frac{\partial}{\partial x_i} [u_i n(t, \boldsymbol{x}, \varphi)] - \frac{\partial}{\partial x_i} \left[D \frac{\partial}{\partial x_i} n(t, \boldsymbol{x}, \varphi) \right]$$
$$= -\frac{\partial}{\partial \varphi_j} \left[n(t, \boldsymbol{x}, \varphi) \dot{\varphi}_j \right] + h(t, \boldsymbol{x}, \varphi)$$

with the components of transport velocity u_i, the internal coordinate of particle properties φ, the diffusion coefficient D and a net production rate h, e.g., from nucleation. Depending on the set of internal coordinates of particle properties, the moments of this distribution function may be e.g. particle number density, particle volume, particle surface area or particle composition etc. There exist many models for solution of the population balance equations, depending on the description of the probability density function. The presently most popular models describe the transformed equations in the moment space with a prescribed probability density function [80] or with an approximated density function [81–83].

For the description of particle formation and growth a method of moments with the so called monodisperse distribution function [84] was applied. The resulting set of equations for steady state transport is:

$$\frac{\partial}{\partial x_i} \left(u_i n_1 - D_1 \frac{\partial}{\partial x_i} n_1 \right) = -J$$
$$\frac{\partial}{\partial x_i} \left(u_i N_\infty - D_p \frac{\partial}{\partial x_i} N_\infty \right) = J - \frac{1}{2} \beta N_\infty^2$$
$$\frac{\partial}{\partial x_i} \left(u_i A_\infty - D_p \frac{\partial}{\partial x_i} A_\infty \right) = J a_1 - \frac{A - A_{\min}}{\tau_C}$$

where u_i is the transport velocity and β is the coagulation coefficient for the free-molecular regime. The monomer concentration n_1 is calculated directly from the rate of formation J of the particle species. N_∞ is the local particle number concentration, A_∞ is the particle surface area, and a_1 is the monomer surface area. The characteristic coalescence time τ_C can be described by

$$\tau_C = A_C T \cdot d_{\mathrm{pp}}^4 \exp\left(-\frac{T_a}{T}\right)$$

In case of $Fe(CO)_5$-doped combustion the parameter A_C and T_a were fitted using the experimental results from [85] as an objective function.

This model is able to describe the particle number density and particle diameter (volume) accounting for the effects of coagulation and coalescence. Under the assumption of a steady state transport and a one-dimensional character of the flow in a flat flame burner, this model can be simplified to a set of ordinary differential equations which are solved using a four-step Runge-Kutta method.In this study we present a comprehensive simulation approach for the precursor reaction and its interaction with the flame, the formation and growth of iron oxide nanoparticles. The results are compared and validated using the large data base compiled by the authors. In a first step the mechanism for combustion of $Fe(CO)_5$ in a H_2/O_2 flame is designed using the experimental results presented in this study and in [46] as an objective function. The resulting detailed mechanism is then automatically reduced for implementation in a multidimensional fluid dynamics calculation for the reactor conditions reported in [85] in the second step. In the third step of our investigation population balances of the particle growth are computed for the calculated velocity, temperature, and concentrations field.

1.6.1 Modeling and Validation of the Fe Chemistry in Flames

IPC is one of the preferred precursors for iron and iron-oxide particle synthesis in various high-temperature processes. The synthesis of metal oxide nanoparticles in the gas-phase is strongly controlled by the kinetics of the precursor decomposition and the following reactions. Most investigations of particle synthesis in flame reactors neglect the interaction of the precursor species with the flame chemistry at low concentrations. In case of $Fe(CO)_5$, nearly any presence of iron species influences the flame significantly. Therefore, the incorporation of the iron species kinetics into the flame chemistry is crucial.

The strong influence of iron-containing compounds on flames was recognized already in the 1960s. In his pioneering work Wagner and co-workers [65, 86] found that $Fe(CO)_5$ has a much stronger influence on the burning velocity of premixed flames than halogenated hydrocarbons or other metal-containing compounds. In their work, Wagner et al. studied not only global burning velocities of doped flames, but also aimed at understanding the effects in detail. They measured flame temperatures by OH absorption spectroscopy, concentrations of OH radicals and iron atoms by absorption spectroscopy as well as the distribution of FeO by emission spectroscopy. Regarding the effect on the flame speed, Rumminger et al. investigated the inhibition mechanism in some detail [64] and suggested that the flame speed reduction is caused by iron-containing gas-phase compounds like FeO and FeOH reacting in catalytic cycles with flame carrier radicals like H and OH. Interestingly, an opposite effect of reduced ignition delay times was reported by Linteris and Babushok [79] in particular for lean H_2/air flames. Dicyclopentadiene iron, also known as ferrocene, has also been shown to have a similar flame speed reducing effect to IPC [87].

The quantitative description of the formation of iron-containing nanoparticles in flames also requires a detailed understanding of the influence of iron compounds on flame chemistry. The flame that is investigated in the present study is the central part of a low-pressure flame reactor for nanoparticle synthesis that has been successfully used for the synthesis of a wide variety of highly specific oxidic nanoparticles with variable sizes and compositions. The formation of iron oxide nanoparticles from an $Fe(CO)_5$-doped flame was investigated by Janzen et al. [85] in similar configurations to the work of Hecht et al. [46]. In his setup the particles were extracted from the flame and characterized using a PMS. In this study also simulations of particle growth were presented using a sectional model of the population balance equations. Nevertheless, the author postulated an instantaneous formation of iron oxide and applied only a strongly simplified mechanism for $Fe(CO)_5$ combustion without any interaction with flame species. The flame temperature was predicted from one dimensional, adiabatic flame assumption using a H_2/O_2 combustion mechanism only.

The experimental setup described in Sect. 1.5 allowed a one-dimensional simulation of the flame with fixed temperature profiles from the corresponding experimental measurements. These temperature profiles originate from the same experiment as the Fe-atom concentration measurements and provide ideal validation data for development and testing of reaction mechanisms. Temperature data were smoothed before their application as reference input profiles. This smoothing was necessary to provide a continuous function of temperature to the equation solver and a smooth objective function for the Fe-atom concentration. Any additional filtering of information from the source data lowered the reliability of the validation. In the present study a good compromise between the uncertainty caused by experimental noise and the uncertainty of the modeled reaction mechanism was found.

The first step in our simulation strategy was performed using methods provided with the chemical kinetics software library Cantera. In order to shorten the computational time, mixture-averaged transport coefficients were used to calculate the diffusive transport. For the first compilation of a reaction mechanism we used the established $Fe(CO)_5$ reaction mechanism of Rumminger et al. [64]. It contains 55 reactions and 12 iron species, including $Fe(CO)_5$. The H_2-O_2-CO system was modeled by the mechanism provided by Mueller et al. [88]. The first simulation efforts with the original mechanism showed much higher concentrations of iron atoms than measured. Also, the initial temperature of the thermal decomposition of $Fe(CO)_5$ contradicted our experimental data. In order to reduce the discrepancies between simulation and measurement in a first attempt we searched for newer, experimentally determined thermochemical data. Data for Fe, FeO and $Fe(OH)_2$ were therefore updated with the current values from the NASA database [89]. The result of this modification is shown in Fig. 1.9. For sake of clarity, only results from the flame doped with 170 ppm $Fe(CO)_5$ are presented here in comparison to the experimental data. The calculated maximum concentration is now nearly identical with the measured one, but the initial decomposition temperature of $Fe(CO)_5$ is still higher than expected from experience (i.e. the measured concentrations increase at lower heights above the burner (HAB) compared to the simulation).

Fig. 1.9 Iron-atom concentrations calculated with the original mechanism (**a**), the original mechanism with modified thermochemical data (**b**) and the reduced mechanism (**c**). The reduced mechanism uses also the modified coefficients for Fe(CO)$_5$ pyrolysis. Simulation results of (**b**) and (**c**) are practically identical

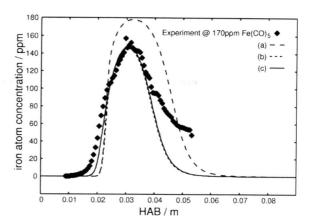

In the work that describes the chemical mechanism used here, Rumminger et al. used bond energies as activation energy in the sequential breaking of the Fe–CO bonds during the decomposition of Fe(CO)$_5$. This sequential process could be replaced by experimental data for the global decomposition reaction

$$Fe(CO)_5 \rightarrow Fe + 5CO$$

proposed by Woiki et al. [24]. They have determined the pre-exponential factor and the activation energy for such a reaction in extensive shock tube experiments to

$$k = 1.93 \times 10^9 \exp(-17400\,\text{cal}/RT)[\text{s}^{-1}].$$

This global decomposition step was incorporated in the iron mechanism. As a result the decomposition of Fe(CO)$_5$ now begins at lower HAB at 450 K fits much better to observations made, e.g., in hot-wall reactors for iron and iron oxide particle synthesis [90]. The earlier decomposition of Fe(CO)$_5$ does not affect the flame chemistry significantly and has only a small (indirect) effect on the flame temperature or the inhibition effects, but is very important for a large set of particle synthesis processes.

The resulting mechanism, already presented in [91, 92] still lacks of the iron oxide formation model, which is necessary for the description of iron oxide particle formation and growth. According to the simulations of the investigated flame, the results are reliable up to an HAB of 50 mm in the 170 ppm case. After this point usually occurrence of iron oxide particles is measured. Therefore, a hypothetical Fe$_2$O$_3$ gaseous species is assumed together with a set of reactions leading to this species from combinatorial considerations. As the reaction parameters of these reactions are unknown, an optimization process is applied with the initially calculated Fe-atom concentration and temperature as objective function and the auxiliary condition to maximize the Fe$_2$O$_3$ concentration. According to this strategy the resulting model should still correctly describe the interaction of the Fe-species with the flame

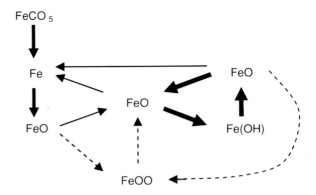

Fig. 1.10 Reaction path in the early flame at 15 mm HAB and 170 ppm Fe(CO)$_5$. The line thickness symbolizes the flux. The FeOOH branch is not included in the final, reduced mechanism

radicals and provide a plausible source term for the nucleation of iron oxide particles. A comparison of the calculations performed with both mechanisms is shown below.

The application of the reaction mechanism into CFD codes for simulation of the laminar reacting flow requires a reduction in number of species and reactions. The specific problem requires a finite rate chemistry model in order to describe the flame correctly. Therefore, the reduced model must also be "detailed" at least in the investigated parameter range of pressure and iron oxide precursor concentrations.

The strategy we chose for the reduction was a brute force method of alternating single reaction exclusion. This process was automated by a script and compared to the simulation based on the original model including the previously discussed modifications. The effect of the modifications was also monitored via an automated reaction path analysis. Reaction paths were calculated in $\Delta x = 1$ mm steps along the spatial coordinate of the 1D flame, to ensure the validity of the currently tested mechanism over a range of temperatures and species compositions. The process was monitored only at fine tuning, like removal of species from the mechanism. In its leanest variant the mechanism consists of 19 species and 23 reactions including the H$_2$/O$_2$- and CO-combustion sub-mechanisms. The H-atom recombination cycle

$$11: FeOH + H \leftrightarrow FeO + H_2$$
$$4: FeO + H_2O \leftrightarrow Fe(OH)_2$$
$$12: Fe(OH)_2 + H \leftrightarrow FeOH + H_2O$$
$$i: H+ \leftrightarrow H_2$$

responsible for the inhibition mechanism was already found by Jensen and Jones [93] and was also identified in the present analysis. There is no further reduction of the mechanism possible for an application in gas-phase synthesis reactors. Figures 1.10 and 1.11 illustrate the reaction paths at different locations in the flame with 170 ppm Fe(CO)$_5$. The FeOOH branch is drawn in the path diagrams, but excluded as a species of low importance from the final, reduced mechanism. The FeH reactions

1 Synthesis of Tailored Nanoparticles in Flames

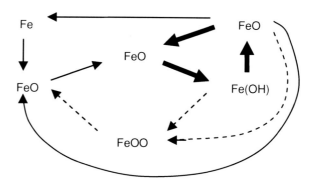

Fig. 1.11 Reaction path in the late flame at 65 mm HAB and 170 ppm Fe(CO)$_5$. The line thickness symbolizes the flux. The FeOOH branch is not included in the final, reduced mechanism

Table 1.1 The reduced mechanism for iron species from Fe(CO)$_5$ in a flame

#	Reaction	A	E_a	Ref
1	Fe(CO)$_5$ = Fe + 5CO	1.93E09	17.4	[24]
2	Fe + O$_2$ = FeO + O	1.26E14	20.0	[94]
3	Fe + O$_2$ + M = FeO$_2$ + M; k_∞	2.0E13	0	[95]
	k_0	1.5E18	4.0	[96]
4	FeO + H$_2$O = Fe(OH)$_2$	1.63E13	0	[93]
5	FeO + H = Fe + OH	1.0E14	6.0	e
6	FeO + H$_2$ = Fe + H$_2$O	1.0E13	5.0	[26]
7	FeO$_2$ + OH = FeOH + O$_2$	1.0E13	12.0	e
8	FeO$_2$ + O = FeO + O$_2$	1.5E14	1.5	e
9	FeOH + O = FeO + OH	5.0E13	1.5	e
10	FeOH + H = Fe + H$_2$O	1.2E12	1.2	e
11	FeOH + H = FeO + H$_2$	1.5E14	1.6	[93]
12	Fe(OH)$_2$ + H = FeOH + H$_2$O	2.0E14	0.6	[93]
13	Fe(OH)$_2$ + FeO = Fe$_2$OOH	1.0E11	0	e
14	2Fe(OH)$_2$ = Fe$_2$O(OH)$_2$ + H$_2$O	5.0E12	0	e
15	Fe$_2$O(OH)$_2$ = Fe$_2$OOOH + H	5.0E12	0	e
16	Fe$_2$OOOH + OH → Fe$_2$O$_3$ + H$_2$O	5.0E12	0	e

$k_f = A \exp(-E_a/RT)$, units: cm-s-mol-kcal, e: estimated in [64], e: estimations in this work

were completely neglected. Table 1.1 lists the set of iron species reactions in the reduced mechanism including the proposed Fe$_2$O$_3$ formation mechanism.

The reaction path analysis (illustrated in Figs. 1.5, 1.6) indicates that the iron conversion after the decomposition of Fe(CO)$_5$ takes place on a path similar to that proposed by Rumminger et al. [64]. The analysis of the model results with respect to flame radicals shows very little change in O and OH concentrations but a significant decrease of H-atom concentration with higher Fe(CO)$_5$ addition. This might be due to the cycle of reactions 11, 4 and 12, leading to the recombination of H radicals. Rather than disrupting the chain reaction and thus reducing the flame speed, the reaction enthalpy of reaction "i" (H + H ↔ H$_2$) is converted into an increased flame temperature under nearly adiabatic conditions. Hence, the sharper rise in the flame temperature with increasing $Fe(CO)_5$ addition may be caused by the exothermic recombination of H atoms. The reaction path analysis also shows that the reaction $FeO + O_2 \rightarrow FeO_2 + O$ that leads to an early increase in O-radical concentration does not play a major role in this flame in contrast to the predictions by Linteris and Babushok [79] who considered this reaction to be important in lean flames.

1.6.2 CFD Simulation of Reactive Flows in Nanoparticle Synthesis

The results of our modeling are presented in three steps according to the mechanism compilation, its implementation into multidimensional CFD simulations and the subsequent modeling of particle formation and growth. The validation of the improved and newly compiled combustion kinetics model of a lean, low pressure H_2/O_2 flame doped with IPC was presented in [91, 92]. These results were used as an objective function for development of a model for Fe_2O_3 formation from the gas phase as previously described. Figure 1.12 displays the calculated iron species concentrations with Fe_2O_3 formation model compared to the original mechanism, lacking the Fe_2O_3 formation. As demanded from the algorithm, the iron species concentrations in the early flame remain unaffected by the newly added set of reactions. The formation of Fe_2O_3 and consumption of $Fe(OH)_x$ species happens at plausible rates.

The reduced mechanism was implemented into the finite rate chemistry model of Ansys-Fluent CFD solver. The simulations were carried out for a pure H_2/O_2 flame and for the $Fe(CO)_5$-doped flame with boundary conditions from [85]. Aim of the numerical experiment was to reconstruct the two-dimensional field of transport velocities, temperature, and species concentrations in the cited experiments for subsequent modeling of particle dynamics. Figure 1.13 shows the result for a rotationally-symmetric calculation of the simplified reactor geometry.

The results from two-dimensional simulations were extracted for a single streamline—in the present case, the symmetry axis. Figure 1.14 shows the temperature and species concentrations extracted from the CFD calculation. Note that the flame front is parallel and close to the burner head. Because the experimental flame may be affected by deposits in the burner matrix and at the reactor walls, this result may not reflect the real conditions perfectly.

The simulation results in Fig. 1.15 from the population balance model show plausible particle diameters. Nevertheless, there ist a significant discrepancy between the experimental results and the model. To our opinion, this difference results mainly from the possible lift of the flame at the burner, which has been frequently observed, while the conditions of the simulations are "perfectly clean" and well defined, but not always well known.

Note that a perfect reconstruction of the velocity, temperature and concentration fields in past experiments is a challenging task. Usually, the boundary conditions are not well defined, i.e., the ambient conditions and the conditions of the burner matrix. Also, the movable skimmer nozzle interacts strongly with the flow. This influence was not considered in our simulation due to the enormous computational time.

1 Synthesis of Tailored Nanoparticles in Flames 31

Fig. 1.12 Concentrations of the main Fe-containing species in the model flame as a function of height above burner (HAB). The upper diagram was calculated without a Fe_2O_3 formation model. The lower diagram displays the consumption of $Fe(OH)_x$ species and formation of (gaseous) iron oxide

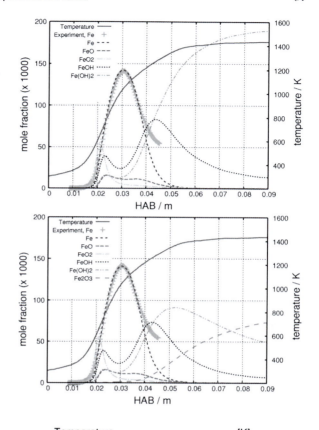

Fig. 1.13 Simulated temperature field for the low-pressure flame. *Lower part* shows the pure flame, *upper part* shows the doped flame. The temperature difference of $\Delta T = 120$ K corresponds well to experimental observations

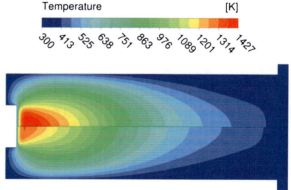

Future work will extend the Fe_2O_3 formation model and implement a more sophisticated model for particle population balances to access the local size distribution function.

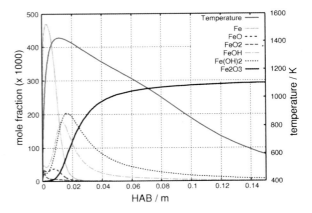

Fig. 1.14 Temperature and gaseous species concentrations at the symmetry axis from the CFD calculation of the reactor conditions described in [85] as a function of HAB

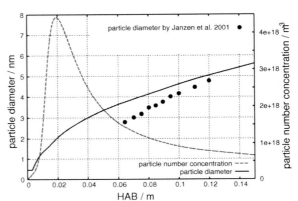

Fig. 1.15 Simulation result from the population balance model. The mean particle diameter and the particle number concentrations are calculated in a one-dimensional geometry at the reactor axis as a function of HAB

1.7 Examples of Flame-Synthesized Particles

Many of the early experimental studies on the gas phase combustion synthesis of particles were initiated and accelerated by industrial needs. As reviewed by Pratsinis [97], the historic evolution was motivated by the industrial importance of fumed silica, titania, and alumina produced in large scale flame reactors. Since the early 1990s combustion scientists became more interested in this subject, and research was intensified towards manufacturing advanced, tailored materials. New routes for the synthesis of non-oxidic ceramic particles in self-sustaining chemical systems were explored by Calcote et al. [98] and realized by, e.g., Axelbaum and coworkers, e.g. [99, 100], by Glassman et al. [101, 102], and by Gerhold and Inkrott [103]. Detailed information can be found in a review article by Brezinsky [104].

The increasing investigation of nanoparticles from the wet phase as well as gas-phase synthesis is accompanied with a more and more detailed understanding of size-dependent nanoparticle properties such as conductivity or optical and catalytic properties. In parallel to the advanced understanding of particle formation processes,

1 Synthesis of Tailored Nanoparticles in Flames

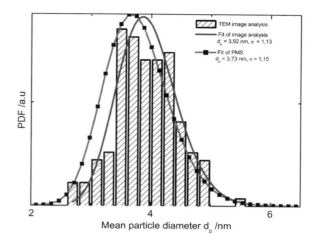

Fig. 1.16 Particle size distribution given as probability density functions (PDF): Comparison between PMS measurement and TEM analysis for the formation of SnO$_2$ nanoparticles taken at 110 mm height above burner. PMS measurement and TEM image analysis were fitted to a log-normal particle size distribution

an increasing demand for directed synthesis of specific nanoparticles with tuned properties has evolved. A main focus of our own work over the past years is also targeted towards a precise understanding of the physics and chemistry of the oxidic particles themselves and methods were developed to synthesize tailor-made nanoparticles. To this goal, high purity conditions are mandatory and impurities that might originate from the combustion process itself were neglected as far as possible. Spicer et al. have studied the formation of SiO$_2$ nanoparticles in a premixed acetylene/oxygen flame [105] and found that under certain conditions silica particles containing some carbon were achieved. We tried to avoid uncertainties resulting from such carbon-containing flames and complex hydrocarbon flame chemistry probably interacting with particle formation. Therefore, most of the results shown are received from experiments carried out in low-pressure premixed H$_2$/O$_2$ flames.

The particle formation and growth within the premixed flame reactor were observed in situ by means of the PMS described above. It enables to measure the initially formed particles evolving within the first few centimeters downstream the flame-stabilizing burner head [32]. A typical result of such a PMS measurement taken during the formation of SnO$_2$ nanoparticles from tetramethyltin Sn(CH$_3$)$_4$ at a reactor pressure of 30 mbar is shown in Fig. 1.16. Both, PMS measurements and TEM investigation of thermophoretically sampled particles are always found in good agreement with respect to mean particle diameter d_P and the geometric standard deviation σ.

Under the given conditions, reasonable particle growth is observed starting from about 80 mm downstream the burner head (Flow coordinate, HAB) as shown in Fig. 1.17. The specification of the PMS enables for the measurement of particles sizing between 2 and 15 nm.

For further investigation, the as-prepared materials were usually sampled by thermophoretic deposition of the particles downstream the reactor either on TEM grids or on a cooled substrate. From Fig. 1.17 it is clear and found as a common result that

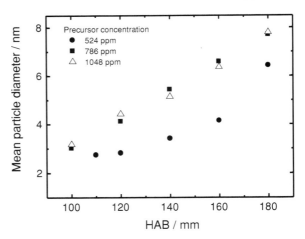

Fig. 1.17 Particle growth of tin oxide (SnO$_2$) nanoparticles from tetratmethyltin (Sn(CH$_3$)$_4$) measured downstream the flame reactor with a PMS

particles grow with increasing distance from the burner head (increasing residence time).

1.7.1 Semi-Conducting Single Oxide Particles

The classical and most extensively studied single oxide, synthesized in nearly every type of flames is SiO$_2$. Typical precursors used are silane (SiH$_4$) and chlorosilanes (SiCl$_{4-x}$H$_x$) or alkylchlorosilanes (SiCl$_{4-x}$(CH$_3$)$_x$); typical flame gases are either H$_2$/O$_2$ or CH$_4$/O$_2$ and we also investigated the formation of primary silica particles in doped, low-pressure premixed H$_2$/O$_2$/Ar flames [61]. As described above, the growth of particles along the flame coordinate was followed by molecular beam sampling and PMS analysis.

Besides silica, many other single oxides have been synthesized in laboratory flames including TiO$_2$, Al$_2$O$_3$, GeO$_2$, PbO, V$_2$O$_5$, Fe$_2$O$_3$, SnO$_2$, ZrO$_2$, ZnO, WO$_3$, see reviews [1–6]. The picture which can be drawn from all these investigations is relatively consistent. The most important parameters determining particle size and morphology are the concentration of precursor and the combined influence of flame temperature (particle temperature) and the residence time of the particles. Burner parameters seem to have only a minor influence as long as sufficient oxygen for precursor decomposition and oxide particle formation is available. The formation of stable and fully oxidized nanoparticles is normally obtained in flames burning under lean conditions. A typical overall equivalence ratio for particle formation in flames is about 0.5, indicating that two times more oxygen is present as necessary to fully oxidize fuel and precursor.

Nevertheless, the properties of the reacting flow in which the particles are synthesized can have a strong effect on the properties of the produced particles. An

1 Synthesis of Tailored Nanoparticles in Flames

early example is the formation of superconducting oxides from $H_2/O_2/Ar$ flames [106, 107]. The stoichiometry, i.e. the oxygen content in the particles, depends on the concentration of oxygen and oxidizing species in the flame and determines the success of the synthesis process. The stoichiometry of oxide particles synthesized in flames affect their crystal structure and lattice parameters. Many metals exist in various oxidation states and can form a variety of sub-oxides. In case of semiconducting oxides, oxygen vacancies in the particles lead to an increased charge carrier density. This is of high interest in the development of functional materials from flame-made nanoparticles for gas sensing and optoelectronics devices. A good example is the formation of the tin oxides SnO and SnO_2 [108, 109]. It has been demonstrated that by varying the combustion properties of the multi-element diffusion flame burner, both stable oxides were obtained [87].

Almost all oxidic semiconductors are n-type conductors, which means, that the majority charge carriers are electrons. The electrons in the oxides originate from oxygen that is being released from the lattice leaving behind two charge carriers. As a result, the adjustment of specific properties with respect to the electrical properties requires a tuning of the particle composition, especially its oxygen content during particle formation.

Laminar, premixed flames burning with various fuel/oxygen ratios seem are an ideal tool for synthesizing metal oxide particles with various stoichiometries [87, 110–112]. The combination of basic combustion parameters with the requirements for particle properties is crucial for the synthesis of well-defined materials. In accordance to [111], the equivalence ratio ϕ is defined as

$$\phi = \frac{\text{mole of oxygen required for complete combustion}}{\text{mole of oxygen supplied}} \tag{1}$$

with $\phi < 1$ and $\phi > 1$ representing lean and rich conditions, respectively.

A 1-D simulation of the gas-phase species that appear in H_2/O_2 flames typically used for nanoparticle synthesis was performed with Chemkin's Premix^{TM}. The underlying gas-phase temperature used for the calculation was measured with laser induced fluorescence of traces of NO added to the burning gas as described above [45]. Despite the concentration of the reactants H_2 and O_2, the diluent Ar, and the product water, only hydrogen, oxygen, and OH radicals show mentionable concentrations along the reaction path while the concentrations of HO_2 as well as H_2O_2 are negligible. Except for molecular oxygen, species with oxidizing and reducing properties for $\phi = 0.5$ and 0.85 are shown in Fig. 1.18.

The simulations shown in Fig. 1.18 indicate that in case of $\phi = 0.5$ the concentration of the oxidizing radicals $O^{\bullet}$ and $OH^{\bullet}$ (filled symbols) reaches the maximum shortly after entering the reaction zone while the portion of the reducing gases H and H_2 vanishes very fast. In case of $\phi = 0.85$, the concentration of O/OH compared to H/H_2 is clearly lower. It is of vital importance that the concentration of $OH^{\bullet}$ can be adjusted by changing ϕ as can be seen for the results found for HAB>25 mm. A high concentration of oxidizing species is vital for the oxidation process and their concentration with respect to height above burner is indicative of the degree of

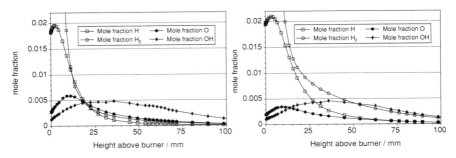

Fig. 1.18 Mole fraction of O•, OH•, H• and H_2 in a $H_2/O_2/Ar$ premixed flame calculated with PREMIX™ for $\phi = 0.5$ (*left*) and 0.85 (*right*)

Table 1.2 Resistivity of as-synthesized tin oxide nanoparticles under ambient conditions synthesized at $0.98 < \phi < 0.25$

Stoichiometry	ϕ	Crystal structure	Color	Resistivity/Ωm
SnO_2	0.25	Tetragonal (Rutile)	White	14.8×10^3
$SnO_{1.74}$	0.63	Tetragonal (Rutile)	Light yellow	7.4×10^3
$SnO_{1.4}$	0.85	Tetragonal (PbO)	Light grey	4.8×10^3
SnO	0.98	Tetragonal (PbO)	Grey	0.98×10^3

oxidation at the position. Hence, the oxidation of the metal oxide precursor is assumed to be highly sensitive to the oxidation potential at the early particle formation. As a result, electrical as well as optical properties change due to the conditions during particle growth [87].

Table 1.2 summarizes some of the results received from electrical measurements on tin oxide powders synthesized between $\phi = 0.98$ and 0.25.

The key parameter which controls the oxygen content in the particles is the H_2/O_2 ratio of the unburned gas. The composition of the particles was carefully determined by various physical methods including Auger electron spectroscopy to measure their exact stoichiometry. As studied for many metals, atomic as well as molecular oxygen play an important role for the oxidation of metal atoms and suboxides, respectively [113, 114]. Therefore, the oxidation process leading to tin dioxide is expected to be incomplete for lack of oxidizing species. Hence, it is understandable that almost no tin dioxide was found at $\phi > 0.85$ due to the fact, that mainly tin monoxide is formed [115].

A further example of stoichiometry tuning by combustion is the synthesis of WO_{3-x} in a premixed flame at ϕ between 0.8 and 0.33 [116]. While the lean conditions lead to bright white stoichiometric WO_3, the decrease in oxidizer helps to form sub-stoichiometric, blue-colored WO_{3-x}. This is especially surprising as fully oxidized tungsten hexafluoride (WF_6) was used as precursor. XPS measurements revealed an amount of about 20% of W^{+V} substituting W^{+VI} for materials

1 Synthesis of Tailored Nanoparticles in Flames

produced at $\phi = 0.8$ and it has been demonstrated by conductivity measurements that the particle resistance depends on x.

1.7.2 Mixed Oxide and Composite Nanoparticles

Crystalline mixed oxides consist of materials that form a common lattice structure with positive and negative ions being homogeneously distributed within the lattice or sitting at defined places within the structure, while nanocomposites consist of nanoparticles with two different crystal phases which are mixed on a nanometer length scale. Mixed oxides as well as nanocomposites are of huge pratical interest, especially for catalytic applications and for tuning of physicochemical properties, e.g. for high-temperature superconductors or super-paramagnetic materials. A highly interesting point with respect to nanocomposite materials is the fact, that an additional functionality can be embedded within a known material without changing its established properties. Flame synthesis has been successfully applied to produce mixed oxide powders both in laboratory, see e.g. [117–120], and industrial quantities [121].

Katz and coworkers [122, 123] are some of the pioneers who synthesized TiO_2/SiO_2 nanocomposites in their counterflow diffusion flame. Friedlander and coworkers [124, 125] later extended the understanding of how mixed particles form. They discussed the influence of precursor chemistry, temperature, and thermodynamics on the intraparticle and interparticle homogeneity. The miscibility/immiscibility of the two compounds, together with intraparticle transport, can limit the approach of the equilibrium phase distribution.

The motivation for the synthesis of mixed and composite oxides is mostly rooted in materials research. Consequently, multiple mixed oxides and nanocomposites like V_2O_5/TiO_2, V_2O_5/Al_2O_3, SiO_2/GeO_2, TiO_2/Al_2O_3, TiO_2/SnO_2, Ta_2O_5/SiO_2 have been synthesized in coflow and counterflow diffusion flames as well as in spray flames, see e.g. [119, 122, 123, 126, 127]. Kinetic and/or temperature control of particle formation in flame reactors not only opens a broad variety with respect to their oxygen content as discussed above, but also enables for the formation of specific nanocomposites. It is known that, depending on flame conditions and temperature, different types of composite materials are accessible [122, 123]. As an example for the change in physico-chemical properties depending on the composition of a mixed oxide we will discuss the systems TiO_2/SnO_2 in more detail.

1.7.2.1 Mixed SnO_2/TiO_2 Particles

The phase diagram of SnO_2/TiO_2 shows a big miscibility gap (cf. Fig. 1.19) and in thermodynamic equilibrium a segregation of SnO_2 and TiO_2 is observed.

A couple of scientists have investigated the properties and main differences of SnO_2/TiO_2 *composite* particles and films compared to $Ti_{1-x}Sn_xO_2$ *mixed oxides* with

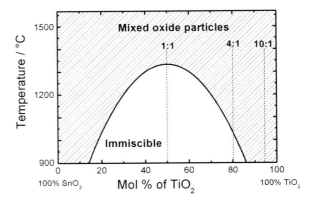

Fig. 1.19 Phase diagram of the SnO$_2$/TiO$_2$ system, adapted from [128]

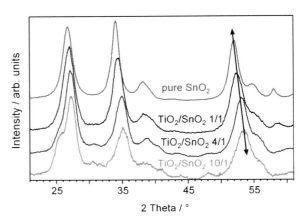

Fig. 1.20 X-ray diffraction pattern of SnO$_2$ and Ti$_{1-x}$Sn$_x$O$_2$ mixed oxides. It is obvious that the signal positions shift with changing phase composition in accordance to Vegard's law

respect to their photo-catalytic potential [129–131]. The nanocomposite materials support a spatial separation of electrons and holes preventing the excitons from recombination and increasing the photo-activity. In comparison to the composite material, the increase in photo-activity of the mixed oxides is expected to originate from an increase in band gap energy of the solid solution. While usually different routes for the formation of composites and mixed oxides are required, flame synthesis has the ability to produce both, composites as well as mixed oxides. Due to the high temperature gradient in flame reactors in the order of 10^4–10^6 K/s [132], phase compositions can be quenched without segregation and "freezing" the actual mixture. As has been shown by Akurati et al. [133], the flame spray synthesis of segregated TiO$_2$/SnO$_2$ nanocomposites is accessible probably due to a relatively low temperature within the particle formation zone, while experiments in our lab for TiO$_2$/SnO$_2$ mixtures ranging from 10/1 to 1/1 always showed complete solid solutions of both oxides [117]. The XRD investigations confirm this finding as the change in lattice constants of the synthesized materials behaves linear in accordance to Vegard's law (Fig. 1.20).

1 Synthesis of Tailored Nanoparticles in Flames 39

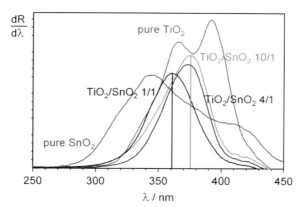

Fig. 1.21 Derivative of the UV-VIS absorption spectra of TiO₂, SnO₂ and several mixed oxides. As can be seen, the absorption of the mixed oxides shifts within the range given by the absorption spectra of SnO₂ and TiO₂, respectively

Additionally, it was found, that with increasing amount of tin substituting titanium, the absorption (and the respective band gap energy) shifts from 375 to 360 nm when changing the composition TiO₂/SnO₂ from 10/1 to 1/1, see Fig. 1.21.

Note that pure TiO₂ exhibits two absorption processes that can be attributed to the anatase structure (left maximum) and the rutile structure (right maximum) while the mixed oxides show only one maximum. This is due to the fact that the addition of tin suppresses the formation of anatase. This is also in accordance to the XRD investigations indicating that the mixed oxides form the rutile structure. As a main result it is found that the synthesis of mixed oxide nanoparticles in flames is possible despite the fact that they are not thermodynamically stable due to kinetic control. Moreover, changes in composition allows for tuning the band gap within the limits given by the respective materials used.

1.7.2.2 Fe₂O₃/SiO₂ Nanocomposites

An interesting example also illustrating the importance of gas-phase kinetics is the formation of Fe₂O₃/SiO₂ nanocomposites. They can be produced in premixed CH₄/O₂ or H₂/O₂/Ar low-pressure flames [120, 134]. If the evolution of different particles along the reaction path and, as a result, different time scales appear for the formation of the respective materials, composites consisting of one material homogeneously dispersed within a second material may result. This principle was used for the formation of Fe₂O₃/SiO₂ nanocomposites.

The thermal decomposition of the iron precursor Fe(CO)₅ with respect to the formation of iron oxide [24] is much faster than the radical driven decomposition of the Si precursor (Si₂(CH₃)₆, Si(CH₃)₄, or SiH₄), see e.g. [135]. While iron silicates like the Fe₂SiO₄ spinel that are found in nature usually have evolved under high pressure and in the presence of further cations like Mg or Ca, it is not expected that iron silicates are formed by clean flame synthesis. Consequently, the iron oxide particles are formed first in the flame and are later encapsulated by SiO₂. The commonly

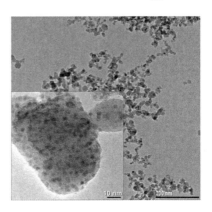

Fig. 1.22 TEM micrograph of γ-Fe$_2$O$_3$/SiO$_2$ nanoparticles. The overview shows the typical fractal geometry of fumed silica, while the inset gives a detailed insight into the homogeneous dispersion of iron oxide (*dark specks*) within the silica matrix

formed nanosized structure is γ-Fe$_2$O$_3$ (maghemite) which exhibits superparamagnetic behavior. This fact was used to engineer a material consisting of iron oxide homogeneously dispersed in silica. The material is of special interest as it allows obtaining particle-size-dependent magnetic properties. Superparamagnetic materials do not retain any magnetization in the absence of an external magnetic field. Therefore, they are of great practical interest for applications that require switchable magnetism, e.g. biomedical applications, such as magnetic resonance imaging, hyperthermia, separation and purification of biomolecules, and drug delivery. In many of these applications, however, the catalytic properties of iron oxides cannot be tolerated. The dilution of Fe$_2$O$_3$ in the silica matrix can be adjusted by means of the corresponding amount of precursor gases, their average spatial arrangement is directly accessible during the synthesis, while the concentration of the precursors is proportional to the particle size. Figure 1.22 shows a representative TEM image of a Fe$_2$O$_3$/SiO$_2$ composite material illustrating the distribution of iron oxide within the silica matrix. The inset illustrates very clearly, that the iron oxide nanoparticles are completely incorporated into the silica matrix. Thus, the chemistry of this "functionalized" silica is not different from that of conventional fumed silica.

SQUID measurements (superconducting quantum interference device) of the samples prove that the iron oxide particles are superparamagnetic and from XRD measurements a crystalline iron oxide phase is confirmed. From XRD investigations alone, it is not possible to distinguish between maghemite and magnetite (Fe$_3$O$_4$). However, Mößbauer spectroscopy at $T = 4.2$ K on pure iron-oxide nanoparticles which are prepared under otherwise comparable conditions had revealed a single sextet with homogeneous line broadening, indicating that the particles are single-phase maghemite [85]. Furthermore, since the composites were synthesized at $\phi = 0.4$, full oxidation of iron resulting in Fe$_2$O$_3$ is expected.

From a series of experiments with various precursor concentrations, Fe$_2$O$_3$/SiO$_2$ nanocomposites with similar loading but different size of Fe$_2$O$_3$ ranging from 9 to 28 nm in diameter were obtained. As expected for these spatially separated γ-Fe$_2$O$_3$

1 Synthesis of Tailored Nanoparticles in Flames

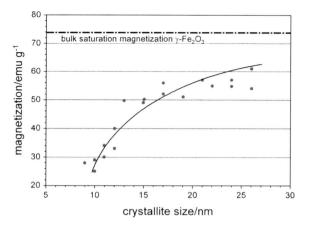

Fig. 1.23 Size-dependent saturation magnetization of γ-Fe_2O_3 nanoparticles dispersed in a silica matrix

nanocrystals, a strong relation between iron oxide particle size and magnetization was found as can be seen from Fig. 1.16.

Within the size range of iron oxide particles that have been prepared, the bulk value for the saturation magnetization of γ-Fe_2O_3 is not reached. Therefore, if a higher magnetization is required, materials with higher saturation are needed. Iron oxides enable this possibility with the aid of nanotechnology. While bulk magnetite Fe_3O_4 exhibits ferromagnetism at room temperature, nanosized magnetite switches to superparamagnetism with a size-depending blocking temperature [136]. In comparison to maghemite, the saturation magnetism of magnetite is significantly higher, achieving 92 emu/g. Hence, the tuning of the oxidation properties within a flame reactor, and therefore tuning the stoichiometry by adjusting the equivalence ratio ϕ during the synthesis becomes again important to force the formation of nanosized magnetite.

The Evonik Degussa company has used this strategy of product design to develop a superparamagnetic iron oxide/silica nanocomposite called MagSilica® [137]. It consists of nanosized magnetite and maghemite with iron oxide nanoparticles sizing in the range of 5–40 nm. The saturation magnetization of this material is higher than that of a pure Fe_2O_3/SiO_2 nanocomposite. Due to the fact that flame reactors enable a cost-effective and versatile industrial process with large production rates, first attempts are made to use this material in large-scale applications like ferrofluids and for the reinforcement of polymers and adhesives. Due to its superparamagnetic properties, the nanocomposite can be processed and manipulated by means of alternating electromagnetic fields enabling heating and hardening of materials that usually contain fumed silica as additives. Contactless hardening, bonding and heating of materials containing the composite instead of classical fumed silica become possible.

1.8 Conclusions

Particle synthesis in flames is an interdisciplinary research field that can be explored from two directions: materials science and combustion science. Materials science is interested in solid state materials, including nanomaterials with tailored new and combined properties. Flame synthesis is from this point of view only one, and often not the most attractive, synthesis route among various other possibilities. Due to the lack of solvents in gas-phase processes, they have an advantage when ultra-clean materials are required. Furthermore, being continuous flow processes, they provide the possibility for scaling production capacities to an industrial scale. Classical combustion science on the other hand considers particle formation mostly from the viewpoint of pollutant formation. Often, interest is limited to characterizing particle sizes, and strategies for using and applying the particles are not addressed properly enough. Pioneering work by Friedlander, Pratsinis, Roth and others helped to bridge the different fields from the viewpoint of chemical engineering.

Understanding the dynamics of larger, highly structured aggregates, which appear later in a flame, is not driven much by processes like chemical reactions. It is therefore obvious that the very early processes in a flame, like precursor decomposition, monomer formation, cluster kinetics, and the early formation of particles should be the natural playground of combustion researchers. As all the processes addressed are strongly controlled by temperature, this property must be carefully measured, together with the concentrations of species in the region of particle inception. This seems to be even more important for those particles with mixed components, where the chemistry becomes critical. Also experimental methods for sizing and collecting particles in the region of their birth are very necessary including shock tube and flow reactor techniques. Because of the diverse materials systems and the multitude of applications with all their specific needs, this research field is far from running out of interesting and important research topics.

Acknowledgments The financial support of this work through the German Research Foundation (DFG) within SFB445 is gratefully acknowledged.

References

1. S.E. Pratsinis, Flame aerosol synthesis of ceramic powders. Prog. Energy Combust. Sci. **24**(3), 197–219 (1998)
2. M.S. Wooldridge, Gas-phase combustion synthesis of particles. Prog. Energy Combust. Sci. **24**(1), 63–87 (1998)
3. P. Roth, Particle synthesis in flames. Proc. Combust. Inst. **31**, 1773–1788 (2007)
4. L. Madler, H.K. Kammler, R. Mueller, S.E. Pratsinis, Controlled synthesis of nanostructured particles by flame spray pyrolysis. J. Aerosol Sci. **33**(2), 369–389 (2002)
5. W.J. Stark, S.E. Pratsinis, Aerosol flame reactors for manufacture of nanoparticles. Powder Technol. **126**(2), 103–108 (2002)

1 Synthesis of Tailored Nanoparticles in Flames

6. D.E. Rosner, Flame synthesis of valuable nanoparticles: recent progress/current needs in areas of rate laws, population dynamics, and characterization. Ind. Eng. Chem. Res. **44**(16), 6045–6055 (2005)
7. B. Giesen, H. Wiggers, A. Kowalik, P. Roth, Formation of Si-nanoparticles in a microwave reactor: comparison between experiments and modelling. J. Nanopart. Res. **7**(1), 29–41 (2005)
8. K. Hitzbleck, H. Wiggers, P. Roth, Controlled formation and size-selected deposition of indium nanoparticles from a microwave flow reactor on semiconductor surfaces. Appl. Phys. Lett. **87**(9), 093105 (2005)
9. L.J. Kecskes, R.H. Woodman, S.F. Trevino, B.R. Klotz, S.G. Hirsch, B.L. Gersten, Characterization of a nanosized iron powder by comparative methods. Kona **21**, 143–150 (2003)
10. D.V. Szabo, D. Vollath, W. Arnold, Microwave plasma synthesis of nanoparticles: application of microwaves to produce new materials. Ceramic Trans. **111**(Microwaves: Theory and Application in Materials Processing V), 217–224 (2001)
11. G.D. Ulrich, Theory of particle formation and growth in oxide synthesis flames. Combust. Sci. Technol. **4**(2), 47–57 (1971)
12. G.D. Ulrich, Flame synthesis of fine particles. Chem. Eng. News **62**(32), 22–29 (1984)
13. J.J. Helble, A.F. Sarofim, Factors determining the primary particle size of flame-generated inorganic aerosols. J. Colloid Interf. Sci. **128**(2), 348–362 (1989)
14. T. Matsoukas, S.K. Friedlander, Dynamics of aerosol agglomerate formation. J. Coll. Int. Sci. **146**(2), 495–506 (1991)
15. D.R. Hardesty, F.J. Weinberg, Electrical control of particulate pollutants from flames. Proc. Combust. Inst. **14**(1), 907–918 (1973)
16. H.K. Kammler, R. Jossen, P.W. Morrison, S.E. Pratsinis, G. Beaucage, The effect of external electric fields during flame synthesis of titania. Powder Technol. **135**, 310–320 (2003)
17. J. Knipping, H. Wiggers, B.F. Kock, T. Hulser, B. Rellinghaus, P. Roth, Synthesis and characterization of nanowires formed by self-assembled iron particles. Nanotechnology **15**(11), 1665–1670 (2004)
18. S. Von Gersum, P. Roth, Oxidation of fullerene C_{60} behind shock waves: tunable diode laser measurements of CO and CO_2. Infrared Phys. Technol. **37**(1), 167–171 (1996)
19. S. Von Gersum, P. Roth, Soot oxidation in high temperature N_2O/Ar and NO/Ar mixtures. Proc. Combust. Inst. **24**(1), 999–1006 (1992)
20. M. Frenklach, L. Ting, H. Wang, M.J. Rabinowitz, Silicon particle formation in pyrolysis of silane and disilane. Isr. J. Chem. **36**(3), 293–303 (1996)
21. J. Herzler, R. Leiberich, H.J. Mick, P. Roth, Shock tube study of the formation of TiN molecules and particles. Nanostructured Mater. **10**(7), 1161–1171 (1999)
22. A. Giesen, J. Herzler, P. Roth, Kinetics of the Fe-atom condensation based on Fe-concentration measurements. J. Phys. Chem. A **107**(26), 5202–5207 (2003)
23. S.K. Friedlander, *Smoke, Dust, and Haze: Fundamentals of Aerosol Dynamics* (Oxford University Press, Oxford, 2000)
24. D. Woiki, A. Giesen, P. Roth, A shock tube study on the thermal decomposition of $Fe(CO)_5$. In: 23rd International Symposium on Shock Waves, Fort Worth, Texas (2001)
25. S.H. Bauer, D.J. Frurip, Homogeneous nucleation in metal vapors. 5. A self-consistent kinetic model. J. Phys. Chem. A **81**(10), 1015–1024 (1977)
26. D.E. Jensen, Condensation modeling for highly supersaturated vapors: application to iron. J. Chem. Soc. Faraday Trans. **76**(11), 1494–1515 (1980).
27. M. Fikri, M. Bozkurt, H. Somnitz, C. Schulz, High temperature shock-tube study of the reaction of gallium with ammonia. Phys. Chem. Chem. Phys. **13**(9), 4149–4154 (2011)
28. M. Fikri, A. Makeich, G. Rollmann, C. Schulz, P. Entel, Thermal decomposition of trimethylgallium $Ga(CH_3)_3$: a shock-tube study and first-principles calculations. J. Phys. Chem. A **112**(28), 6330–6337 (2008)
29. J. Herzler, P. Roth, C. Schulz, Kinetics of the reaction of Ga atoms with O_2 in The International Shock Wave Symposium, Bangalore, 2005
30. G.P. Smith, D.M. Golden, M. Frenklach, N.W. Moriarty, B. Eiteneer, M. Goldenberg, C.T. Bowman, R.K. Hanson, S. Song, W. C. Gardiner Jr., V.V. Lissianski, Z. Qin, http://www.me.berkeley.edu/gri_mech/

31. A. Giesen, D. Woiki, J. Herzler, P. Roth, Oxidation of Fe atoms by O_2 based on Fe- and O-concentration measurements. Proc. Combust. Inst. **29**, 1345–1352 (2002)

32. P. Roth, A. Hospital, Design and test of a particle mass-spectrometer (Pms). J. Aerosol Sci. **25**(1), 61–73 (1994)

33. J. Griesheimer, K.H. Homann, Large molecules, radicals ions, and small soot particles in fuel-rich hydrocarbon flames: Part II. Aromatic radicals and intermediate PAHs in a premixed low-pressure naphthalene/oxygen/argon flame. Proc. Combust. Inst. **27**, 1753–1759 (1998)

34. R. Humpfer, H. Oser, H.-H. Grotheer, T. Just, The reaction system CH_3+OH at intermediate temperatures. Appearance of a new product channel. Proc. Combust. Inst. **25**, 721–731 (1994)

35. I.K. Lee, M. Winterer, Aerosol mass spectrometer for the in situ analysis of chemical vapor synthesis processes in hot wall reactors. Rev. Sci. Instrum. **76**(9), 095194 (2005)

36. N.A. Fuchs, On the stationary charge distribution on aerosol particles in a bipolar ionic atmosphere. Pure Appl. Geophys. **56**(1), 185–193 (1963)

37. A. Hospital, P. Roth, In-situ mass growth measurements of charged soot particles from low pressure flames. Proc. Combust. Inst. **23**, 1573–1579 (1991)

38. T.P. Huelser, A. Lorke, P. Ifeacho, H. Wiggers, C. Schulz, Core and grain boundary sensitivity of tungsten-oxide sensor devices by molecular beam assisted particle deposition. J. Appl. Phys. **102**(12), 124305 (2007)

39. H. Maetzing, W. Baumann, M. Hauser, H.R. Paur, H. Seifert, A. Van Raaij, P. Roth, A mass spectrometer for nanoparticles. VDI-Berichte **1803** (Nanofair 2003: New Ideas for Industry, 2003), 327–330 (2003)

40. M.G.D. Strecker, P. Roth, in *Fine Solid Particles*, ed. by J. Schwedes, S. Bernotat (Shaker Verlag, Aachen, 1997)

41. A.P. George, R.D. Murley, E.R. Place, Formation of titanium dioxide aerosol from the combustion supported reaction of titanium tetrachloride and oxygen. Faraday Symposia Chem. Soc. **7**, 63–71 (1973)

42. R.A. Dobbins, C.M. Megaridis, Morphology of flame-generated soot as determined by thermophoretic sampling. Langmuir **3**(2), 254–259 (1987)

43. C. Janzen, H. Kleinwechter, J. Knipping, H. Wiggers, P. Roth, Size analysis in low-pressure nanoparticle reactors: comparison of particle mass spectrometry with in situ probing transmission electron microscopy. J. Aerosol Sci. **33**(6), 833–841 (2002)

44. K. Kohse-Höinghaus, R.S. Barlow, M. Aldén, J. Wolfrum, Combustion at the focus: laser diagnostics and control. Proc. Combust. Inst. **30**, 89–123 (2005)

45. H. Kronemayer, P. Ifeacho, C. Hecht, T. Dreier, H. Wiggers, C. Schulz, Gas-temperature imaging in a low-pressure flame reactor for nano-particle synthesis with multi-line NO-LIF thermometry. Appl. Phys. B **88**, 373–377 (2007)

46. C. Hecht, H. Kronemayer, T. Dreier, H. Wiggers, C. Schulz, Imaging measurements of atomic iron concentration with laser-induced fluorescence in a nano-particle synthesis flame reactor Appl. Phys. B **94**, 119–125 (2009)

47. B.F. Kock, C. Kayan, J. Knipping, H.R. Orthner, P. Roth, Comparison of LII and TEM sizing during synthesis of iron particle chains. Proc. Combust. Inst. **30**, 1689–1697 (2005)

48. M. Tamura, J. Luque, J.E. Harrington, P.A. Berg, G.P. Smith, J.B. Jeffries, D.R. Crosley, Laser-induced fluorescence of seeded nitric oxide as a flame thermometer. Appl. Phys. B **66**(4), 503–510 (1998)

49. W.G. Bessler, F. Hildenbrand, C. Schulz, Two-line laser-induced fluorescence imaging of vibrational temperatures of seeded NO. Appl. Opt. **40**, 748–756 (2001)

50. W.G. Bessler, C. Schulz, Quantitative multi-line NO-LIF temperature imaging. Appl. Phys. B **78**, 519–533 (2004)

51. R.W. Dibble, R.E. Hollenbach, Laser Rayleigh thermometry in turbulent flames. Proc. Combust. Inst. **18**, 1489–1499 (1981)

52. D. Hoffman, K.-U. Münch, A. Leipertz, Two-dimensional temperature determination in sooting flames by filtered Rayleigh scattering. Opt. Lett. **21**, 525–527 (1996)

53. K. Kohse-Höinghaus, J.B. Jeffries (eds.), *Applied Combustion Diagnostics* (Taylor and Francis, New York, 2002)

1 Synthesis of Tailored Nanoparticles in Flames

54. F. Beyrau, A. Bräuer, T. Seeger, A. Leipertz, Gas-phase temperature measurement in the vaporizing spray of a gasoline direct-injection injector by use of pure rotational coherent anti-Stokes Raman scattering. Opt. Lett. **29**(3), 247–249 (2003)
55. M.G. Allen, Diode laser absorption sensors for gas dynamic and combustion flows. Meas. Sci. Technol. **9**(4), 545–562 (1998)
56. A.O. Vyrodow, J. Heinze, M. Dillmann, U.E. Meier, W. Stricker, Laser-induced fluorescence thermometry and concentration measurements on NO A-X (0,0) transitions in the exhaust gas of high pressure CH_4/air flames. Appl. Phys. B. **61**, 409–414 (1995)
57. T. Lee, W.G. Bessler, H. Kronemayer, C. Schulz, J.B. Jeffries, R.K. Hanson, Quantitative temperature measurements in high-pressure flames with multi-line NO-LIF thermometry. Appl. Opt. **44**(31), 6718–6728 (2005)
58. I. Düwel, H.W. Ge, H. Kronemayer, R. Dibble, E. Gutheil, C. Schulz, J. Wolfrum, Experimental and numerical characterization of a turbulent spray flame. Proc. Combust. Inst. **31**, 2247–2255 (2007)
59. H. Kronemayer, W.G. Bessler, C. Schulz, Gas-phase temperature imaging in spray systems using multi-line NO-LIF thermometry. Appl. Phys. B **81**(8), 1071–1074 (2005)
60. W.G. Bessler, C. Schulz, V. Sick, J.W. Daily, A versatile modeling tool for nitric oxide LIF spectra (www.lifsim.com). In: 3rd Joint meeting of the US sections of The Combustion Institute, Chicago, 01–06 2003, pp. PI05
61. D. Lindackers, M.G.D. Strecker, P. Roth, C. Janzen, S.E. Pratsinis, Formation and growth of SiO_2 particles in low pressure $H_2/O_2/Ar$ flames doped with SiH_4. Combust. Sci. Technol. **123**(1–6), 287–315 (1997)
62. H. Kunieda, T.J. Turner, A. Hisamitsu, K. Koyama, R. Mushotzky, Y. Tsusaka, Rapid variabilita of the iron fluorescence line from the Seyfert 1 galaxy NGC6814. Nature **345**, 786–788 (1990)
63. Y. Ralchenko, NIST Atomic spectra Darabase (Version 3.1.2), [Online]. Available: http:// physics.nist.gov/asd3 [2007, August 1]. National Institute of Standards and Technology (2007)
64. M.D. Rumminger, D. Reinelt, V. Babushok, G.T. Linteris, Numerical study of the inhibition of premixed and diffusion flames by iron pentacarbonyl. Combust. Flame **116**(1–2), 207–219 (1999)
65. U. Bonne, W. Jost, H.G. Wagner, Iron pentacarbonyl in methane-oxygen (or air) flames. Fire Res. Abstr. Rev. **4**, 6–18 (1962)
66. K. Tian, Z.S. Li, S. Staude, B. Li, Z.W. Sun, A. Lantz, M. Aldén, B. Atakan, Influence of ferrocene addition to a laminar premixed propene flame: Laser diagnostics, mass spectrometry and numerical simulations. Proc. Combust. Inst. **32**, 445–452 (2009)
67. M.J. Dyer, D.R. Crosley, Two-dimensional imaging of OH laser-induced fluorescence in a flame. Opt. Lett. **7**(8), 382–384 (1982)
68. H.W. Kim, M. Choi, In situ line measurement of mean aggregate size and fractal dimension along the flame axis by planar laser light scattering. J. Aerosol Sci. **34**(12), 1633–1645 (2003)
69. R. Starke, P. Roth, Soot particle sizing by LII during shock tube pyrolysis of C_6H_6. Combust. Flame **127**(4), 2278–2285 (2001)
70. A. D'Anna, A. Rolando, C. Allouis, P. Minutolo, A. D'Alessio, Nano-organic carbon and soot particle measurements in a laminar ethylene diffusion flame. Proc. Combust. Inst. **30**, 1449–1456 (2005)
71. L.A. Melton, Soot diagnostics based on laser heating. Appl. Opt. **23**(13), 2201–2208 (1984)
72. C. Schulz, B.F. Kock, M. Hofmann, H.A. Michelsen, S. Will, B. Bougie, R. Suntz, G.J. Smallwood, Laser-induced incandescence: recent trends and current questions Appl. Phys. B **83**(3), 333–354 (2006)
73. A.V. Filippov, M.W. Markus, P. Roth, In situ characterization of ultrafine particles by laser-induced incandescence: sizing and particle structure determination. J. Aerosol Sci. **30**(1), 71–87 (1999)
74. A.V. Filippov, D.E. Rosner, Energy transfer between an aerosol particle and gas at high temperature ratios in the Knudsen transition regime. Int. J. Heat Mass Transf. **43**(1), 127–138 (2000)

75. R.L. van der Wal, Soot precursor material: Visualization via simultaneous lIF-LII and characterization via tem. Proc. Combust. Inst. **26**(2), 2269–2275 (1996)
76. S. Will, S. Schraml, K. Bader, A. Leipertz, Performance characteristics of soot primary particle size measurements by time-resolved laser-induced incandescence. Appl. Opt. **37**(24), 5647–5658 (1998)
77. C. Schulz, Laser-induced incandescence: quantitative interpretation, modeling, application. in Proceedings International Bunsen Discussion Meeting and Workshop, Duisburg, Germany, Sept 25–28, 2005, ISSN 1613–0073, Vol. 195, CEUR Workshop Proceedings, (http://CEUR-ws.org/Vol-195/)
78. H.A. Michelsen, F. Liu, B.F. Kock, H. Bladh, A. Boiarciuc, M. Charwath, T. Dreier, R. Hadef, M. Hofmann, J. Reimann, S. Will, P.-E. Bengtsson, H. Bockhorn, F. Foucher, K.-P. Geigle, C. Mounaïm-Rousselle, C. Schulz, R. Stirn, B. Tribalet, S. Suntz, Modeling laser-induced incandescence of soot: a summary and comparison of LII models. Appl. Phys. B **87**, 503–521 (2007)
79. G.T. Linteris, V.I. Babushok, Promotion or inhibition of hydrogen-air ignition by iron-containing compounds. Proc. Combust. Inst. **33**, 2535–2542 (2009)
80. A.D. Randolph, M.A. Larson, *Theory of Particulate Processes: Analysis and Techniques of Continuous Crystallization* (Academic Press, New York, 1971)
81. R. McGraw, Description of aerosol dynamics by the quadrature method of moments. Aerosol Sci. Technol. **27**, 255–265 (1997)
82. D.L. Marchisio, R.D. Virgil, R.O. Fox, Quadrature method of moments for aggregation-breakage processes. J. Coll. Int. Sci. **258**, 322–334 (2003)
83. D.L. Marchisio, R.O. Fox, Solution of population balance equations using the direct quadrature method of moments. J. Aerosol Sci. **36**(1), 43–73 (2005)
84. F.E. Kruis, K.A. Kusters, S.E. Pratsinis, B. Scareltt, A simple model for the evolution of the characteristics of aggregate particles undergoing coagulation and sintering. Aerosol Sci. Technol. **19**, 514–526 (1993)
85. C. Janzen, P. Roth, Formation and characteristics of Fe_2O_3 nano-particles in doped low pressure $H_2/O_2/Ar$ flames. Combust. Flame **125**(3), 1150–1161 (2001)
86. G. Lask, H.G. Wagner, Influence of additives on the velocity of laminar flames. Proc. Combust. Inst. **8**, 432–438 (1962)
87. P. Ifeacho, T. Huelser, H. Wiggers, C. Schulz, P. Roth, Synthesis of SnO_{2-x} nanoparticles tuned between $0 < x < 1$ in a premixed low pressure $H_2/O_2/Ar$ flame. Proc. Combust. Inst. **31**, 1805–1812 (2007)
88. M.A. Mueller, R.A. Yetter, F.L. Dryer, Flow reactor studies and kinetic modeling of the $H_2/O_2/NO_x$ and $CO/H_2O/O_2/NO_x$ reactions. Int. J. Chem. Kin. **31**(10), 705–724 (1999)
89. B.J. McBride, S. Gordon, M.A. Reno, Coefficients for Calculating Thermodynamic and Transport Properties of Individual Species. NASA, Report TM-4513 (1993)
90. Y. Sawada, Y. Kageyama, M. Iwata, A. Tasdaki, Synthesis and magnetic properties of ultrafine iron particles prepared by pyrolysis of carbonyl iron. Jap. J. Appl. Phys. **31**(12A), 3858 (1992)
91. S. Staude, C. Hecht, I. Wlokas, C. Schulz, B. Atakan, Experimental and numerical investigation of $Fe(CO)_5$ addition to a laminar premixed hydrogen/oxygen/argon flame. Z. Phys. Chem. **223**, 639–649 (2009)
92. I. Wlokas, C. Schulz, Model for the formation of Fe_2O_3 in premixed $Fe(CO)_5$-doped low-pressure H_2/O_2 flames. European Combustion Meeting (2011)
93. D.E. Jensen, G.A.J. Jones, Chem. Phys. **60**, 3421 (1974)
94. A. Fontijn, S.C. Kurzius, J.J. Houghton, Proc. Combust. Inst. **14**, 167 (1973)
95. M. Helmer, J.M.C. Plane, Experimental and theoretical study of the reaction $Fe + O_2 + N_2 - > FeO_2 + N_2$. J. Chem. Soc. Faraday Trans. **90**(3), 395–402 (1994)
96. U.S. Akhamodov, I.S. Zaslonko, V.N. Smirnov, Kinet. Catal. **29**(2), 291 (1988)
97. M.D. Allendorf, J.R. Bautista, E. Potkay, Temperature measurements in a vapor axial deposition flame by spontaneous Raman spectroscopy. J. Appl. Phys. **66**(10), 5046–5051 (1989)
98. H.F. Calcote, W. Felder, A new gas-phase combustion synthesis process for pure metals, alloys, and ceramics. Proc. Combust. Inst. **24**(1), 1869–1876 (1992)

1 Synthesis of Tailored Nanoparticles in Flames

99. R.L. Axelbaum, Synthesis of stable metal and non-oxide ceramic nanoparticles in sodium/halide flames. Powder Metallurgy **43**(4), 323–325 (2000)
100. R.L. Axelbaum, D.P. Dufaux, C.A. Frey, S.M.L. Sastry, A flame process for synthesis of unagglomerated, low-oxygen nanoparticles: application to Ti and TiB_2. Metall. Mat. Trans. B **28**(6), 1199–1211 (1997)
101. K. Brezinsky, J.A. Brehm, C.K. Law, I. Glassman, Supercritical combustion synthesis of titanium nitride. Proc. Combust. Inst. **26**(2), 1875–1881 (1996)
102. I. Glassman, K.A. Davis, K. Brezinsky, A gas-phase combustion synthesis process for non-oxide ceramics. Proc. Combust. Inst. **24**(1), 1877–1882 (1992)
103. B.W. Gerhold, K.E. Inkrott, Nonoxide ceramic powder synthesis. Combust. Flame **100**(1–2), 146–152 (1995)
104. K. Brezinsky, Gas-phase combustion synthesis of materials. Proc. Combust. Inst. **26**(2), 1805–1816 (1996)
105. P.T. Spicer, C. Artelt, S. Sanders, S.E. Pratsinis, Flame synthesis of composite carbon black-fumed silica nanostructured particles. J. Aerosol Sci. **29**(5–6), 647–659 (1998)
106. B.D. Merkle, R.N. Kniseley, F.A. Schmidt, I.E. Anderson, Superconducting yttrium barium copper oxide ($YBa_2Cu_3O_x$) particulate produced by total consumption burner processing. Mat. Sci. Eng. A **124**(1), 31–38 (1990)
107. M.R. Zachariah, S. Huzarewicz, Aerosol processing of yttrium barium copper oxide super-conductors in a flame reactor. J. Mat. Res. **6**(2), 264–269 (1991)
108. D.L. Hall, A.A. Wang, K.T. Joy, T.A. Miller, N.S. Wooldridge, Combustion synthesis and characterization of nanocrystalline tin and tin oxide (SnO^x, x=0-2) particles. J. Am. Ceramic Soc. **87**(11), 2033–2041 (2004)
109. T.A. Miller, S.D. Bakrania, C. Perez, M.S. Wooldridge, A new method for direct preparation of tin dioxide nanocomposite materials. J. Mat. Res. **20**(11), 2977–2987 (2005). doi:10.1557/jmr.2005.0375
110. M. Huber, W.J. Stark, S. Loher, M. Maciejewski, F. Krumeich, A. Baiker, Flame synthesis of calcium carbonate nanoparticles. Chem. Commun. **5**, 648–650 (2005)
111. R.N. Grass, W.J. Stark, Gas phase synthesis of fcc-cobalt nanoparticles. J. Mat. Chem. **16**(19), 1825–1830 (2006)
112. V. Simanzhenkov, H. Wiggers, P. Roth, Properties of flame synthesized germanium oxide nanoparticles. J. Nanosci. Nanotechnol. **5**(3), 436–441 (2005)
113. D.P. Belyung, A. Fontijn, The $AlO+O_2$ reaction system over a wide temperature-range. J. Phys. Chem. A **99**(32), 12225–12230 (1995)
114. A. Fontijn, Wide-temperature range observations on reactions of metal atoms and small radicals. Pure Appl. Chem. **70**(2), 469–476 (1998)
115. J. Herzler, P. Roth, High temperature gas phase reaction of SnO_g with O_2. Phys. Chem. Chem. Phys. **5**(8), 1552–1556 (2003)
116. A. Gupta, P. Ifeacho, C. Schulz, H. Wiggers, Synthesis of tailored WO_3 and WO_x (2.9 < x < 3) nanoparticles by adjusting the combustion conditions in a $H_2/O_2/Ar$ premixed flame reactor. Proc. Combust. Inst. **33**(2), 1883–1890 (2011)
117. P. Ifeacho, H. Wiggers, P. Roth, SnO_2/TiO_2 mixed oxide particles synthesized in doped premixed $H_2/O_2/Ar$ flames. Proc. Combust. Inst. **30**, 2577–2584 (2005)
118. B.K. McMillin, P. Biswas, M.R. Zachariah, In situ characterization of vapor phase growth of iron oxide-silica nanocomposites.1. 2-D planar laser-induced fluorescence and Mie imaging. J. Mat. Res. **11**(6), 1552–1561 (1996)
119. S. Vemury, S.E. Pratsinis, Dopants in flame synthesis of titania. J. Am. Ceramic Soc. **78**(11), 2984–2992 (1995)
120. M.R. Zachariah, M.I. Aquino, R.D. Shull, E.B. Steel, Formation of superparamagnetic nanocomposites from vapor-phase condensation in a flame. Nanostructured Mater. **5**(4), 383–392 (1995)
121. A. Gutsch, J. Averdung, H. Muehlenweg, From technical development to successful nanotechnological product. Chemie Ingenieur Technik **77**(9), 1377–1392 (2005)

122. C.H. Hung, J.L. Katz, Formation of mixed oxide powders in flames: Part I. Titania-silica. J. Mat. Res. **7**(7), 1861–1869 (1992)
123. C.H. Hung, P.F. Miquel, J.L. Katz, Formation of mixed-oxide powders in flames Part 2: SiO_2-GeO_2 and Al_2O_3-TiO_2. J. Mat. Res. **7**(7), 1870–1875 (1992)
124. S.H. Ehrman, S.K. Friedlander, M.R. Zachariah, Characteristics of SiO2/TiO2 nanocomposite particles formed in a premixed flat flame. J. Aerosol Sci. **29**(5–6), 687–706 (1998)
125. S.H. Ehrman, S.K. Friedlander, M.R. Zachariah, Phase segregation in binary SiO_2/TiO_2 and SiO_2/Fe_2O_3 nanoparticle aerosols formed in a premixed flame. J. Mat. Res. **14**(12), 4551–4561 (1999)
126. P.F. Miquel, J.L. Katz, Formation and characterization of nanostructured V-P-O particles in flames: a new route for the formation of catalysts. J. Mat. Res. **9**(3), 746–754 (1994)
127. H. Schulz, L. Mädler, S.E. Pratsinis, P. Burtscher, N. Moszner, Transparent nanocomposites of radiopaque, flame-made Ta_2O_5/SiO_2 particles in an acrylic matrix. Adv. Funct. Mat. **15**(5), 830–837 (2005)
128. N.N. Padurow, Mischbarkeit im System Rutil-Zinnstein. Naturwissenschaften **43**(17), 395–396 (1956)
129. J. Shang, W.Q. Yao, Y.F. Zhu, N.Z. Wu, Structure and photocatalytic performances of glass/SnO_2/TiO_2 interface composite film. Appl. Catalysis A **257**(1), 25–32 (2004)
130. J. Yang, D. Li, X. Wang, X.J. Yang, L.D. Lu, Rapid synthesis of nanocrystalline TiO_2/SnO_2 binary oxides and their photoinduced decomposition of methyl orange. J. Solid State Chem. **165**(1), 193–198 (2002)
131. J. Lin, J.C. Yu, D. Lo, S.K. Lam, Photocatalytic activity of rutile $Ti_{1-x}Sn_xO_2$ solid solutions. J. Catalysis **183**(2), 368–372 (1999)
132. S. Tsantilis, S.E. Pratsinis, Soft- and hard-agglomerate aerosols made at high temperatures. Langmuir **20**(14), 5933–5939 (2004)
133. K.K. Akurati, A. Vital, R. Hany, B. Bommer, T. Graule, M. Winterer, One-step flame synthesis of SnO_2/TiO_2 composite nanoparticles for photocatalytic applications. Int. J. Photoenergy **7**(4), 153–161 (2005)
134. C. Janzen, J. Knipping, B. Rellinghaus, P. Roth, Formation of silica-embedded iron-oxide nanoparticles in low-pressure flames. J. Nanopart. Res. **5**(5–6), 589–596 (2003)
135. C.J. Butler, A.N. Hayhurst, E.J.W. Wynn, The size and shape of silica particles produced in flames of H_2/O_2/N_2 with a silicon-containing additive. Proc. Combust. Inst. **29**, 1047–1054 (2002)
136. M. Stjerndahl, M. Andersson, H.E. Hall, D.M. Pajerowski, M.W. Meisel, R.S. Duran, Superparamagnetic Fe_3O_4/SiO_2 nanocomposites: enabling the tuning of both the iron oxide load and the size of the nanoparticles. Langmuir **24**(7), 3532–3536 (2008)
137. M. Kroell, M. Pridoehl, G. Zimmermann, L. Pop, S. Odenbach, A. Hartwig, Magnetic and rheological characterization of novel ferrofluids. J. Magn. Magn. Mater. **289**, 21–24 (2005)

Chapter 2
Chemical Vapor Synthesis of Nanocrystalline Oxides

Ruzica Djenadic and Markus Winterer

Abstract The generation of nanoparticles in the gas phase by Chemical Vapor Synthesis (CVS) may be described from the point of view of chemical engineering as a sequence of unit operations among which reactant delivery, reaction energy input, and product separation are key processes which determine the product characteristics and quality required by applications of nanoparticles and powders. In case of CVS, the volatility of the reactants (precursors) may severely limit the possible type of products as well as the production rate. It is shown that these limits can be lifted by use of a laser flash evaporator which also enables the use of precursor mixtures for the production of complex oxides as shown for Co-doped ZnO and the pulsed operation to influence powder characteristics. The mode in which energy is supplied to the particle synthesis reactor has also substantial influence on particle and powder characteristics as is shown for TiO_2 using different time-temperature profiles.

2.1 Introduction

Nanocrystalline oxides are used in ceramics, catalysts, fuel cells, photovoltaic devices, or gas sensors, etc. and can be prepared by a large number of methods each with different advantages and disadvantages [1]. Fumed silica (SiO_2) is the first industrially produced nanostructured material [2] using a gas phase process that has been transferred to other oxides, such as titania (TiO_2), alumina (Al_2O_3), or zirconia (ZrO_2) which allows the tailoring of their properties based on process simulations [3].

In this chapter, we report on new developments of Chemical Vapor Synthesis (CVS), a modified Chemical Vapor Deposition (CVD) method in which the process

---

R. Djenadic · M. Winterer (✉)
Nanoparticle Process Technology, Faculty of Engineering and CENIDE
(Center for Nanointegration Duisburg-Essen), University of Duisburg-Essen,
Lotharstraße 1, 47057 Duisburg, Germany
e-mail: markus.winterer@uni-due.de

A. Lorke et al. (eds.), *Nanoparticles from the Gas Phase*, NanoScience and Technology,
DOI: 10.1007/978-3-642-28546-2_2, © Springer-Verlag Berlin Heidelberg 2012

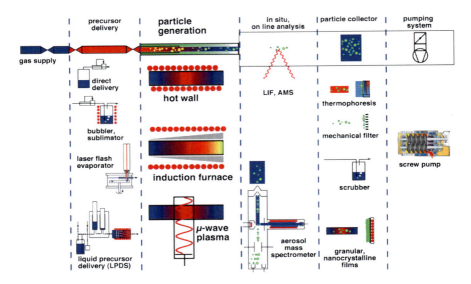

Fig. 2.1 Overview of the modular CVS reactor at the University Duisburg-Essen with different options for the unit operations: precursor delivery, particle generation, process analysis, and particle collection (drawing of screw pump courtesy of Dr.-Ing. K. Busch GmbH)

parameters are adjusted such that (nano-) particles are formed instead of films [4]. In this contribution novel modules for unit operations, relevant for the CVS process such as precursor delivery and energy input are described (Fig. 2.1), as well as their influence on particle respectively powder characteristics.

It will be discussed how the process parameters influence particle characteristics, such as size distribution, microstructure, morphology, and crystallinity which are important for the generation of functional nanomaterials.

2.2 Influence of Pulsed Precursor Delivery on Particle Size and Size Distribution

Pulsed flows in reactors are so far mostly known for processes in the liquid and less for the gas phase. Pulsed plasmas were used to generate FePt nanoparticles from the gas phase [5]. Pulsed reactant flows are used in Atomic Layer Deposition (ALD [6]). A pulsed flow of liquid reactants was used to produce carbon fibers in a continuous process [7]. For the liquid phase it could be shown that the particle size distribution (PSD) for droplets in an oscillating reactor becomes narrower [8]. Nanoparticles of very narrow size distribution can be generated in segmented liquid flows [9, 10].

In case of CVS a pulsed flow of reactants—depending on pulse length and frequency—can lead to a reduction in reactant and particle concentration by axial dispersion and thereby suppress dynamically the particle growth by chemical

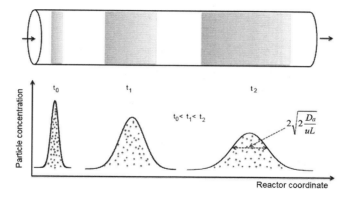

Fig. 2.2 The spreading of a precursor pulse according to the axial dispersion model (after Levenspiel [12])

reaction and coagulation during their residence time. In this way, the microstructure of the particles (degree and type of agglomeration and consequently PSD) can be influenced. A pulsed flow of reactants can be implemented experimentally using a laser flash evaporator [11] by pulsing the CO_2-laser power. It is of great advantage for a systematic investigation that all other process parameters remain unchanged. In this way, it should be possible to reduce the particle number density and to achieve narrow PSD in CVS. Figure 2.2 schematically illustrates this idea. When the pulse of precursor is delivered into the reactor (time: t_0) due to the molecular diffusion it starts to spread (times: t_1 and t_2) along the reactor. This can be described by using the axial dispersion coefficient, D_a [12]:

$$\frac{\partial C}{\partial t} = \left(\frac{D_a}{uL}\right) \frac{\partial^2 C}{\partial z^2} - \frac{\partial C}{\partial z} \quad (2.1)$$

where (D_a/uL) is the parameter measuring the extent of the axial dispersion and is called 'dispersion number' (dimensionless, the inverse of the Bodenstein number, Bo [13]), u is the velocity, L is the reactor length, C is the concentration, and $z = (ut + x)$. The solution of Eq. (2.1) is a symmetrical curve similar to a Gaussian distribution described by mean value and variance ($\sigma^2 = 2D_a/uL$). The main challenge is to keep pulses apart from each other in order to avoid their intermixing and to retain a pulsed flow for the total residence time. As discussed later, this will be achieved by varying the laser pulse repetition frequency and duty cycle.

2.2.1 Experimental Methodology

Particle synthesis. Figure 2.3 shows a schematic drawing of the CVS setup used for the TiO_2 particle synthesis from titanium diisopropoxide bis (tetramethylheptan

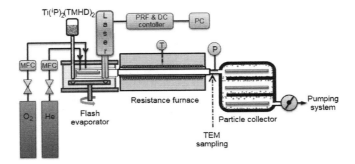

Fig. 2.3 The CVS setup for synthesis of TiO$_2$ nanoparticles using laser flash evaporation as precursor delivery method

edionate), Ti(^{i}P)$_2$(TMHD)$_2$ (95 %, ABCR, Germany). The solid precursor was continuously (manually) fed into the flash evaporator and evaporated using a CO$_2$ laser (Coherent GEM, operated at 95 W). After the precursor is evaporated, its vapor was carried by a helium stream (1,020 sccm) into the hot-wall reactor where the TiO$_2$ particles are formed in reaction with 1,000 sccm of oxygen at a temperature of 1,273 K and a pressure of 20 mbar. Particles are collected in the thermophoretic particle collector.

In order to study the influence of the particle number concentration on the width of the PSD, precursor has to be delivered in pulses to the reactor. This was utilized by changing the CO$_2$ laser pulse repetition frequency and the duty cycle. Each state when the laser is "ON" defines the amount of the precursor in the reactor and consequently the particle number concentration. In the experiments, the laser repetition frequency and duty cycle were varied ($f = 0.05$–25 kHz; DC $= 20$–100 %). A schematic illustration of some frequency and duty cycle combinations are shown in Fig. 2.4. When the laser is operated at a frequency of 25 kHz or duty cycle of 100 %, precursor delivery can be considered as continuous.

Particle characterization. Crystal structure and phase composition of the as synthesized powders were analyzed from the X-ray diffraction (XRD) patterns obtained using the Panalytical X'Pert Pro powder diffractometer. The XRD patterns were refined by the Rietveld method using the MAUD software [14]. The XRD patterns were recorded using CuK$_\alpha$ radiation in an 2θ interval from 20–120° with steps of 0.03° and counting time 200 s/step.

Transmission electron microscopy (TEM) was used to determine the particle size and size distribution of TiO$_2$ nanoparticles. For this purpose thermophoretic (in-situ) particle deposition on Cu grids with carbon film was used. Images were recorded using a Philips Tecnai TEM-F20 Super Twin operating at 200 kV (0.23 nm point resolution) with a Gatan Multiscan CCD (794IF) camera. The PSD was obtained by the log-normal size distribution (fit using the geometric mean as particle size and the geometric standard deviation, σ_g, to describe the width of the distribution) of

Fig. 2.4 Schematic illustration of some frequency and duty cycle combinations used (the *gray areas* represent the 'ON' state of the laser in which a defined amount of precursor is evaporated)

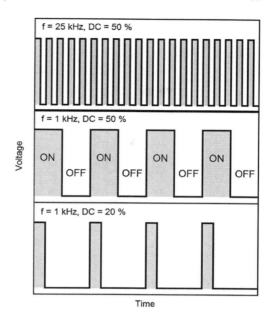

binned data obtained from the TEM images by measuring the size of several hundred particles.

In order to determine the degree of particle agglomeration, photon correlation spectroscopy (PCS) was used to measure the particle size by a Zetasizer Nano S (Malvern) equipped with a green laser ($\lambda = 532$ nm). Prior to the measurement 12.5 mg nanoparticles were dispersed in 25 ml of 0.01 M HCl (pH $\approx$ 2) and ultrasonically treated.

2.2.2 Results and Discussion

Figure 2.5 shows diffraction patterns of two selected powders synthesized by evaporating the precursor using a CO_2 laser operated at higher (25 kHz) and lower (1 kHz) frequencies. The as-synthesized TiO_2 powders are highly crystalline nanoparticles consisting of anatase with a very small contribution of rutile.

The powders synthesized varying the laser duty cycle (DC = 20–100 %) showed high crystallinity as well. It is interesting to mention that only powders synthesized by evaporating the precursor with the laser operating at 25 kHz and 50 % of duty cycle were grayish, while all other powders were white, clearly indicating a difference in the precursor concentration in the reactor. At high frequency (25 kHz), due to the high precursor concentration, the process conditions were not optimal for complete precursor decomposition.

Fig. 2.5 XRD patterns of TiO$_2$ particles synthesized using a CO$_2$ laser operating at 50% of duty cycle and two frequencies: 1 and 25 kHz (as an example)

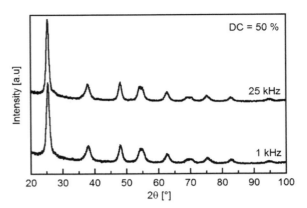

2.2.2.1 Influence of Laser Pulse Repetition Frequency

For lower laser pulse repetition frequencies it is expected that the particle number concentration in the reactor is lower and thus the formation of smaller particles with a narrower PSD was expected. Figure 2.6 shows TEM images of TiO$_2$ particles synthesized using a high (25 kHz) and a low (0.05 kHz) laser frequency. From the images it can be already seen that the particle size does not differ much. Detailed analysis of particle size from the TEM images for the frequency range 0.05–25 kHz showed that the particle size is not affected by a change of the laser pulse repetition frequency (Fig. 2.7). The slight fluctuation in particle size observed at lower frequency is most likely related to a variation of the amount of precursor available for evaporation due to the manual feed of the precursor powder. Although it was expected that with lowering the laser pulse repetition frequency, the PSD will become narrower, only an insignificant narrowing of the PSD from σ_g of 1.25(3) at 25 kHz to 1.23(2) at 0.05 kHz was observed (Fig. 2.8).

Friedlander [15] has shown that the PSD can reach a limiting 'self-preserving' size distribution. The self-preserved size distribution is characterized by a geometric standard deviation of 1.40 and 1.46 for the continuum and free molecule regime, respectively. As the width of TiO$_2$ PSD presented here is about 1.25, the self-preserved size distribution has not been reached and is already rather narrow. Therefore, variations in other process parameters mask probably changes in the size distribution originating from the pulsed operation of the flash evaporator.

Kodas and Friedlander [16] investigated the monodisperse aerosol production in tubular flow reactors and found that in order to produce monodisperse aerosols, nucleation and growth should be separated, the residence time distribution should be narrow and particles should be exposed to a similar monomer concentration. Therefore, the manual precursor feeding, which causes the fluctuations in the evaporated amount of precursor, could be one reason for the insignificant change in the width of the PSD.

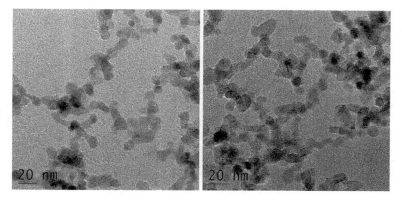

Fig. 2.6 TEM images of TiO$_2$ nanoparticles synthesized using a laser operating at 25 kHz (*left*) and 0.05 kHz (*right*) with 50% of duty cycle

Fig. 2.7 Particle size as a function of laser pulse repetition frequency at 50% duty cycle obtained by the TEM image analysis

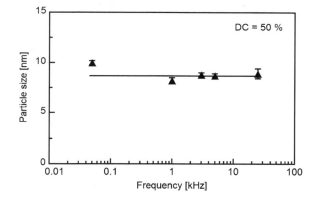

Fig. 2.8 Geometric standard deviation as a function of laser pulse repetition frequency at 50% duty cycle obtained by the TEM image analysis

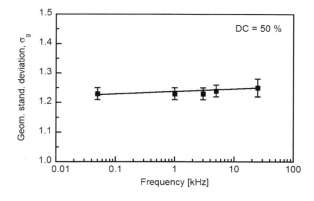

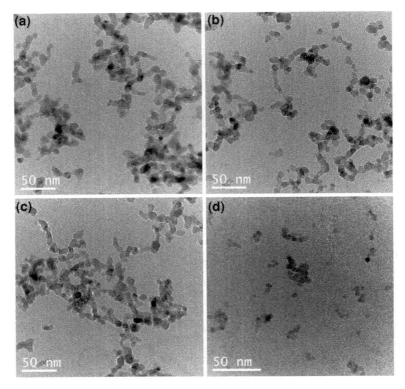

Fig. 2.9 TEM images of TiO$_2$ nanoparticles synthesized using the laser operating at 1 kHz and different duty cycles: 100 (**a**), 80 (**b**), 60 (**c**), and 20 % (**d**)

2.2.2.2 Influence of Laser Duty Cycle

Figure 2.9 shows the TEM images of TiO$_2$ particles synthesized using a frequency of 1 kHz and varying the duty cycles. It is clear that the duty cycle has a significant influence on the particle size, as it is also observed in Fig. 2.10. Therefore, choosing a lower duty cycle, the amount of precursor, and consequently the number concentration of particles was evidently reduced, leading to the formation of smaller particles.

Not only the particles size decreased but also the degree of agglomeration showed a significant decrease, as it can be seen from the cubed ratio of particle size obtained from the PCS and XRD (Fig. 2.11). Concerning the width of the PSD, results showed that at 1 kHz, the duty cycle has no influence on the PSD (Fig. 2.12). Similar conclusions can be drawn for the experiments performed at lower frequency (0.2 kHz), shown in Figs. 2.13, 2.14, and 2.15. The duty cycle had only an impact on the particle size (Figs. 2.13, 2.14), while no influence on the width of the PSD was observed (Fig. 2.15).

Fig. 2.10 Particle size obtained from the TEM images as a function of laser duty cycle at 1 kHz

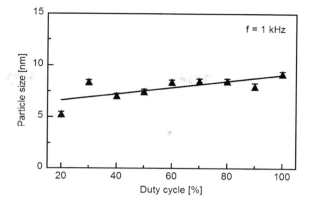

Fig. 2.11 Degree of agglomeration (cubed ratio of particle size obtained from the PCS and XRD) as a function of duty cycle at frequency of 1 kHz (line is only a guide to the eyes)

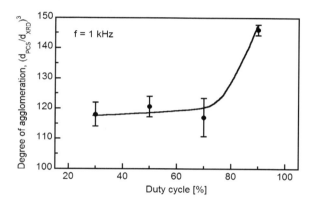

Fig. 2.12 Geometric standard deviation (obtained from the log-normal fit of the TEM data) as a function of the laser duty cycle at 1 kHz

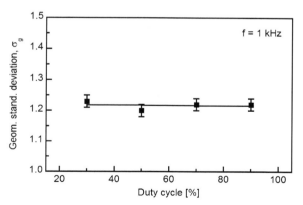

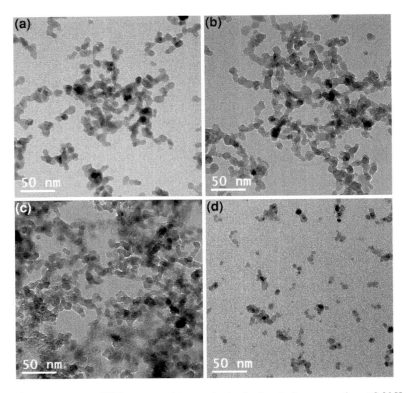

Fig. 2.13 TEM images of TiO$_2$ nanoparticles synthesized using the laser operating at 0.2 kHz and different duty cycles: 100 (**a**), 80 (**b**), 60 (**c**), and 20% (**d**)

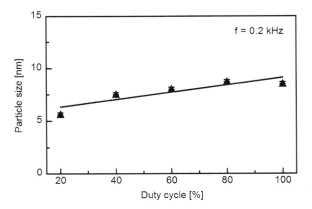

Fig. 2.14 Particle size (obtained from the TEM images) as a function of laser duty cycle at 0.2 kHz

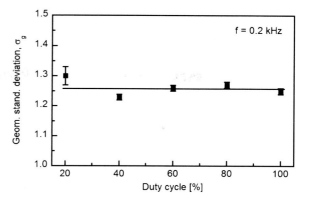

Fig. 2.15 Geometric standard deviation (obtained from the log-normal fit of the TEM data) as a function of laser duty cycle at 0.2 kHz

2.3 Influence of the Time-Temperature Profiles on Powder Characteristics

The quality and application of nanostructured materials are strongly related to the particle and powder characteristics. Powders of small grain size, narrow size distribution, low degree of agglomeration, and high purity are required for the fabrication of solid nanocrystalline materials and the exploitation of size effect in applications. A large variety of methods [17] based on solid, liquid, or gas phase processes are in use for nanocrystalline powder synthesis. For the production of nanocrystalline materials from nanopowders produced by gas phase methods not only the grain size, but also the particle size, i.e. the control of the particle morphology or agglomeration, is important [18]. The objective in most nanoparticle synthesis is to grow dense, non-agglomerated nanoparticles. Therefore, it is requisite to understand the process of particle formation in order to have control upon the powder characteristics.

In the following a simple reaction-coagulation-sintering model (CVSSIN) [4] is used to describe the CVS process. As a model system titania (TiO_2) was used. Nanocrystalline titania, especially anatase, is used in applications such as photocatalysis and photovoltaics [19, 20], thus substantial experimental and theoretical information are available. Additionally, the product of the process is in all relevant cases crystalline, consisting mostly of anatase. This enables not only a detailed model description of the CVS process, but also an extensive characterization of the generated particles and the product powders. The model assumes stationary, ideal, 1D plug flow, without axial dispersion. Processes, which influence the formation of particles from molecular precursors and powder characteristics, such as particle microstructure, morphology, size distribution, and crystallinity are included in the model: conversion of the precursor into monomers (growth species), formation of clusters (primary particles or grains) from monomers, coagulation of primary particles and formation of agglomerates, sintering of the primary particles within the agglomerates, heat exchange (with the hot wall), and heat production by the above processes. In this study, we demonstrate the influence of the key CVS-process

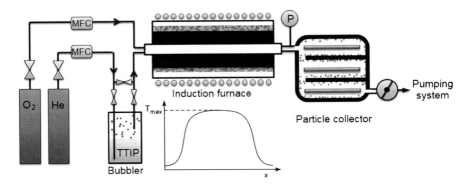

Fig. 2.16 The CVS setup used for the synthesis of TiO$_2$ nanoparticles

parameter, the time-temperature profile in the CVS-reactor, on the particle generation and the corresponding powder characteristics, especially the degree of agglomeration.

2.3.1 Experimental Methodology

Powder synthesis. Figure 2.16 shows a schematic illustration of the CVS setup used for the synthesis of TiO$_2$ particles from titanium-tetraisopropoxide, TTIP (98 %, ABCR, Germany). It consists of a precursor delivery system (bubbler), a hot-wall reactor with an alumina tube, a thermophoretic powder collector, and a pumping system (Busch Cobra NC 0600A). Helium carrier gas (150 sccm) is used to transport the precursor vapor from the bubbler (the temperature is maintained at 333 K using an oil thermostat) to the reaction zone, and oxygen (1,000 sccm) is used as a reactant. The mass flow of the gases is controlled by thermal mass flow controllers (MKS Instruments). In the hot zone of the reactor the precursor vapor decomposes and reacts to form oxide particles. The particles are then transported by the gas stream to the particle collector, where they are separated from the carrier gas flow and byproducts by thermophoresis. A hot-wall reactor is assembled using an induction furnace. The reactor tube (alumina) was inductively heated using a graphite susceptor and a high frequency generator (Hüttinger BIG 20/100). Reactor wall temperatures from 873 up to 2,023 K are investigated and measured using a pyrometer (Sensotherm Metis MI16, $\lambda = 1.6\,\mu$m). The process pressure is held constant at 20 mbar for all experiments using a capacitive absolute pressure gauge (MKS Baratron) and a butterfly valve with adjusting the effective pumping speed of the vacuum pump at the reactor exit.

Powder characterization. Powders have been characterized using X-ray diffraction photon correlation spectroscopy and high-resolution transmission microscopy as described in Sect. 2.2.1.

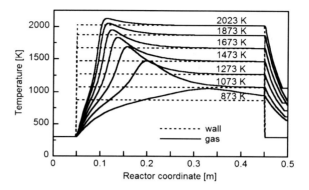

Fig. 2.17 Gas temperature profiles (*full lines*) as a function of reactor coordinate for hypothetical 'flat' wall temperature profiles (*dashed lines*)

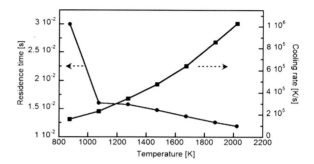

Fig. 2.18 Residence time and maximum cooling rate in the CVS reactor as a function of the maximum wall temperature obtained from the model

2.3.2 Results and Discussion

Results of the CVSSIN model. The CVSSIN model revealed that a simple increase of the wall temperature not only increases the gas temperature in the reactor, but changes the time-temperature profile completely (Fig. 2.17). As the conversion of the TTIP precursor into TiO_2 particles is a highly exothermic reaction, a hot spot in the reactor develops. The difference between the maximum gas temperature and the wall temperature decreases with increasing wall temperature and the position of the hot spot moves closer to the reactor entrance. The total residence time in the reactor decreases and the cooling rate at the exit of the reactor increases with increasing wall temperature (Fig. 2.18). As the temperature increases the primary and agglomerate particle size increases (Fig. 2.19). The agglomerate size reaches a maximum at 1,273 K (Fig. 2.19). Above these temperatures, due to a faster chemical reaction rate, the number density of smaller particles increases, and sintering is enhanced, leading to a decrease of agglomerate size.

The degree of agglomeration (cubed ratio of agglomerate and primary particle size) displays a minimum after a first maximum which is formed due to the particle nucleation burst by chemical reaction (Fig. 2.20). At that minimum the gas temperatures are sufficiently high to completely coalesce the very small primary particles

Fig. 2.19 Primary and agglomerate size obtained from the model as a function of the maximum wall temperature

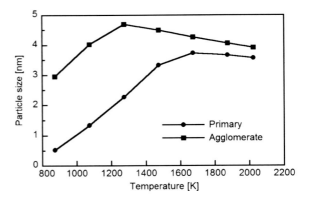

Fig. 2.20 Degree of agglomeration obtained from the model as a function of reactor coordinate

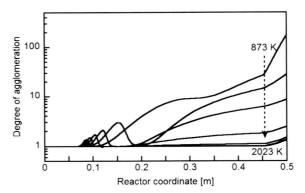

inside of the agglomerates. However, at the reactor exit, where the particles are collected, agglomeration increases and shows a maximum for particles produced at 873 K (Fig. 2.20). At the lowest process temperature the sintering kinetic of the primary particles is too slow, while at higher process temperatures the sintering kinetics is enhanced and coalescence is possible, leading to a reduction of the agglomeration (Fig. 2.21).

Experimental results. Figure 2.22 displays a typical X-ray diffraction pattern of a TiO_2 powder synthesized at 1,273 K together with the Rietveld refinement. The XRD patterns of all as-synthesized TiO_2 powders [21] show strong reflections of the anatase phase indicating that the powders are well crystalline. The Rietveld analysis of the XRD data shows very good agreement between experiment and refinement. The TEM image of TiO_2 particles synthesized at 1,273 K (Fig. 2.23a) shows that particles are spherical with a size of about 7 nm (Fig. 2.23b). The crystallite size obtained from Rietveld refinement of the XRD data (volume weighted) and particle size obtained from the PCS (volume weighted) as a function of process temperature are shown in Fig. 2.24. The crystallite size (as a measure for the primary particle size) increases with process temperature, and ranges from about 2 nm at 873 K up

2 Chemical Vapor Synthesis of Nanocrystalline Oxides

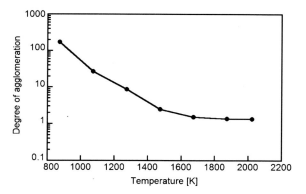

Fig. 2.21 Degree of agglomeration as a function of maximum wall temperature according to the CVSSIN model

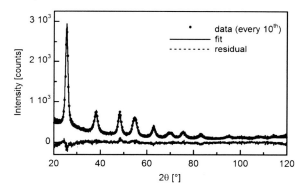

Fig. 2.22 The XRD pattern of TiO_2 nanoparticles synthesized at 1,273 K using Rietveld refinement

to 11 nm at 2,023 K, while the agglomerate size (as measured by PCS) decreases. The cubed ratio of the particle size determined from PCS to the primary particle size obtained from XRD is a measure for the degree of agglomeration and it is shown in Fig. 2.25. The experimental data follows the trend of the CVSSIN model showing that at the highest process temperature (2,023 K) the lowest degree of agglomeration is achieved as it was predicted by CVSSIN model.

However, the absolute values of the model are considerably smaller than the experimentally determined data. It is known [22] that surface reactions dominate particle growth at lower temperatures.

As surface reactions were not included in the CVSSIN model, this could be the reason for the observed discrepancy between model and experimental results at lower temperatures. On the other hand, the preparation of colloidal particle dispersions (ultrasonic treatment in water, no surfactant added) and the measurement of the agglomerate size by PCS may not be the optimum procedure to compare experimental with simulational results. It was shown by Grass et al. [23] that non-agglomerated particles can be formed at low precursor flow rates and high quenching rates. Materials properties (solid state diffusion coefficients) and the time-temperature profile

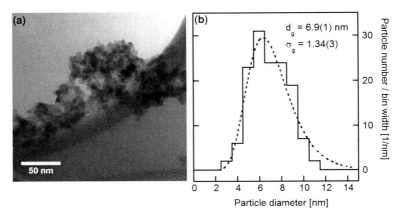

Fig. 2.23 TEM image (**a**) and corresponding PSD (**b**) of TiO$_2$ nanoparticles synthesized at 1,273 K

Fig. 2.24 Particle size obtained from XRD and PCS (volume weighted) as a function of process temperature (error bars are smaller than symbols)

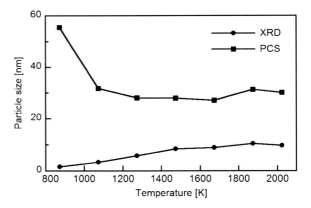

control the agglomeration. This is in agreement with the results of titania generated by CVS presented here.

2.4 Control of Composition by Laser Flash Evaporation: Crystal and Local Structure of Cobalt Doped Zinc Oxide Nanoparticles

Nanostructured magnetic semiconductor materials are promising materials for novel concepts in information technology. Inspired by theoretical predictions, at first by Dietl [24] later by Sato and Katayama-Yoshida [25], extensive research efforts have been carried out by many groups on searching for room temperature ferromagnetism

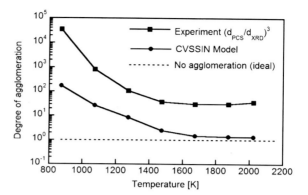

Fig. 2.25 Comparison of the experimentally determined degree of agglomeration (calculated from the particle size obtained from PCS and XRD) with the CVSSIN model

in transition metal (TM) doped ZnO. The key requirement for practical applications is the generation of intrinsic ferromagnetism with Curie temperatures above room temperature. The coupling between magnetism and charge carriers is possible, if the magnetic dopants are located on (substitutional) lattice sites. As chemical composition, crystal and local structure play an important role in the particle properties, a detailed structural characterization is essential for their understanding. Therefore, the crystal and local structure of the Co-doped ZnO, $Zn_{1-x}Co_xO$ ($x = 0.25, 0.30$, and 0.50) were studied in detail.

2.4.1 Experimental Methodology

Particle synthesis. Nanocrystalline Co-doped ZnO particles ($Zn_{1-x}Co_xO$, $x = 0.25, 0.30$, and 0.50) are synthesized in the CVS reactor which consists of two sequential resistive furnaces and a ceramic (alumina) tube (Fig. 2.26). Anhydrous solid zinc acetate, $Zn(OAc)_2$ (Sigma Aldrich, 99.9% purity), and cobalt acetate, $Co(OAc)_2$ (Sigma Aldrich, 99.9% purity), powders are thoroughly mixed in a mortar under inert conditions inside a glovebox corresponding to a nominal Co content, x. The precursor mixture was transferred under inert conditions to the laser flash evaporator. The radiation of a CO_2 laser (95 W, $f = 25$ kHz, DC = 50%) is used to evaporate the precursor mixture. The precursor vapors are transported into the hot-wall reactor using helium (1,020 sccm) as carrier gas where they react with oxygen (1,000 sccm) to form particles.

A hot-wall temperature of 1,373 K and a total pressure of 20 mbar are used for all experiments. The particles are thermophoretically separated from the gas flow in a particle collector.

Particle characterization. Atomic absorption spectroscopy (AAS) is used to determine the Co concentration in Co-doped ZnO nanoparticles. The measurement was carried out using a Thermo Scientific Atomic Absorption Spectrometer (M Series).

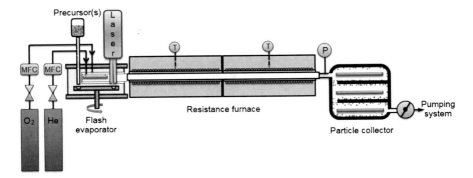

Fig. 2.26 The CVS setup for the synthesis of Co-doped ZnO nanoparticles from mixtures of precursor powders by laser flash evaporation

Fourier transform infrared (FTIR) spectroscopy was used to compare absorption lines of the precursors with the emission line of the CO_2 laser used for their evaporation. Measurements were performed on a Bruker IFS 66v/s instrument.

As described in Sect. 2.2.1 XRD measurements were used in order to analyze the phase composition, crystal, and microstructure of the samples.

In order to investigate the local structure around Zn and Co atoms in Co-doped ZnO nanoparticles, X-ray absorption spectra (XAS) were measured at the Co K-edge and Zn K-edge at the HASLAB (DESY) beamline X1. The absorption of the samples is optimized by diluting appropriate amounts homogeneously into starch powder and pressing a pellet (13 mm diameter) uniaxially (25 kN/10 min). Transmission spectra are collected at ambient temperature. The spectra of commercial CoO and Co_3O_4 powders (Sigma Aldrich) are also recorded as a reference. The X-ray absorption fine structure (XAFS) data were reduced using the program *xafsX* [26]. The extracted extended X-ray absorption fine structure (EXAFS) data are then analyzed by the Reverse Monte Carlo method (RMC) using the *rmcxas* program [27]. Initial atomic configurations are generated from results of the Rietveld refinements of XRD data of the corresponding samples and contain an appropriate number of Co atoms. Zn and Co EXAFS spectra for $x = 0.25$ and 0.30 are analyzed simultaneously using a single atomic configuration, where an appropriate number of Zn atoms in a wurtzite lattice are substitutionally replaced by Co atoms. The sample with $x = 0.5$ was analyzed assuming a mixture of CoO (67 vol %) and Co_3O_4 (33 vol%) consistent with the XRD analysis as initial configurations. Theoretical amplitude and phase functions for RMC analysis are obtained using the program *FEFF* (version 8) [28].

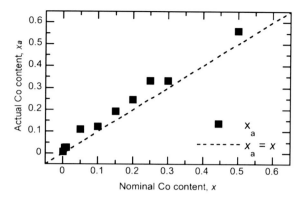

Fig. 2.27 Correlation between actual (x_a) and nominal (x) Co content (error bars are smaller than the symbols) [29]

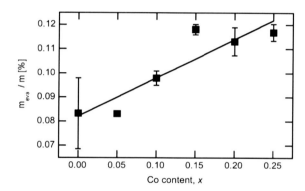

Fig. 2.28 Evaporated mass fraction of the precursor mixture as a function of nominal Co content [29]

2.4.2 Results and Discussion

Chemical composition. As shown in a previous study [29] the actual Co concentration (x_a) in the as-synthesized $Zn_{1-x}Co_xO$ nanoparticles is systematically higher compared to the nominal concentration (x) of the precursor mixture used (Fig. 2.27). It is also observed that the evaporation rate of the precursor mixture as determined from its weight loss is enhanced with increasing Co concentration (Fig. 2.28) which may explain the systematic deviation of the actual Co content from the nominal. The increasing evaporation rate of the precursor mixture with increasing Co content can be explained by the higher absorption of laser light by Co-acetate compared to Zn-acetate as can be seen from a comparison of the FTIR spectra of the precursors with the laser emission line (Fig. 2.29). However, the chemical composition of the product can be easily adjusted knowing the evaporation rates of the used precursor materials. This is of great advantage for the generation of complex oxide materials.

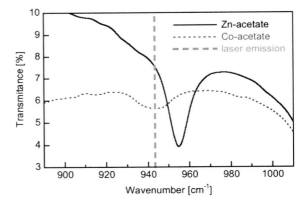

Fig. 2.29 FTIR spectra of the Zn- and Co-acetate precursors and the emission line of the CO_2 laser (wavelength of 10.6 μm) [29]

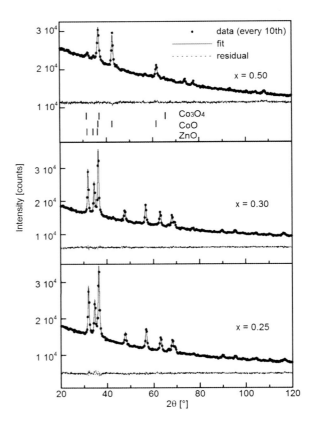

Fig. 2.30 Rietveld refinement of X-ray diffractograms of as-synthesized $Zn_{1-x}Co_xO$ nanoparticles with nominal Co contents of $x = 0.25, 0.30$, and 0.50 (vertical bars correspond to the three most pronounced Bragg reflections for wurtzite ZnO, CoO, and Co_3O_4) refinement results are compiled in Table 2.1

Table 2.1 Results of Rietveld refinement of the XRD data of $Zn_{1-x}Co_xO$ nanoparticles ($x = 0.25$, 0.30, and 0.50)

Nominal Co content x	Actual Co content x_a	Phase composition (vol%)			ZnO (wurtzite)		
		ZnO	CoO	Co_3O_4	Lattice parameters		Crystallite size
		P63mc	Fm-3m	Fd-3m	a (Å)	c (Å)	d (nm)
0.25	0.30	100	–	–	3.2584(2)	5.2121(4)	20.0(1)
0.30	0.33	97.5	2.5(3)	–	3.2572(2)	5.2098(4)	23.2(1)
0.50	0.56	25	52(2)	23(2)	3.259(1)	5.214(3)	14.3(5)

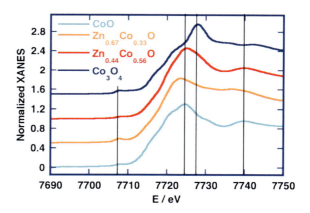

Fig. 2.31 Co K-edge XANES spectra of $Zn_{1-x}Co_xO$, CoO, and Co_3O_4 the actual Co contents are indicated as obtained by chemical analysis

Crystal structure. Figure 2.30 displays the XRD patterns and the result of the Rietveld refinement for as-synthesized $Zn_{1-x}Co_xO$ nanoparticles ($x = 0.25, 0.30$, and 0.50). The Rietveld refinement of the XRD data shows that the sample with $x = 0.25$ consists purely of wurtzite phase. In samples with a Co concentration of $x \geq 0.30$, CoO, and Co_3O_4 are present as additional phases. The maximum Co solubility in wurtzite-type ZnO, for which the particles are still single-phase, varies greatly for samples of the same nominal composition but prepared by different methods [30–34]. The Co solubility limit in nanoparticles studied in this work, is about $x_a = 0.33$ (nominal content $x = 0.25$) (Fig. 2.30, Table 2.1).

Further Co addition causes the generation of second phases: CoO in the sample with nominal Co content of $x = 0.30$ ($x_a = 0.332$) and CoO and Co_3O_4 in the sample with $x = 0.50$ ($x_a = 0.56$). This is consistent with other literature reports [35, 36].

Local structure. XAFS provides information about the distribution and location of Co in the wurtzite lattice complementary to X-ray diffraction. Figure 2.31 shows normalized Co K-edge XANES spectra of the $Zn_{1-x}Co_xO$ ($x = 0.25$ and 0.50) together with reference spectra of CoO and Co_3O_4. Comparing the XANES spectra of $Zn_{1-x}Co_xO$ with the spectra of the CoO reference, reveals that the Co is present

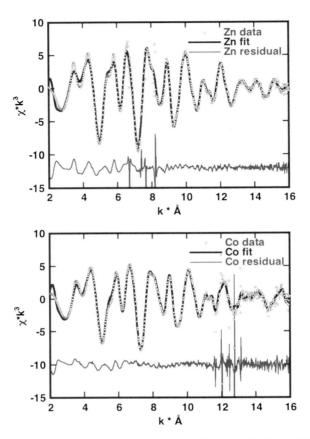

Fig. 2.32 Zn (*top*) and Co K-edge (*bottom*) EXAFS spectra for $Zn_{1-x}Co_xO$ $x = 0.25$ fit by Reverse Monte Carlo modeling. The *sharp peaks* in the spectra are due to monochromator glitches. For the initial atomic configuration substitutional Co atoms on Zn sites were assumed

in the Co^{2+} valence state. The small pre-edge peak at about 7,708 eV appears due to the transition of Co 1s electron to 4p–3d hybridized states in tetrahedral symmetry [37]. This indicates that Co is either located on Zn sites or tetrahedral interstitial sites.

The EXAFS spectra of $Zn_{1-x}Co_xO$ samples with $x = 0.25$ (Fig. 2.32) and $x = 0.30$ are very similar as well as the Zn and Co K-edge spectra of each of these samples in contrast to the sample with $x = 0.5$ (Fig. 2.33). This indicates that Co atoms in the samples with lower Co content are on Zn lattice sites, whereas for the highest content investigated, Co occupies sites with different local structure. This observation is consistent with the XRD results described above. Therefore, the samples with $x = 0.25$ and $x = 0.30$ where fit by RMC modeling using a wurtzite lattice with Zn atoms randomly substituted by Co, according to the composition, as initial configuration. The results of the simultaneously fit Zn- and Co- spectra are

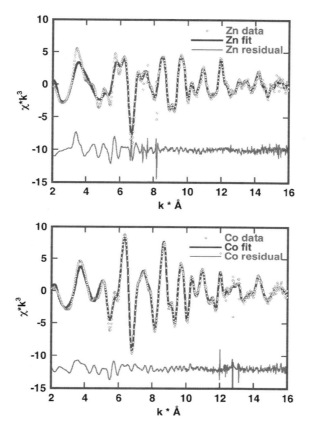

Fig. 2.33 Zn (*top*) and Co K-edge (*bottom*) EXAFS spectra for $Zn_{1-x}Co_xO$ $x = 0.5$ fit by Reverse Monte Carlo modeling. The *sharp peaks* in the spectra are due to monochromator glitches

shown in Fig. 2.32. The agreement between model and data is very good. The main deviations are due to monochromator glitches, insufficient background subtraction, and multiple scattering for which higher than pair correlation functions need to be considered. The corresponding partial pair distribution functions (PDFs) are shown in Fig. 2.34. All PDFs show essentially only a broadening of each peak due to thermal (and structural) disorder in the material.

In contrast to the samples with lower Co content the sample with $x = 0.5$ could not be fit well using a pure wurtzite model. Therefore, according to the XRD results the Co spectrum was fit with a mixture of CoO and Co_3O_4 and the Zn spectrum using pure ZnO in wurtzite structure (Fig. 2.34). The Co spectrum is fit quite well, whereas the fit of the Zn spectrum shows significant deviations from the data. Probably, Co is not segregated completely and is still present as dopant up to the solubility limit of about 33 mol% in the wurtzite lattice. This may also explain additional peaks in the

Fig. 2.34 Partial PDFs for $Zn_{1-x}Co_xO$ $x = 0.25$ derived from RMC. For the initial atomic configuration substitutional Co atoms on Zn sites were assumed. *Top* cationic partial PDFs, *bottom* anionic partial pair distribution functions (PDFs generated by the initial atom configuration: *thin lines*, PDFs of the RMC fit: *thick lines*)

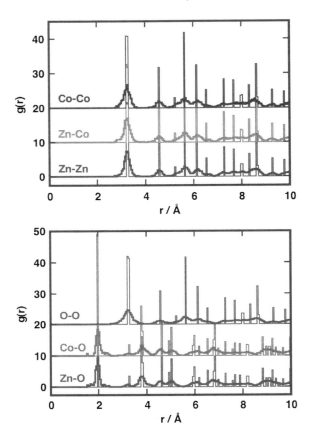

Co-O (at 1.93 Å) and Co-Co (at 3.98 Å) PDF's of CoO (Fig. 2.35) and in the Zn-Zn PDF of the pure ZnO model for the Zn K-edge spectrum (Fig. 2.36).

The moment analysis of the partial PDFs (Table 2.2) shows that—within statistical error—the local structure samples with a lower Co content is identical. The phase separation visible in the XRD data at $x = 0.30$ could not be detected. The atom distances and the second moments which are a measure for the static and dynamic respectively structural and thermal disorder is identical for both samples of lower Co content as well as the second moment of the Co-Co, Zn-Co and Zn-Zn and the Co-O and Zn-O PDFs indicating that the majority of the Co atoms is on Zn lattice sites.

2 Chemical Vapor Synthesis of Nanocrystalline Oxides

Fig. 2.35 Partial PDFs for $Zn_{1-x}Co_xO$ $x = 0.5$ derived from RMC simulation of the Co K-edge spectrum. A complete segregation of Co to a mixture of CoO (67 vol%) and Co_3O_4 (33 vol%) phases is assumed. *Top* partial PDFs for the CoO and *bottom* for the Co_3O_4 configuration (after the RMC fit: *thick lines*, initial configuration: *thin lines*)

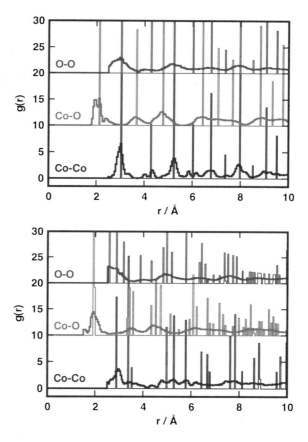

Fig. 2.36 Partial PDFs for $Zn_{1-x}Co_xO$ $x = 0.5$ derived from RMC. Partial PDFs obtained from the Zn K-edge spectrum using pure ZnO as model (after the RMC fit: *thick lines*, initial configuration: *thin lines*)

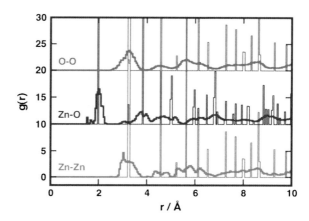

Table 2.2 Results of EXAFS data analysis of $Zn_{1-x}Co_xO$ nanoparticles using *rmcxas*

x (Co)		Shell	N	R (Å)	p_2 (10^{-3} Å^2)
0.25		Co-O	4.0(8)	2.02(3)	18(8)
		Co-Co	4.2(7)	3.23(3)	24(6)
		Co-Zn	na	3.23(3)	25(7)
		Zn-O	3.9(7)	1.96(3)	18(9)
		Zn-Co	4.2(7)	3.23(3)	24(6)
		Zn-Zn	7.6(12)	3.23(3)	20(7)
0.30		Co-O	3.9(8)	2.03(3)	17(8)
		Co-Co	4.2(7)	3.24(3)	27(7)
		Co-Zn	na	3.23(3)	25(8)
		Zn-O	3.9(7)	1.97(3)	14(7)
		Zn-Co	4.4(7)	3.23(3)	25(8)
		Zn-Zn	7.6(12)	3.22(3)	22(8)
0.50	(CoO)	Co-O	6.1(11)	2.12(4)	51(17)
		Co-Co	12(2)	3.01(3)	30(10)
	(Co$_3$O$_4$)	Co-O	5.1(10)	2.01(4)	33(12)
		Co-Co	6.9(17)	2.89 (5)	33(12)
	(ZnO)	Zn-O	3.9(8)	2.01(3)	14(5)
		Zn-Zn	12(2)	3.22(4)	50(11)

N the coordination number, R the interatomic distance, p_2 the second moment, *na* not analzyed

2.5 Summary and Conclusions

It has been shown that variations of precursor delivery and time-temperature profile in Chemical Vapor Synthesis have substantial influence on the particle and powder characteristics. The chemical engineering of CVS has been further developed, materials that can be produced by CVS have been extended to more complex systems and particles using a laser flash evaporator as precursor delivery method and powder characteristics have been optimized.

A novel method, pulsed precursor delivery using a laser flash evaporator, was developed to feed reactant material into a CVS reactor. A systematic study showed so far only limited influence on the width of the PSD, but a significant impact on the reduction of the particle size and the degree of agglomeration. The laser pulse repetition frequency had only little influence on the particle size and the size distribution. On the other hand, it is possible to lower the particle number concentration, thus to reduce the final particle size with rather narrow size distribution by decreasing the laser duty cycle. A low duty cycle leads also to a significant decrease in the degree of particle agglomeration.

The characteristics of nanocrystalline powders were studied using TiO_2 as a model material produced by chemical vapor synthesis. The time-temperature history of the gas phase in the reactor has considerable influence. With increasing wall temperatures the gas temperature becomes sufficiently high for a complete coalescence of the primary particles inside the agglomerates and the quenching rate at the reactor exit

2 Chemical Vapor Synthesis of Nanocrystalline Oxides

fast enough to prevent extensive formation of hard agglomerates. Therefore, the fast quenching at the end of the reactor is essential for the formation of weakly agglomerated particles.

A high solubility of cobalt in ZnO is achieved using chemical vapor synthesis, as $Zn_{1-x}Co_xO$ nanoparticles for Co contents of up to $x_a = 0.33$ (according to chemical analysis) are of single wurtzite phase. Instrumental is here that the evaporation of mixtures of precursors is possible generating solid solutions. Detailed analysis of both crystallographic (from XRD analyzed by Rietveld refinement) and local structure (from EXAFS analyzed by Reverse Monte Carlo simulations) provides evidence that it is possible to prepare nanocrystalline ZnO particles by chemical vapor synthesis where Co is substituting Zn in the wurtzite lattice which is the key requirement for the development of dilute magnetic semiconductors based on Co-doped ZnO. Sample preparation procedures and conditions play a very important role in obtaining Co-doped ZnO samples of high homogeneity and chemical vapor synthesis, as a nonequilibrium process where precursors are mixed on the molecular level in the gas phase, provides the possibility to prepare such samples.

Acknowledgments The authors gratefully acknowledge the generous support by the Collaborative Research Center SFB 445 'Nanoparticles from the Gas Phase' funded by the German Research Foundation. We are also very thankful to Dr. Marina Spasova (team of Prof. Farle) and Dr. Ralf Theissmann (team of Prof. Schmechel) for providing the TEM imaging, Kerstin Brauner (team of Prof. Epple) for the determination of the chemical composition by atomic adsorption spectroscopy, Andreas Gondorf (team of Prof. Lorke) for the FTIR spectroscopy measurements and Dr. Adam Webb at HASYLAB/DESY for supporting us at beamline X1.

References

1. J.A. Rodriguez, M. Fernandez-Garcia, *Synthesis, Properties and Applications of Oxide Nanomaterials* (Wiley-Interscience, New York, 2007)
2. G.W. Kriechbaum, P. Kleinschmidt, Adv. Mater. **1**, 330 (1989)
3. A. Gutsch, H. Mühlenweg, M. Krämer, Small **1**, 31 (2005)
4. M. Winterer, *Nanocrystalline Ceramics—Synthesis and Structure* (Springer, Heidelberg, 2002)
5. I. Matsui, J. Nanopart. Res. **8**, 429 (2006)
6. M. Leskelä, M. Ritala, Angew. Chem. Int. Ed. **42**, 5548 (2003)
7. T. Masuda, S.R. Mukai, H. Fujikawa, Y. Fujikata, K. Hashimoto, Mat. Manufact. Proc. **9**, 237 (1994)
8. N.E. Pereira, X.W. Ni, Chem. Eng. Sci. **56**, 735 (2001)
9. N. Jongen, M. Donnet, P. Bowen, J. Lemaitre, H. Hofmann, R. Schenk, C. Hofmann, M. Aoun-Habbache, S. Guillemet-Fritsch, J. Sarrias, A. Rousset, M. Viviani, M.T. Buscaglia, V. Buscagkia, P. Nanni, A. Testino, J.R. Herguijuela, Chem. Eng. Technol. **26**, 303 (2003)
10. B.K.H. Yen, A. Günther, M.A. Schmidt, K.F. Jensen, M.G. Bawendi, Angew. Chem. Int. Ed. **44**, 5447 (2005)
11. M. Winterer, V.V. Srdic, R. Djenadic, A. Kompch, T.E. Weirich, Rev. Sci. Instrum. **78**, 123903 (2007)
12. O. Levenspiel, *Chemical Reaction Engineering* 3rd edn. (Wiley, New York, 1999)
13. M. Jakubith, *Chemische Verfahrenstechnik—Einführung in Reaktionstechnik und Grundoperationen* (VCH, New York, 1991)
14. L.Lutterotti, P. Scardi, J. Appl. Cryst. **23**, 246 (1990)

15. S.K. Friedlander, *Smoke, Dust, and Haze—Fundamentals of Aerosol Dynamics*, 2nd edn. (Oxford University Press, Oxford, 2000)
16. T.T. Kodas, S.K. Friedlander, AIChE J. **34**, 551 (1998)
17. G.-M. Chow, K.E. Gonsalves, *Nanotechnology—Molecular Designed Materials*, ACS Symphosium Series, vol. 662 (American Ceramic Society, Washington, DC, 1996)
18. R.C. Flagan, M.M. Lunden, Mat. Sci. Eng. A **204**, 113 (1995)
19. X. Chen, S.S. Mao, J. Nanosci. Nanotechnol. **6**, 906 (2006)
20. G. Li, L. Li, J. Boerio-Goates, B. Woodfield, J. Am. Chem. Soc. **127**, 8659 (2005)
21. R. Djenadic, S.R. Chowdhury, M. Spasova, M. Gondorf, E. Akyildiz, M. Winterer, Mater. Res. Soc. Symp. Proc. **1056**, 1056-HH08-07 (2008)
22. K. Nakaso, K. Okuyama, M. Shimada, S.E. Pratsinis, Chem. Eng. Sci. **58**, 3327 (2003)
23. R.N. Grass, S. Tsantilis, S.E. Pratsinis, AIChE J. **52**, 1318 (2006)
24. T. Dietl, H. Ohno, F. Matsukura, J. Cibert, D. Ferrand, Science **287**, 1019 (2000)
25. K. Sato, H. Katayama-Yoshida, Phys. E **10**, 251 (2001)
26. M.J. Winterer, Phys. IV **7**, C2–243 (1997)
27. M. Winterer, J. Appl. Phys. **88**, 5635 (1988)
28. A.L. Ankudinov, B. Ravel, J.J. Rehr, R.C. Albers, S.D. Conradson, Phys. Rev. B **58**, 7565 (1998)
29. R. Djenadic, G. Akgül, K. Attenkofer, M. Winterer, J. Phys. Chem. C **114**, 9207 (2010)
30. S. Kolesnik, B. Dabrowski, J. Mais, J. Appl. Phys. **95**, 2582 (2004)
31. A.S. Risbud, N.A. Spaldin, Z.Q. Chen, S. Stemmer, R. Seshadri, Phys. Rev. B **68**, 205202 (2003)
32. L.B. Duan, W.G. Chu, J. Yu, Y.C. Wang, L.N. Zhang, G.Y. Liu, J.K. Liang, G.H. Rao, J. Magn. Magn. Mater. **320**, 1573 (2008)
33. V. Jayaram, J. Rajkumar, B.S. Rani, J. Am. Ceram. Soc. **82**, 473 (1999)
34. T.A. Schaedler, A.S. Gandhi, M. Saito, M. Rühle, R. Gambino, C.G. Levi, J. Mater. Res. **12**, 791 (2006)
35. B.B. Straumal, A.A. Mazilkin, S.G. Protasova, A.A. Myatev, P.B. Straumal, B. Baretzky, Acta Mater. **56**, 6246 (2008)
36. B. Straumal, B. Baretzky, A. Mazilkin, S. Protasova, A. Myatev, P. Straumal, J. Eur. Ceram. Soc. **29**, 1963 (2009)
37. Z. Sun, W. Yan, G. Zhang, H. Oyanagi, Z. Wu, Q. Li, W. Wu, T. Shi, Z. Pan, P. Xu, S. Wei, Phys. Rev. B **77**, 245208 (2008)

Chapter 3
Nucleation, Structure and Magnetism of Transition Metal Clusters from First Principles

Sanjubala Sahoo, Markus E. Gruner, Alfred Hucht, Georg Rollmann and Peter Entel

Abstract Properties of transition metal (TM) clusters such as structural stability, growth and magnetic properties are studied using the density functional theory (DFT). We find that for both elemental and binary clusters, different morphologies are stable for different ranges of cluster sizes. We discuss possible structural transformations namely Jahn-Teller (JT) and Mackay transformation (MT) occurring in TM clusters. While the JT-distorted cluster is stable for a Fe_{13} icosahedron, the MT-distorted structure is stable for Co_{13}. For Ni_{13}, however, both distortions lead to similar energies. In larger clusters, both JT and MT compete with each other, and as a result we find a higher stability for large Fe clusters with a shell wise Mackay transformation. Studies on binary Fe-Pt clusters show a segregation tendency of Pt atoms to the surfaces of the clusters. The ordered Fe-Pt icosahedral structures show enhanced stability compared to the $L1_0$ cuboctahedron. From the studies on magnetocrystalline anisotropy (MAE) for clusters, we find that relaxed Fe_{13} and Ni_{13} have several orders of magnitude larger MAE as compared to the corresponding bulk values. However, Co_{13} does not follow this trend.

3.1 Introduction

Understanding the physics of clusters (nanoparticles) is of immense interest because clusters possess novel properties, which can be harnessed for technological applications. While semiconducting nanoparticles are of interest for electronics and logic gates, transition metal (TM) nanoparticles are projected to play a role in data storage devices [1]. For this, it is desirable to have a large magnetization and a large magnetocrystalline anisotropy (MAE). Therefore, thorough studies of TM

S. Sahoo · M. E. Gruner · A. Hucht · G. Rollmann · P. Entel (✉)
Faculty of Physics and Center for Nanointegration CENIDE, University of Duisburg-Essen,
Lotharstraße 1, 47057 Duisburg, Germany
e-mail: entel@thp.uni-duisburg.de

A. Lorke et al. (eds.), *Nanoparticles from the Gas Phase*, NanoScience and Technology,
DOI: 10.1007/978-3-642-28546-2_3, © Springer-Verlag Berlin Heidelberg 2012

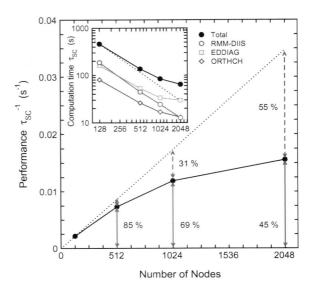

Fig. 3.1 Scaling of VASP with respect to the number of processors on the IBM Blue Gene/L for a large Fe cluster of size 561 atoms. The performance shown on the ordinate is the inverse of the average time (τ_{SC}^{-1}) for an electronic self consistency (SC) step. The *inset* shows the double logarithmic plot for the number of processors versus computation time (τ_{SC}) for several subroutines in an electronic SC step. The *dotted line* is a theoretical extrapolation of the scaling to linear behavior. Figure taken from [14, 15]

nanoparticle morphologies and their magnetic properties are essential. It is well known that the size and the structure of the nanoparticles play an important role in determining their properties [2–5]. The early attempts of studying the physics of nanoparticles have been through model studies or semi-empirical calculations. However, since the electronic structure is essential for a thorough understanding of material properties especially at very small scales, a first-principles approach is required. Ab initio methods demand exceedingly more computational resources in terms of memory capacity as well as simulation time, thus using conventional desktop computer hardware, one is currently capable of handling systems consisting up to about hundreds of atoms. In order to simulate larger systems, one has to make use of massively parallel computing. In the course of our research, structural optimizations of TM clusters with up to 1,415 atoms were carried out from first principles [6], which is only possible on world leading supercomputing hardware as the IBM Blue Gene/P installation located at the Jülich Supercomputing Center. It is important to note that such calculations also allow atomistic design of ultrahard nanomagnets, which may be important in actual devices [7].

Our total energy investigations of the electronic structure of materials were carried out within the framework of density functional theory (DFT) [8, 9]. For our calculations, we took advantage of the Vienna ab initio simulation package (VASP) [10], which uses plane wave basis sets and the projector augmented wave method [11] to deal with the interaction between core and valence electrons. For the vast majority of our calculations we used the generalized gradient approximation (GGA) for the exchange and correlation parameterized by Perdew and Wang [12] in combination with the spin interpolation formula of Vosko et al. [13]. Due to the approximation of the wavefunctions by a plane wave basis, the clusters had to be placed in a cubic

3 Nucleation, Structure and Magnetism of Transition Metal Clusters

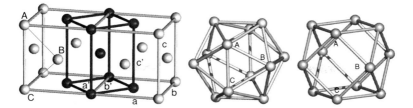

Fig. 3.2 *Left* the face-centered cubic lattice (*grey balls*) with the inscribed body-centered tetragonal Bain cell (*black balls*). The Bain transformation takes place by the variation of c with respect to a and b. The *prime letters* correspond to the bct structure. *Middle and right* icosahedron and cuboctahedron, respectively. Due to the Mackay transformation, the bond $\overline{AC}$ elongates resulting in the formation of a square with the two adjacent triangles turning into the same plane. The *capital letters* represent the same atomic sites in each structure. Figure originally published in [27]

supercell being surrounded by a sufficient amount of vacuum to prevent interactions with the periodic images, while k-space integration was restricted to the Γ-point.

The VASP code has proven in many examples to provide an excellent compromise between speed and accuracy, especially on massively parallel supercomputing hardware [14, 15]. This is demonstrated in Fig. 3.1 for the case of a 561 atom Fe-cluster which was simulated on a IBM Blue Gene/L using up to CPU-2048 cores. While a performance decrease with increasing parallelization is expected due to the communication overhead and serial parts of the code, we could demonstrate that the VASP code can achieve on 1,024 cores agreeable 70 % of the ideal performance. Especially, the optimization of the trial wavefunctions according to the residual vector minimization scheme (labeled 'RMM-DIIS') scales nearly perfectly, even at the largest processor numbers under consideration. However, with increasing number of processors, its computation time gets outweighed by other routines ('EDDIAG' and 'ORTCH') providing the calculation of the electronic eigenvalues, subspace diagonalization and orthonormalization of the wavefunctions, making strong use of linear algebra, and fast Fourier transform library routines.

3.2 Magnetism and Structural Transformations in Nanoparticles

The nature of chemical bonding depends on the separation of atoms and their geometrical orientations. In binary clusters, it also depends on the composition and distribution of elements in the clusters. Their properties can be tailored by changing the composition and chemical ordering [16–18]. In addition to above, for example, in TM clusters, there is a close interlink between magnetism and structure of clusters. It is known that clusters of certain sizes are more stable compared to other sizes. The size dependent stability of clusters can be studied in experiments through mass spectroscopy, where stable structures would yield a higher probability of occurrence.

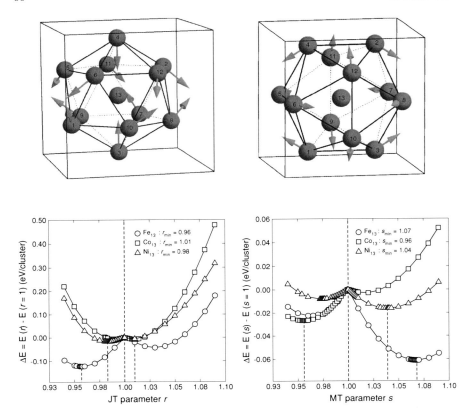

Fig. 3.3 *Top panel* the JT-distorted and the MT Fe$_{13}$ cluster are illustrated in the *left* and *right* panels, respectively. *Arrows* represent the direction of relative shift of atoms with respect to the ideal positions. For the JT-distorted and MT Fe cluster, the displacements marked by the *arrows* have been scaled up by factors of 20 and 30, respectively. The *box* is drawn to support the 3-D impression. The actual simulation box size is 15^3 Å^3. *Bottom panel* the values of r and s corresponding to the minimum energy (shown as *dashed lines*) for 13-atom Fe, Co and Ni clusters. The perfect ICO corresponds to $s = 1$ and $r = 1$

In a similar experiment for Co and Ni, Pellarin et al. [19] have shown regular peaks in the mass spectra which have been interpreted as magic cluster sizes with icosahedral geometry. Similar observations for fivefold icosahedral geometry have been predicted from theoretical calculations for very small clusters. Experimentally, however, a small cluster of 6 nm size has already been shown to have a bcc structure [20]. This suggests hints for size-dependent competition of structures in iron clusters, which goes hand-in-hand with our theoretical studies.

Structural transformations in solids are a relatively old subject. Starting from a statistical model to describe a structural phase transition, there have been studies on more complicated structural transformations achieved by application of pressure or by external (magnetic) fields. Most notable among them is the series of Heusler

3 Nucleation, Structure and Magnetism of Transition Metal Clusters

Table 3.1 The bondlengths (Å) for the minimum energy JT-distorted ($r = 0.965$ for Fe_{13}, 1.01 for Co_{13}, 0.98 for Ni_{13}) and Mackay-transformed ($s = 1.07$ for Fe_{13}, 0.96 for Co_{13}, 1.04 for Ni_{13}) clusters

Bonds	Fe_{13} (JT)	Co_{13} (JT)	Ni_{13} (JT)
2 × center shell	2.34	2.35	2.28
10 × center shell	2.42	2.32	2.33
Bonds	Fe_{13} (MT)	Co_{13} (MT)	Ni_{13} (MT)
Center shell	2.39	2.33	2.32
24 × bond length (surface)	2.50	2.46	2.43
6 × bond length (surface)	2.58	2.41	2.47

alloys [21, 22], which is an active research area in itself. For Fe and related bulk alloys, it is known that bulk can transform from fcc to bcc-phase. According to the Bain transformation [23] the fcc-bcc transformation takes place passing through an intermediate body-centered tetragonal (bct) phase, by simple elongation of the [001] axis with respect to the other two perpendicular crystal axes as shown in Fig. 3.2 (left). There are also other competing realizations of the above-mentioned transformation path, named Nishiyama-Wassermann (N-W) and Kurdjumow-Sachs (K-S). They are more realistic for bulk alloys since they allow for a compatible interface between austenite and martensite during the process [24]. Such restrictions do not persist for small clusters. Nevertheless, for the same reasons, other more complicated structural transformations may take place. For the systems under consideration, we consider two other important structural transformations. The Jahn-Teller transformation (JT), which relates to the tendency of a system to avoid an enhanced degeneracy of occupied electronic states arising from a more symmetrical geometry through a distortion of the structure. For 13 atom icosahedral clusters (Fe, Co, Ni), JT can be characterized by the parameter r, defined as $r = |\underline{r}_3 - \underline{r}_4|/|\underline{r}_1 - \underline{r}_2|$ (see top left panel of Fig. 3.3 for the labeling of atoms). The Mackay transformation (MT) describes a specific transformation path of the cluster from icosahedron to cuboctahedron geometry. The MT can be characterized by s, which is the square of the ratio of the stretched to the unstretched edges (see Fig. 3.2) being equal to 1 and 2 for the perfect icosahedron (ICO) and cuboctahedron (CUBO), respectively [25]. That is, $s = |\underline{r}_4 - \underline{r}_2|^2/|\underline{r}_4 - \underline{r}_{12}|^2$, as shown in Fig. 3.3.

It is found that for Fe_{13}, the JT-distorted structure is lower in energy. The gain in energy due to the relaxation from the perfect ICO for Fe_{13} is 125 meV/cluster along the JT distortion and 61 meV/cluster along the MT, with a total moment of 44 μ_B. The existence of two different metastable structures has also been reported earlier [26–28]. On the other hand, Co_{13} stabilizes MT distorted cluster, where the JT-distorted and the Mackay-distorted cluster is 7 and 27 meV lower in energy with respect to the perfect ICO. The energy differences are nearly the same (16 meV/cluster) for the JT and MT Ni_{13} clusters. In order to understand the role of each type of distortion, we have performed a systematic study of energy versus

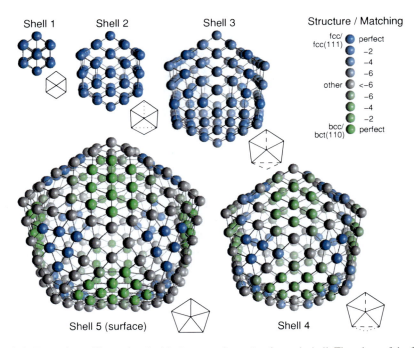

Fig. 3.4 Fe$_{561}$ cluster illustrating the Mackay transformation for each shell. The edges of the faces are shown in the *lined-sketch* below each shell. The *color gradient* represents the number of nearest neighbors through common neighbor analysis (CNA). The signatures of perfect face centered cubic and body centered coordination are shown in *blue* and *green*, respectively. One can observe a gradual evolution from a fcc-like to a predominately body centered neighboring, while going from interior to the surface of the nanocluster. Figure taken from [27]

the distortion parameters. The results are shown in Fig. 3.3 (bottom panels). The structural parameters r and s for the minimum energy are listed in Table 3.1.

With increasing size of clusters there is a competition of various structural motifs, in addition to the two described above. Recently, the authors addressed this problem within a large-scale ab initio approach which revealed that already above a crossover size of about 150 atoms, the bulk-like bcc structure is energetically favored for Fe clusters [27]. However, there is even the possibility for partial transformation along a prescribed transformation path, with varying degree from the inside to the surface of the cluster. A relevant example for such a hybrid morphology is the so-called shell wise Mackay Transformation (SMT), which is a strong candidate for the ground state of Fe$_{55}$. In a strict sense, it is only applicable to so-called magic number clusters, consisting of an integer number n of closed geometric shells. Such structures obey the following formation law for the number of atoms N:

$$N = (10n^3 + 15n^2 + 11n + 3)/3 . \tag{3.1}$$

Even up to 561 atoms, the energy difference between the bcc and the SMT isomer remains—in contrast to the CUBO and the ICO—in the range of thermal energies; thus, the occurrence of the SMT structure appears realistic at finite temperatures.

The search for the origin of this unusual transformation requires an inspection at the atomistic level, which may become a tedious or even impossible task for large system sizes. In our case, help comes from a widely used tool for structure identification in classical molecular dynamics studies of large systems, the so-called common neighbor analysis (CNA) [29]. The CNA characterizes the local environment of an atom by a set of signatures, which can be compared with the characteristic result for an ideal bulk or surface structure. In the first neighbor shell, signatures are obtained for each pair of neighbors, containing information on the number of nearest neighbors both atoms have in common, the number of bonds between these neighbors and the longest chain connecting them. While the bcc and fcc clusters (not shown) are uniform in structure, one finds for the icosahedron the typical mixture of fcc and hcp environments and typical signatures indicating the presence of fivefold symmetry axes. The SMT isomer, however, which is shown in Fig. 3.4 in a shell wise decomposed fashion, possesses a largely icosahedral outer shape and a core with a nearly face centered cubic coordination. However, we find no trace of the typical icosahedral signatures anywhere in the cluster. Instead, the signatures corresponding to the surface and subsurface atoms show typical signs of a bcc-like coordination, which is the energetically most favorable configuration for bulk iron and can thus be assumed as the driving force for this unusual transformation. Such a process results in a pair distribution function which is rather typical for amorphous materials [15, 30].

The calculated magnetic moments can be directly compared with the measurements on small TM nanoparticles by Billas et al. [31–33]. The size dependent Stern-Gerlach measurements of the magnetic moments of Fe_N, Co_N and Ni_N clusters clearly confirm the expected strong enhancement of the average magnetic moment of small clusters relative to bulk. With increase in cluster size up to 700 atoms, the average magnetic moments exhibit an oscillatory variation while approaching the corresponding bulk values. The convergence of the magnetic moment is faster for Co and Ni clusters than for the Fe clusters. In order to investigate such a trend, we have calculated the magnetic moments for relaxed Fe clusters. Figure 3.5 compares the magnetic moments of Fe clusters with up to 641 atoms for different geometries as a function of the cluster size with the experimental data and previous theoretical work [27, 34]. The bcc Fe clusters show good agreement with the experimental results [31, 32] up to 400 atoms but sizeable differences beyond, although a bcc ground state can be safely expected. In fact, for $N = 561$, the SMT isomer shows the best agreement with the experimental magnetization data, owing to a shell wise antiferromagnetic configuration of the cluster core, i.e., while the outermost shells generally couple ferromagnetically, there is an alternation of the orientation of the Fe magnetic moments from shell to shell in the interior. This may be seen as an indication for the thermodynamical relevance of the SMT structures for small cluster sizes, In the previous study of Tiago et al. [35], the magnetic moments refer to unrelaxed Fe clusters, while in our calculations, the clusters are optimized

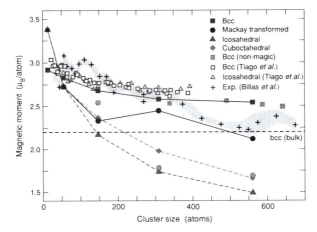

Fig. 3.5 Evolution of magnetic moments as a function of cluster size for Fe clusters. *Lines* are only guide to the eyes. The filled symbols such as *squares, circles, triangles, diamonds* denote the moments for bcc, Mackay transformed, icosahedral and cuboctahedral clusters. The *shaded circles* represent the SMT structures with magnetic configuration higher in energy and the *shaded squares* represent the relaxed bcc clusters. The *crosses* denote the experimental data taken from Refs. [31, 32] and the bcc bulk data are taken from Ref. [34]. The *solid region* are the extension of error bars. Figure taken from [27]

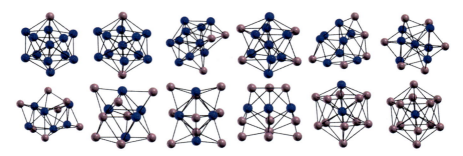

Fig. 3.6 The lowest energy structures for 13-atom Fe$_{13-n}$Pt$_n$ clusters. The *blue* and *brown* balls represent the Fe and Pt atoms, respectively

without symmetry constraints in order to minimize the interatomic forces. This leads to significant deviations between the data for the icosahedral clusters, while moments of the bcc isomers can be said to agree within the methodological accuracy. The central difference is the neglect of the compression of the cluster core due to the tension imposed by the surface shells which leads to the above-mentioned shell wise antiferromagnetic configuration, which triggers an instability between ferro- and antiferromagnetic arrangements present in the more close packed structures [36].

3.3 Structure and Magnetism in Binary Nanoparticles

There have been several studies on binary clusters (for a recent review, see, e.g., [37]). Using EAM model potentials, Montejano-Carrizales et al. [38] have shown surface segregation of Cu atoms on Cu-Ni clusters. On the other hand, using molecular dynamics simulations, Baletto et al. [39] observed well-defined single layer shells of Ag on Pd and Cu clusters. Experimentally, Rousset et al. [40] have also observed the segregation tendency for Pd-Pt binary clusters.

In the following, we will discuss the segregation and magnetic properties of Fe-Pt clusters. It has been found that the alloying of $4d$ or $5d$ elements to the $3d$ host affects their magnetic properties due to intermixing of d-orbitals. For example, Zitoun et al. [41] have experimentally observed an enhanced magnetic moment for Co-Rh binary clusters in comparison to their corresponding pure clusters. Similar enhancement is also observed in experiments in Fe-Pt binary clusters by Lu et al. [42]. For our studies, we have considered the closed shell icosahedron as the starting geometry for $Fe_{13-n}Pt_n$ clusters, where n is an integer between 1 and 13 describing the composition of the binary cluster. For each cluster composition several selected configurations are geometrically optimized. The lowest energy structures are shown in Fig. 3.6. We find that the icosahedral structure is stable only for very low and high concentrations of Pt. The structures for intermediate compositions tend to deform. To quantify the phenomena, we plot in Fig. 3.7 the number of nearest neighbor bonds (for homo species, hetero species, and total number of bonds) versus the number of Pt atoms. The dotted lines show the linear variation that would be achieved in an infinite system. We find that the total number of bonds (black squares) is not constant with respect to Pt concentration suggesting structural deformation. The number of Pt-Pt bonds is small in the range from $n = 1$ to $n = 7$. For higher n, it increases strongly. From the above analysis and Fig. 3.6, we can conclude that the number of Fe-Fe bonds tend to maximize in the ground state structures for all compositions. This is a clear hint for the segregation behavior which can be backed by the observation of the nonsymmetrical nature of the number of Fe-Fe (blue circle) and Pt-Pt bonds (red circle) in Fig. 3.7 with respect to the 50 % composition. For a pure mixing, the number of bonds should be symmetrical with respect to the 50 % composition line.

The mixing energy for the ground state structures for each composition is shown in Fig. 3.8. It is obtained as follows,

$$E^{\text{Mix}} = \frac{1}{N} \left[E_{A_n B_{N-n}} - \frac{n_A}{N} E_{A_N} - \frac{(N - n_A)}{N} E_{B_N} \right] \qquad (3.2)$$

where, E^{Mix} is the mixing energy per atom, A, B denote the different species of atoms in the cluster, N refers to the total number of atoms in the bimetallic cluster, n_A is the number of atoms of species A, $E_{A_n B_{N-n}}$ describes the total energy of the bimetallic cluster.

Concerning the magnetic properties of Fe-Pt clusters, we found ferromagnetic ordering for all lowest energy structures and all compositions. The magnetic moments

Fig. 3.7 The variation of nearest neighbor bonds Pt-Pt (*red circle*), Fe-Fe (*blue circle*), Fe-Pt (*diamond*) as a function of the number of Pt atoms. The *black squares* represent the total number of bonds in the cluster. For a perfect icosahedral cluster the total number of bonds should be 42. Because of the structural deformation for the intermediate cluster compositions, there is a reduced number of total bonds

Fig. 3.8 Mixing energy (eV/atom) with respect to Pt composition for the lowest energy structures of $Fe_{13-n}Pt_n$ clusters. See Fig. 3.6 for the energetically preferable structures

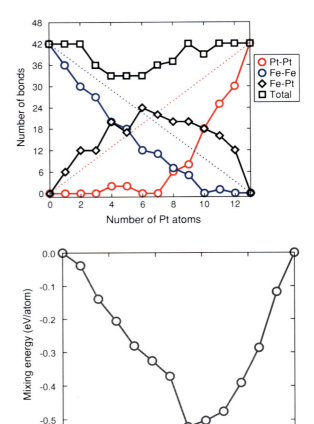

on Fe and Pt atoms for each composition are listed in Table 3.2. It is observed that the total moment of the Fe-Pt clusters decrease with increasing number of Pt atoms (see Table 3.2). This is understandable as Pt is nonmagnetic in the bulk, while an induced moment on the Pt atoms appears due to the presence of Fe atoms. For instance, we obtain a maximum induced moment (0.54 μ_B) on Pt atoms for the Fe_1Pt_{12} cluster.

For technologically relevant applications, particles with larger diameters in the range of at least a few nanometers are required. Near-stoichiometric Fe-Pt particles represent a widely discussed example as they were envisaged as medium for future high-end magnetic data storage devices [43]. This owes to the observation that stoichiometric FePt alloys in the thermodynamically stable $L1_0$ phase, which is characterized by a stacking of Fe and Pt layers along the [001] direction, exhibit a magnetic anisotropy constant which is about one order of magnitude larger than that

3 Nucleation, Structure and Magnetism of Transition Metal Clusters

Table 3.2 The average magnetic moments in units of μ_B on Fe (M_{Fe}) and Pt (M_{Pt}) atoms for the lowest energy structures of $Fe_{13-n}Pt_n$ clusters

Cluster	M_{Fe}	M_{Pt}	M_{Tot}
Fe_{13}	3.00		44.00
$Fe_{12}Pt_1$	2.86	0.39	38.00
$Fe_{11}Pt_2$	2.82	0.37	34.34
$Fe_{10}Pt_3$	2.95	0.56	34.00
Fe_9Pt_4	2.89	0.42	30.00
Fe_8Pt_5	3.01	0.41	28.00
Fe_7Pt_6	3.09	0.50	26.00
Fe_6Pt_7	3.17	0.52	24.00
Fe_5Pt_8	3.21	0.57	22.00
Fe_4Pt_9	3.30	0.41	18.00
Fe_3Pt_{10}	3.34	0.33	14.00
Fe_2Pt_{11}	3.10	0.32	9.91
Fe_1Pt_{12}	3.15	0.33	7.30
Pt_{13}		0.15	2.03

M_{Tot} the total cluster magnetic moment in μ_B

of current memory devices [44]. This helps to overcome the so-called superparamagnetic limit, which threatens the long-time stability of the information stored, essentially described by the Néel relaxation time. The latter can be expressed by an exponential dependence on the product of anisotropy constant and volume divided by temperature and thus imposes a lower boundary for the possible size above which the magnetization of a grain is (for a sufficiently long time) not affected by thermal relaxation processes. Considering the bulk values, particle sizes of 4 nm or even smaller appear feasible for perfectly ordered $L1_0$ particles. This, however, has not been achieved in practice, so far. See [45, 46] for a recent discussion of the coercive field and the MAE of experimentally fabricated Fe-Pt clusters. As an example, Fig. 3.9 shows the variation of coercive field as a function of the particle size for these clusters at 0 and 300 K, where a crossover from low to high coercivity takes place for particle size with ≈ 7 nm in diameter. One possible explanation for the low coercive field of the smaller particles (among several others) is the presence of twin structures or other morphologies at the relevant particle sizes, which do not possess the desired hard magnetic properties.

Figure 3.10 illustrates the structural stabilities of Fe-Pt clusters for corresponding different morphologies as a function of the cluster size up to 561 atoms (≈ 2.5 nm in diameter) [47]. Although the largest particles are still too small to avoid superparamagnetism, general trends can be formulated from the results, as the largest clusters already possess a balanced surface-to-volume ratio (45 % at 2.5 nm as compared to 32 % at 4 nm). It is obvious from the figure that disordered phases are considerably higher in energy than the desired $L1_0$ phase reflecting the strong ordering tendencies present in the bulk alloys. However, ordered multiply twinned morphologies are substantially more favorable throughout the investigated size range. These

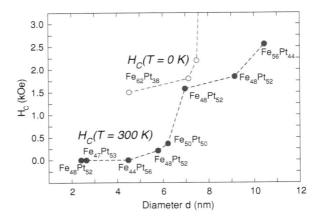

Fig. 3.9 The coercive field as a function of diameter for Fe-Pt clusters at two different temperatures. The *open* and *filled circles* denote the coercive fields at 0 and 300 K, respectively (figure plotted from the data in [45, 46])

results explain, at least in part, the experimental difficulties to obtain particles with large MAE at small sizes. The calculated energy differences can be as large as $\Delta E = E - E_{L1_0} = -30$ meV/atom in the case of the radially ordered $Fe_{265}Pt_{296}$ icosahedron, which is still close to thermal energies. Instead of a $L1_0$-type arrangement of the atoms in the individual twins forming the icosahedron, the shell wise alternating arrangement of the Fe and Pt-rich shells perpendicular to the (111) facets of the twins corresponds to an effective realization of the $L1_1$-type of order. In the bulk, this structure is found unstable for FePt alloys [48], but in small particles it is stabilized by the extremely favorable surface energy of purely Pt-covered (111) surfaces [49]. From the bulk and surface energies, the crossover to a $L1_0$ order should be expected for diameters of a few nanometers. This is in agreement with recent DFT total energy comparison of large clusters with up to 1,415 atoms (seven closed geometric shells), where the structural stabilities of ordered icosahedra and alternate icosahedra are compared with respect to the $L1_0$ cuboctahedron as a function of Fe-Pt cluster size (See Fig. 3.11). Figure 3.10 suggests that up to certain cluster size, different isomers of icosahedra are energetically more preferable compared to the $L1_0$ structure, while beyond that range, a crossover to $L1_0$ order occurs. Figure 3.11 is an extension to larger cluster sizes of the Fe-Pt nanoparticles (shown in Fig. 3.10) confirming this crossover. This yields an estimate of 3.5 to 4 nm for the critical diameter [6, 15]. The inset shows the energetics of the same morphologies for Co-Pt nanoparticles with up to 561 atoms. It shows a similar energy trend as found for Fe-Pt clusters up to 2.5 nm range, but multiply twinned and segregated structures are significantly more preferred than the Fe-Pt case.

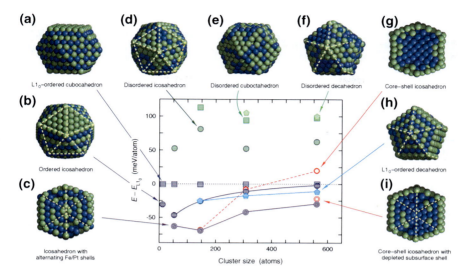

Fig. 3.10 The energies of Fe-Pt nanoparticles as a function of cluster size. The energy difference is with respect to the L1$_0$ cuboctahedron (*a*). The morphologies are shown only for Fe$_{265}$Pt$_{296}$ cluster. The *blue* and *green balls* denote the Fe and Pt atoms. Options *c*, *g*, and *i* show the cross sections of the icosahedral structures. The *yellow dashed lines* represent the interfaces of visible twins. The *squares*, *circles* and *pentagons* represent the cuboctahedra, icosahedra, and decahedra, respectively. The *shaded green* and *hatched blue* symbols refer to the disordered and ordered structures. *Thick and nested symbols*: core-shell icosahedra (*red*) and shell wise ordered icosahedra (*violet*). The *lines* are guide to the eyes. Figure originally published in [47]

3.4 MAE of Clusters

TM clusters are projected as possible candidates for future recording media, where hypothetically, each cluster can store one bit of information by means of their magnetic configuration. The magnetism of an ensemble of small clusters is not stable due to superparamagnetism induced by thermal fluctuation, which is a major obstacle for the miniaturization of the storage devices. Therefore, materials with a large MAE are needed to block the thermal fluctuations. For this purpose, MAE, which manifests itself in a significant dependence of the internal energy on the direction of the spontaneous magnetization, is a primary requirement. It has been observed that TM clusters may give rise to novel magnetic anisotropic properties compared to the bulk. For example, XMCD studies by Gambardella et al. [50] predict a large MAE of a single Co atom (9.0 meV/atom) deposited on a Pt(111) surface. Experimentally, Balashov et al. [51], through inelastic tunneling spectroscopy, have found very large value of MAE for Fe and Co clusters up to 3 atoms compared to the bulk. Theoretically, there exist abundant studies on the MAE of clusters deposited on substrates, e.g., [52–58] however, studies related to MAE of gas phase clusters are still limited [59–62]. Most theoretical studies are based on semi-empirical techniques.

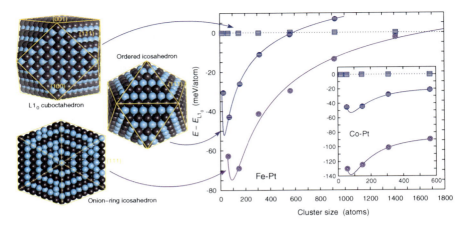

Fig. 3.11 The energies of Fe-Pt nanoparticles as a function of cluster size with up to 1,415 atoms. The energy difference is with respect to the L1$_0$ cuboctahedron shown in the *top left panel*. The clusters on the *left* show the morphologies of the Fe$_{679}$Pt$_{736}$ (L1$_0$ and onion-ring) and Fe$_{440}$Pt$_{483}$ (ordered icosahedron) clusters. The *blue* and *black balls* represent the Fe and Pt atoms, respectively. The *hatched square* represents the reference structure (L1$_0$ cuboctahedron). The *hatched* and *nested circles* indicate the ordered icosahedron (constructed from the L1$_0$ cuboctahedron through a full transformation along the Mackay path) and onion ring icosahedron (the cross section is shown on the *left*). The *yellow dashed lines* represent the interfaces of visible twins. The *inset* shows the energetics for Co-Pt clusters with up to 561 atoms. Figure is originally published in [6]

For example, using the tight binding technique, Pastor et al. [59] have calculated the MAE of small Fe clusters up to 7 atoms, where they have found large MAE compared to the corresponding bulk values as well as thin films. Using the same method, Guirado-Lopez et al. [60] have calculated the MAE for Co nanoparticles. They have also obtained large MAE for Co clusters compared to the bulk. Calculation of the MAE for isolated clusters based on ab initio methods is mostly available for TM dimers [62–64]. For larger clusters, there exist only few reports [7, 65, 66]. Using DFT, Kortus et al. [65] have calculated the MAE of 5- and 13-atom Co and Fe clusters, where they have obtained the MAEs for Fe$_5$, Co$_5$ and Fe$_{13}$ as 0.2, 0.1, and 0.27 meV/atom, respectively. However, for Co$_{13}$, they obtain nearly zero MAE. Calculations for the orbital moments as well as MAE for Co clusters up to 6 atoms based on full potential linearized augmented plane wave method by Hong et al. [66] have shown significantly enhanced values of orbital moments relative to the hcp bulk Co. However, they obtain a rather small value of MAE for Co clusters (<1 meV/atom).

In our work, we have studied the MAE for 13-atom Fe, Co, and Ni (M$_{13}$) clusters. It is well known that the MAE in cubic bulk TM is in the order of $\sim$1 μeV/atom. MAE in bulk depends mainly on dipolar and spin-orbit interactions [67]. However, since the dipolar-interactions are feeble in spherical clusters, only spin-orbit coupling plays a major role. As a result, calculation of MAE requires much careful attention in order to overcome minute computational noise, which can easily interfere with

3 Nucleation, Structure and Magnetism of Transition Metal Clusters

the estimated quantities. Therefore, one must have a very refined convergence with respect to charge density, as well as a proper choice of the energy cutoff and a large Fourier mesh. For our studies of the MAE of clusters, we have used a Gaussian broadening parameter of 0.01 eV for the energy levels and a very high value of plane-wave cutoff of 1,000 eV with a 0.046 Å^{-1} Fourier grid spacing. The results discussed in the following are published in Ref. [68].

3.4.1 MAE for Perfect Clusters

Figure 3.12 shows the perfect icosahedral cluster and the definition of the (x, z)-plane with angle θ for the magnetization directions used for the MAE calculations. For $\theta = 0$, the magnetization is directed parallel to an axis passing through the central atom and the center of a bond connecting two outer atoms; with increasing θ, the magnetization direction passes through the center of a triangular facet, through an outer atom and finally arrives again at a bond center for $\theta = \pi/2$.

The minimum energy center-shell distances for each of the perfect icosahedral M_{13} clusters are calculated, which are found to be 2.39, 2.33 and 2.32 Å for Fe$_{13}$, Co$_{13}$ and Ni$_{13}$, respectively with total magnetic moments 44 μ_B (Fe$_{13}$), 31 μ_B (Co$_{13}$), and 8 μ_B (Ni$_{13}$). For Fe$_{13}$ with 44 μ_B, we find difficulties in achieving convergence in the calculation of MAE at the low smearing of 0.01 eV because of dense energy levels near the Fermi level. Therefore, we have slightly enlarged the center-shell distance to 2.57 Å (from 2.39 Å), where we obtain a magnetic moment of 46 μ_B. Figure 3.13 shows the comparison between the θ-dependent energy differences $\Delta E(\theta)$ of a perfect Fe$_{13}$ ICO (with total magnetic moment 46 μ_B) obtained from ab initio calculations with the Néel model. We obtain the global minima for the magnetization directed parallel to the line from the central atom to the circumcenter of the triangular facets and the global maxima for the magnetization parallel to the line from center atom to one of the surface atoms. The magnitude of $\Delta E(\theta)$ for Fe$_{13}$ cluster is found to be 1.7 μeV/atom, which is comparable to that of the bulk bcc Fe. The MAE of perfect Co$_{13}$ and Ni$_{13}$ is found to be 0.31, and 0.77 μeV/atom, respectively. $\Delta E(\theta)$ shows a similar qualitative trend for Co$_{13}$ and Ni$_{13}$ as obtained for perfect Fe$_{13}$.

The total energy obtained as a function of θ from the ab initio calculations for Fe$_{13}$ are fitted to the anisotropy expansion of the Néel surface anisotropy model [69, 70] for a nearly spherical cluster, which is defined as

$$\Delta E_{\text{Néel}} = \sum_n E_n = -\sum_n \sum_{i=1}^{N} D_n (\mathbf{e}_i \cdot \mathbf{S}_i)^n. \tag{3.3}$$

where, D_n is the anisotropy constant of order n (an even integer), $\mathbf{e}_i$ the normalized position vector of atom i along the radial directions and $\mathbf{S}_i$ is the corresponding magnetic moment. The D_n is assumed to be θ-independent. From the symmetry

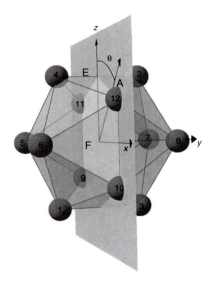

Fig. 3.12 Variation of θ along the x–z plane passing through the perfect icosahedral cluster. The letters E, F, and A are the middle of an edge, the middle of a facet, and an outer atom, respectively

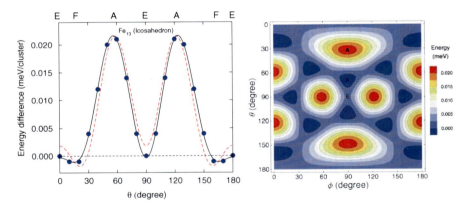

Fig. 3.13 *Left* the θ-dependent energy (in meV per cluster) for perfect Fe_{13} icosahedral cluster showing a comparison between the ab-initio data (*filled circles* connected by *solid line* as guide to the eyes) and the fit to the anisotropy term of Néel model (*dashed line*). The energy difference is defined as $\Delta E = E(\theta) - E(0)$. The letters A, E, and F refer to the positions defined in Fig. 3.12. *Right* the (θ, ϕ) for the MAE of the perfect Fe_{13} cluster

considerations of perfect ICO, the second and fourth-order contributions have no θ-dependence. The major contribution to the anisotropy energy is from the next order contribution, i.e., the sixth order. Still higher order contributions are negligible and the $\Delta E(\theta)$ could be successfully fitted to the sixth order fit.

3 Nucleation, Structure and Magnetism of Transition Metal Clusters

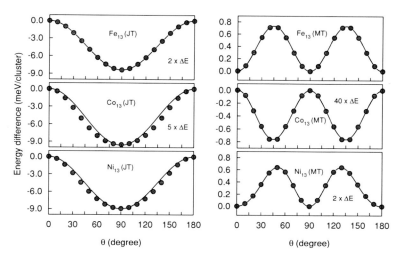

Fig. 3.14 The energy differences $\Delta E = E(\theta) - E(0)$ in meV/cluster as a function of θ for the JT (*left*) and Mackay-distorted (*right*) M_{13} clusters. The magnetization direction varies in the $x - z$ plane. For the JT case, the $x - z$ plane passes through atoms 3, 4, 12, and middle of the bond 8–10 (see *left panel* of Fig. 3.2). For the MT case, θ passes through all atoms shown in Fig. 3.12. For the JT-distorted clusters, the energy difference for the Co_{13} and Fe_{13} clusters is multiplied by factors of 5 and 2, respectively, whereas for the MT-distorted clusters, the energy difference for Co_{13} and Ni_{13} is multiplied by factors of 40 and 2, respectively

3.4.2 MAE for Relaxed Clusters

As already discussed in the previous section, the M_{13} clusters show tendencies for structural distortions. We have calculated the MAE for the JT-distorted and Mackay-distorted clusters (see Fig. 3.3 and Table 3.1 for the structural parameters). In order to calculate the MAE of relaxed clusters, the perfect ICO structure is relaxed along two different paths. These paths are, on one hand, the JT type distortion and on the other hand, the distortion along the Mackay transformation path, which are discussed in Sect. 3.3. The relaxation patterns for both distortions are illustrated in Fig. 3.3. Note that the clusters have different spatial orientations in order to underline the remaining symmetry in each case. Therefore, the θ path for the calculation of MAE is different for the two distortions. For the JT distortion, the θ path is A-E-A-F-E-F-A (see Fig. 3.3, top left panel) and for the MT distortion, it follows the path E-F-A-E as shown in Fig. 3.12.

Figure 3.14 shows the θ-dependent energy differences for the JT-distorted (left panel) and MT-distorted (right panel) M_{13} clusters. For the JT-distorted M_{13} clusters, a large second-order contribution of D_2 is obtained because of the symmetry breaking. All higher order terms are significantly small and hence can be safely neglected. On the other hand, for the MT clusters, due to their cubic symmetry, no second order contribution is found, and the lowest order contribution is only D_4 for this case. The

Table 3.3 Anisotropy constants D_n (in units of meV) according to Eq. (3.3), obtained by fitting the GGA results for perfect and relaxed M_{13} clusters

Cluster	Moment (μ_B)	D_2	D_4	D_6	$r-1$	$s-1$	ΔE^{DFT}
Fe$_{13}$ (ICO)	46			−0.04	0	0	0.020
Co$_{13}$ (ICO)	31			−0.01	0	0	0.004
Ni$_{13}$ (ICO)	8			−0.02	0	0	0.010
Fe$_{13}$ (JT)	44	15.0			−0.04		4.200
Co$_{13}$ (JT)	31	−16.0			0.01		1.900
Ni$_{13}$ (JT)	8	44.1			−0.02		8.900
Fe$_{13}$ (MT)	44		−11.5			0.07	0.710
Co$_{13}$ (MT)	31		−0.4			−0.04	0.020
Ni$_{13}$ (MT)	8		−10.1			0.04	0.320

In the icosahedral symmetry, second- and fourth-order contributions do not depend on θ for any value of $D_{2,4}$. Thus, the related data are omitted. The same holds for the second-order terms in cubic symmetry (MT clusters). In all cases, only the leading order terms are essential and all higher order terms can be neglected. r and s are parameters, which describe the JT and the Mackay-transformation, respectively (see text). The DFT energy differences (in meV) are shown in the last column of the table

higher orders can be neglected. Table 3.3 reports the corresponding values. For all cases, the MAE of the JT-distorted clusters is found to be larger than that of MT M_{13} clusters. For instance, for Fe$_{13}$, the MAE for the JT-distorted cluster is calculated to be 322 μeV/atom, which is approximately six times larger in comparison to the MT Fe$_{13}$ cluster (55 μeV/atom). Similarly, for Co$_{13}$, the MAE for the JT-distorted cluster has a value of 147 μeV/atom. This is approximately 100 times larger than the MT-distorted (1.42 μeV/atom) cluster. However, among all JT and MT clusters, the JT-distorted Ni$_{13}$ cluster has the largest value (688 μeV/atom), which is about 30 times larger than that of the MT-distorted one. The reason behind the large values of the MAE for the JT-distorted clusters relative to the MT ones is the lower symmetry of the JT-distorted cluster.

We obtain a larger value of MAE per atom for JT- and MT-distorted Fe$_{13}$ and Ni$_{13}$ clusters relative the bulk, where LSDA calculations found 1.4 μeV/atom (bcc Fe), 2.7 μeV/atom (fcc Ni) [71]. For Co$_{13}$, a different trend is observed. Though the JT-distorted Co$_{13}$ has a large MAE value $\sim$100 times larger than that of the bulk (1.3 μeV/atom for fcc Co), for MT Co$_{13}$, it is similar to the bulk.

3.5 Summary and Outlook

We have investigated the ground state structural and magnetic properties of elemental and binary nanoclusters. We have considered two types of structural transformations in TM clusters, namely Jahn-Teller and Mackay transformation. Our calculations suggest an enhanced stability for the JT-distorted Fe$_{13}$ cluster, while Mackay distortion is observed for Co$_{13}$. For Ni$_{13}$, both JT- and MT geometries are found to be

3 Nucleation, Structure and Magnetism of Transition Metal Clusters

similar in energy. For large icosahedral Fe clusters, we observe a shell wise Mackay transformation with decreasing order as we go from center to shell. Thus, the interior of the cluster tends towards a cuboctahedral geometry. The magnetic moment as a function of the Fe cluster size has been shown to agree well with the experimental results. Our studies on binary Fe-Pt clusters show segregation behavior with a tendency for the Pt atoms to remain on the surface of the clusters. From energetics, we find that for binary Fe-Pt clusters, with diameters less than 3.5–4 nm icosahedral structures are more stable. We have studied systematically the MAE of 13 atom clusters of Fe, Ni, and Co. As expected, we obtain large MAEs for relaxed Fe_{13} and Ni_{13} clusters compared with the symmetric clusters, which are by several orders of magnitude larger than the corresponding values of the bulk. However, Co_{13} does not comply to this above trend.

A future issue not discussed in this paper the continuation of the zero-temperature calculations to finite temperatures, which, in principle, is possible by either VASP using ab initio molecular dynamics simulations of smaller clusters or by directly computing the free energies and retaining the most important elementary excitations such as electronic, vibrational, and magnetic ones as was recently done by calculating the structural transformations in bulk Ni_2MnGa [72]. The thermodynamic properties of TM clusters are an important issue, especially, regarding the role of extrinsic effects not discussed here like growth of the particles in the gas phase (H_2) or the impact of a substrate on the cluster properties. Extrinsic effects can, for example, be responsible for the crossover from negative to positive thermal expansion with increasing size, which has been recently discussed for Pt clusters [73, 74]. Although the discussion of the thermodynamic properties of magnetic TM clusters is a natural extension of the ground state properties discussed here, this subject is beyond the scope of the present paper and is left for future computations using parallel computer platforms.

Acknowledgments We thank the John von Neumann Institute for Computing, the Jülich supercomputing Center and the Center for Computational Sciences and Simulation (CCSS), University of Duisburg-Essen for computation time and support. Also Financial support was granted by the Deutsche Forschungsgemeinschaft through SFB 445.

References

1. S. Sun, C.B. Murray, D. Weller, L. Folks, A. Moser, Monodisperse FePt nanoparticles and ferromagnetic FePt nanocrystal superlattices. Science **287**, 1989 (2000)
2. Y. Xie, J.A. Blackman, Magnetism of iron clusters embedded in cobalt. Phys. Rev. B **66**, 085410 (2002)
3. O. Diéguez, M.M.G. Alemany, C. Rey, P. Ordejón, L.J. Gallego, Density-functional calculations of the structures, binding energies, and magnetic moments of Fe clusters with 2 to 17 atoms. Phys. Rev. B **63**, 205407 (2001)
4. A.V. Postnikov, P. Entel, J.M. Soler, Density functional simulation of small Fe nanoparticles. Eur. Phys. J. D **25**, 261 (2003)
5. H.M. Duan, Q.Q. Zheng, Symmetry and magnetic properties of transition metal clusters. Phys. Lett. A **280**, 333 (2001)

6. M.E. Gruner, P. Entel, Competition between ordering, twinning, and segregation in binary magnetic $3d$-$5d$ nanoparticles: a supercomputing perspective. Int. J. Quant. Chem. **112**, 277 (2012)
7. C. Antoniak, M.E. Gruner, M. Spasova, A.V. Trunova, F.M. Römer, A. Warland, B. Krumme, K. Fauth, S. Sun, P. Entel, M. Farle, H. Wende, A guideline for atomistic design and understanding of ultrahard nanomagnets. Nat. Commun. **2**, 528 (2011)
8. P. Hohenberg, W. Kohn, Inhomogeneous electron gas. Phys. Rev. **136**, B864 (1964)
9. W. Kohn, L.J. Sham, Self-consistent equations including exchange and correlation effects. Phys. Rev. **140**, A1133 (1965)
10. G. Kresse, J. Furthmüller, Efficient iterative schemes for ab initio total-energy calculations using a plane-wave basis set. Phys. Rev. B **54**, 11169 (1996)
11. P.E. Blöchl, Projector augmented-wave method. Phys. Rev. B **50**, 17953 (1994)
12. J.P. Perdew, in *Electronic Structure of Solids'91*, ed. by P. Ziesche, H. Eschrig (Akademie, Berlin, 1991), pp. 11–20
13. S.H. Vosko, L. Wilk, M. Nusair, Accurate spin-dependent electron liquid correlation energies for local spin density calculations: a critical analysis. Can. J. Phys. **58**, 1200 (1980)
14. M.E. Gruner, G. Rollmann, P. Entel, Large-scale first-principles calculations of magnetic nanoparticles, in *Proceedings of the NIC Symposium 2008*, vol. 39, ed. by G. Münster, D. Wolf, M. Kremer, NIC Series (John von Neumann Institute for Computing, Jülich, 2008), p. 161
15. M.E. Gruner, P. Entel, Simulating functional magnetic materials on supercomputers. J. Phys. Condens. Matter *21*, 293201 (2009)
16. R. Ferrando, A. Fortunelli, G. Rossi, Quantum effects on the structure of pure and binary metallic nanoclusters. Phys. Rev. B. **72**, 085449 (2005)
17. S. Heinrichs, W. Dieterich, P. Maass, Modeling epitaxial growth of binary alloy nanostructures on a weakly interacting substrate. Phys. Rev. B. **75**, 085437 (2007)
18. F.R. Negreiros, Z. Kuntová, G. Barcaro, G. Rossi, R. Ferrando, A. Fortunelli, Structures of gas-phase Ag-Pd nanoclusters: a computational study. J. Chem. Phys. **132**, 234703 (2010)
19. M. Pellarin, B. Baguenard, J.L. Vialle, J. Lerme, M. Broyer, J. Miller, A. Perez, Evidence for icosahedral atomic shell structure in nickel and cobalt clusters. Comparison with iron clusters. Chem. Phys. Lett. **217**, 349 (1994)
20. T. Vystavel, G. Palasantzas, S.A. Koch, J.T.M. De Hosson, Nanosized iron clusters investigated with in situ transmission electron microscopy. Appl. Phys. Lett. **82**, 197 (2003)
21. M.I. Katsnelson, Y.V. Irkhin, L. Chioncel, A.I. Lichtenstein, R.A. de Groot, Half-metallic ferromagnets: from band structure to many-body effects. Rev. Mod. Phys. **80**, 315 (2008)
22. M.E. Gruner, P. Entel, I. Opahle, M. Richter, Ab initio investigation of twin boundary motion in the magnetic shape memory Heusler alloy Ni_2MnGa. J. Mater. Sci. **43**, 3825 (2008)
23. E.C. Bain, The nature of martensite. Trans. Am. Inst. Min. Metall. Pet. Eng. **70**, 25 (1924)
24. Z. Nishiyama, *Martensitic Transformation* (Academic Press, New York, 1978)
25. A.L. Mackay, A dense non-crystallographic packing of equal spheres. Acta Cryst. **15**, 916–918 (1962)
26. G. Rollmann, S. Sahoo, P. Entel, Structural and magnetic properties of Fe-Ni clusters. Phys. Status Solid. **201**, 3263 (2004)
27. G. Rollmann, M.E. Gruner, A. Hucht, P. Entel, Shellwise Mackay transformation in iron nanoclusters. Phys. Rev. Lett. **99**, 083402 (2007)
28. G. Rollmann, P. Entel, S. Sahoo, Competing structural and magnetic effects in small iron clusters. Comput. Mater. Sci. **35**, 275 (2006)
29. H. Jónsson, H.C. Andersen, Icosahedral ordering in the Lennard-Jones liquid and glass. Phys. Rev. Lett. **60**, 2295 (1988)
30. M.E. Gruner, G. Rollmann, A. Hucht, P. Entel, Structural and magnetic properties of transition metal nanoparticles from first principles, in *Advances in Solid State Physics*, vol. 47, ed. by R. Haug (Springer, Berlin, 2008), p. 117
31. I.M.L. Billas, J.A. Becker, A. Châtelain, W.A. de Heer, Magnetic moments of iron clusters with 25 to 700 atoms and their dependence on temperature. Phys. Rev. Lett. **71**, 4067 (1993)

3 Nucleation, Structure and Magnetism of Transition Metal Clusters

32. I.M.L. Billas, A. Châtelain, W.A. de Heer, Magnetism from the atom to the bulk in iron, cobalt, and nickel clusters. Science **265**, 1682 (1994)
33. I.M.L. Billas, A. Châtelain, W.A. de Heer, Magnetism of Fe, Co and Ni clusters in molecular beams. J. Magn. Mat. **168**, 64 (1997)
34. H.C. Herper, E. Hoffmann, P. Entel, Ab initio full-potential study of the structural and magnetic phase stability of iron. Phys. Rev. B **60**, 3839 (1999)
35. M.L. Tiago, Y. Zhou, M.M.G. Alemany, Y. Saad, J.R. Chelikowsky, Evolution of magnetism in iron from the atom to the bulk. Phys. Rev. Lett. **97**, 147201 (2006)
36. M.E. Gruner, G. Rollmann, S. Sahoo, P. Entel, Magnetism of close packed Fe_{147} clusters. Phase Trans. **79**, 701 (2006)
37. R. Ferrando, J. Jellinek, R.L. Johnston, Nanoalloys: from theory to applications of alloy clusters and nanoparticles. Chem. Rev. **108**, 845 (2008)
38. J.M. Montejano-Carrizales, M.P. Iñiguez, J.A. Alonso, Embedded-atom method applied to bimetallic clusters: the Cu-Ni and Cu-Pd systems. Phys. Rev. B **49**, 16649 (1994)
39. F. Baletto, C. Mottet, R. Ferrando, Growth simulations of silver shells on copper and palladium nanoclusters. Phys. Rev. B **66**, 155420 (2002)
40. J.L. Rousset, A.M. Cadrot, F.J. Cadete Santos Aires, A. Renouprez, P.Mélinon, A. Perez, M. Pellarin, J.L. Vialle, M. Broyer, Study of bimetallic PdPt clusters in both free and supported phases. J. Chem. Phys. **102**, 8574 (1995)
41. D. Zitoun, M. Respaud, M. Fromen, M.J. Casanove, P. Lecante, C. Amiens, B. Chaudret, Magnetic enhancement in nanoscale CoRh particles. Phys. Rev. Lett. **89**, 037203 (2002)
42. L.Y. Lu, D. Wang, X.G. Xu, Q. Zhan, Y. Jiang, Enhancement of magnetic properties for FePt nanoparticles by rapid annealing in a vacuum. J. Chem. Phys. **113**, 19867 (2009)
43. M.L. Plumer, E.J. Van, D. Weller (eds.), *The physics of ultra-high-density magnetic recording* (Springer, Berlin, 2001)
44. D. Weller, A. Moser, Thermal effect limits in ultrahigh-density magnetic recording. IEEE Trans. Magn. **35**, 4423 (1999)
45. U. Wiedwald, L. Han, J. Biskupek, U. Kaiser, P. Ziemann, Preparation and characterization of supported magnetic nanoparticles prepared by reverse micelles. Beilstein J. Nanotechnol. **1**, 24 (2010)
46. B. Rellinghaus, S. Stappert, M. Acet, E.F. Wassermann, Magnetic properties of FePt nanoparticles. J. Magn. Magn. Mater. **266**, 142 (2003)
47. M.E. Gruner, G. Rollmann, P. Entel, M. Farle, Multiply twinned morphologies of FePt and CoPt nanoparticles. Phys. Rev. Lett. **100**, 087203 (2008)
48. A. Dannenberg, M.E. Gruner, P. Entel, First-principles study of the structural stability of $L1_1$ order in Pt-based alloys. J. Phys. Conf. Ser. **200**, 072021 (2010)
49. A. Dannenberg, M.E. Gruner, A. Hucht, P. Entel, Surface energies of stoichiometric FePt and CoPt alloys and their implications for nanoparticle morphologies. Phys. Rev. B **80**, 245438 (2009)
50. P. Gambardella, S. Rusponi, M. Veronese, S.S. Dhesi, C. Grazioli, A. Dallmeyer, I. Cabria, R. Zeller, P.H. Dederichs, K. Kern, C. Carbone, H. Brune, Giant magnetic anisotropy of single cobalt atoms and nanoparticles. Science **300**, 1130 (2003)
51. T. Balashov, T. Schuh, A.F. Takács, A. Ernst, S. Ostanin, J. Henk, I. Mertig, P. Bruno, T. Miyamachi, S. Suga, W. Wulfhekel, Magnetic anisotropy and magnetization dynamics of individual atoms and clusters of Fe and Co on Pt(111). Phys. Rev. Lett. **102**, 257203 (2009)
52. R. Félix-Medina, J. Dorantes-Dávila, G.M. Pastor, Ground-state magnetic properties of Co_N clusters on Pd(111): spin moments, orbital moments, and magnetic anisotropy. Phys. Rev. B **67**, 094430 (2003)
53. S. Rohart, C. Raufast, L. Favre, E. Bernstein, E. Bonet, V. Dupuis, Magnetic anisotropy of $Co_x Pt_{1-x}$ clusters embedded in a matrix: influences of the cluster chemical composition and the matrix nature. Phys. Rev. B **74**, 104408 (2006)
54. A.N. Andriotis, M. Menon, Orbital magnetism: pros and cons for enhancing the cluster magnetism. Phys. Rev. Lett. **93**, 026402 (2004)

55. Š. Pick, V.S. Stepanyuk, A.L. Klavsyuk, L. Niebergall, W. Hergert, J. Kirschner, P. Bruno, Magnetism and structure on the atomic scale: small cobalt clusters in Cu(001). Phys. Rev. B **70**, 224419 (2004)
56. J. Hafner, D. Spišák, Morphology and magnetism of Fe_n clusters ($n = 1 - 9$) supported on a Pd(001) substrate. Phys. Rev. B **76**, 094420 (2007)
57. S. Bornemann, J. Minár, J.B. Staunton, J. Honolka, A. Enders, K. Kern, H. Ebert, Magnetic anisotropy of deposited transition metal clusters. Eur. Phys. J. D **45**, 529 (2007)
58. B. Nonas, I. Cabria, R. Zeller, P.H. Dederichs, T. Huhne, H. Ebert, Strongly enhanced orbital moments and anisotropies of adatoms on the Ag(001) surface. Phys. Rev. Lett. **86**, 2146 (2001)
59. G.M. Pastor, J. Dorantes-Dávila, S. Pick, H. Dreyssé, Magnetic anisotropy of $3d$ transition-metal clusters. Phys. Rev. Lett. **75**, 326 (1995)
60. R.A. Guirado-López, J.M. Montejano-Carrizales, Orbital magnetism and magnetic anisotropy energy of Co nanoparticles: role of polytetrahedral packing, polycrystallinity, and internal defects. Phys. Rev. B **75**, 184435 (2007)
61. L. Fernández-Seivane, J. Ferrer, Magnetic anisotropies of late transition metal atomic clusters. Phys. Rev. Lett. **99**, 183401 (2007)
62. P. Błoński, J. Hafner, Magnetic anisotropy of transition-metal dimers: density functional calculations. Phys. Rev. B **79**, 224418 (2009)
63. T.O. Strandberg, C.M. Canali, A.H. MacDonald, Transition-metal dimers and physical limits on magnetic anisotropy. Nat. Mater. **6**, 648 (2007)
64. D. Fritsch, K. Koepernik, M. Richter, H. Eschrig, Transition metal dimers as potential molecular magnets: a challenge to computational chemistry. J. Comp. Chem. **29**, 2210 (2008)
65. J. Kortus, T. Baruah, M.R. Pederson, Magnetic moment and anisotropy in Fe_nCo_m clusters. Appl. Phys. Lett. **80**, 4193 (2002)
66. J. Hong, R.Q. Wu, First principles determinations of magnetic anisotropy energy of Co nanoclusters. J. Appl. Phys. **93**, 8764 (2003)
67. J.H. van Vleck, On the anisotropy of cubic ferromagnetic crystals. Phys. Rev. B **52**, 1178 (1937)
68. S. Sahoo, A. Hucht, M.E. Gruner, G. Rollmann, P. Entel, A. Postnikov, J. Ferrer, L. Fernández-Seivane, M. Richter, D. Fritsch, S. Sil, Magnetic properties of small Pt-capped Fe, Co., and Ni clusters: a density functional theory study. Phys. Rev. B **82**, 054418 (2010)
69. L. Néel, The surface anisotropy of ferromagnetic substances. C. R. Acad. Sci. Paris **237**, 1468 (1953)
70. L. Néel, Surface magnetic anisotropy and orientational superstructures. J. Phys. Radium **15**, 225 (1954)
71. S.V. Halilov, A. Ya, Perlov, P. M. Oppeneer, A. N. Yaresko, and V. N. Antonov, Magnetocrystalline anisotropy energy in cubic Fe, Co., and Ni: applicability of local-spin-density theory reexamined. Phys. Rev. B **57**, 9557 (1998)
72. M.A. Uijttewaal, T. Hickel, J. Neugebauer, M.E. Gruner, P. Entel, Understanding the phase transitions of the Ni_2MnGa magnetic shape memory system from first principles. Phys. Rev. Lett. **102**, 035702 (2009)
73. S.I. Sanchez, L.D. Menard, A. Bram, J.H. Kang, M.W. Small, R.G. Nuzzo, A.I. Frenkel, The Emergence of nonbulk properties in supported metal clusters: negative thermal expansion and atomic disorder in Pt nanoclusters supported on γ-Al_2O_3. J. Am. Chem. Soc. **131**, 7040 (2009)
74. B. Roldan Cuenya, A.I. Frenkel, S. Mostafa, F. Behafarid, J.R. Croy, L.K. Ono, Q. Wang, Anomalous lattice dynamics and thermal properties of supported size- and shape-selected Pt nanoparticles. Phys. Rev. B **82**, 155450 (2010)

Chapter 4
Synthesis and Film Formation of Monodisperse Nanoparticles and Nanoparticle Pairs

Shubhra Kala, Marcel Rouenhoff, Ralf Theissmann and Frank Einar Kruis

Abstract The use of well-defined nanoparticles for functional film applications is described. The advantages of applying size-fractionation, e.g. by means of mobility analysis, are described together with the technological obstacles which have to be overcome. The synthesis of Au and Ge nanoparticles by means of spark discharge is described. To prepare alloy nanoparticles, two different approaches have been utilized. Au-Ge pair nanoparticles are formed by bipolar mixing after separate size selection of both materials. The synthesis of AuGe alloyed nanoparticles is also performed by co-sparking from two different electrodes. The development of an electrostatic precipitator for functional film formation is described.

4.1 Introduction

In the last ten years, the use of nanoparticles for functional applications has seen a tremendous development. Examples are the use of quantum dots in electronic applications, magnetic dots for high-density magnetic recording, metal oxide nanoparticles for thin gas-sensitive films, thermoelectric films and hybrid inorganic-organic films for solar cells. In most of these functional applications, thin films are used. The most important nanoparticle films can be distinguished according to the following classification:

(1) Sub-monolayers of single, non-contacting nanoparticles on a substrate
(2) Several monolayers of nanoparticles on top of each other
(3) Micrometer thick nanoparticle layers, usually showing a clearly porous structure
(4) Composite layers of nanoparticles imbedded in a continuous (e.g. polymer) matrix

---

S. Kala · M. Rouenhoff · R. Theissmann · F. E. Kruis(✉)
Faculty of Engineering and CENIDE, University of Duisburg-Essen, 47057 Duisburg, Germany
e-mail: einar.kruis@uni-due.de

A. Lorke et al. (eds.), *Nanoparticles from the Gas Phase*, NanoScience and Technology,
DOI: 10.1007/978-3-642-28546-2_4, © Springer-Verlag Berlin Heidelberg 2012

In our research, we strongly propagate the use of well-defined nanoparticles for functional film formation. First demonstrated for producing films from monodisperse, monocrystalline PbS nanoparticles [11], the basic concept is (1) to synthesize the particles by means of a physical synthesis step forming a polydisperse nanoaerosol, (2) to select a small portion of the size spectrum by means of electrical mobility-based size fractionation in a Differential Mobility Analyzer (DMA), (3) to convert the aggregated and irregularly shaped nanoparticles into quasi-spherical and monocrystalline particles by means of in-flight thermal annealing and (4) to deposit the nanoparticles on suitable substrates. Clearly, this approach has the disadvantage of losing a large fraction of the originally synthesized particles. This has the following causes:

(1) Only a certain portion of the nanoparticles can be electrically charged, e.g. when radioactive sources are used, the bipolar charge equilibrium poses a limit to the charging probability.
(2) The electrical mobility window is much narrower than the original size distribution. This is inherent to fractionation techniques.
(3) The multi-step process leads to particle losses.

In practise, only some 1 % of the evaporated material is finally converted into the well-defined nanoparticles sought for. It has to be realized, however, that the cost of the evaporation material is far less than the cost of the equipment, electrical power and inert gas required.

Then, why do we choose this technologically complicated and time-consuming experimental strategy? The arguments are the following.

(1) Many of the functional applications are strongly dependent on the size of the nanoparticles. In conventional processing for material science, the process conditions are changed to vary the mean particle size. A first problem is that control over the width of the distribution is not possible or only rather minimal, whereas the width of the size distribution will have a major impact on the properties. A second problem is that changing the process conditions not only changes the mean size but also other particle properties such as crystallinity and particle form, so that it will never be clear what will have been the real cause of the observed change in properties. Use of narrowly distributed particles allows to perform more fundamental scientific investigation towards structure-property relationships.
(2) In conventional gas-phase synthesis approaches, aggregation is unavoidable, as Brownian coagulation is unavoidable. Although the aggregates might be weak, their existence makes it virtual impossible to obtain monocrystalline, single, quasi-spherical particles. In-flight sintering of these aggregates is necessary to improve the product quality, but the strong size dependence of sintering as well as the unavoidable Brownian coagulation makes this method very problematic. When in-flight sintering size-selected aggregates, however, the initial aggregate size is uniform so that a more or less uniform sintering time is needed. Furthermore, the number concentration due to the size selection step decreased to

4 Synthesis and Film Formation of Monodisperse Nanoparticles

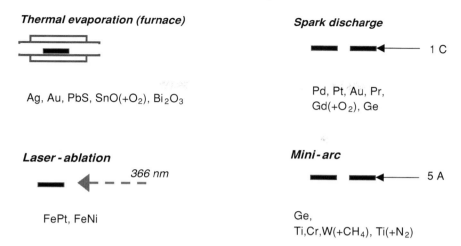

Fig. 4.1 Different physical synthesis techniques used in the authors' laboratory. The elements and compounds successfully synthesized are also indicated

10^4–10^6 cm^{-3}, so that Brownian coagulation during sintering is no longer a problem.

(3) For commercial applications of nanoparticulate films, it is essential that the films can be produced in a reproducible manner. Although nanoparticles can be dispersed in liquids and films can be formed from the dispersion, the solvent and dispersant aids will have an unclear effect on the films formed. Applying direct deposition from nanoaerosols on the substrate has the advantages of reducing these uncertainties and avoiding the use of potential harmful substances. Measurements on nanoparticulate films deposited from polydisperse nanoaerosols show a large sample-to-sample variation in properties such as electrical conductivity of the thin films. Use of exactly defined and very reproducible sizes is essential to obtain reproducible properties of the films. The subject of film formation from nanoaerosols is an unexplored domain of engineering full of potential problems. The most economic deposition techniques are nonvacuum ones, so that the deposition strongly depends on the nanoparticle trajectories between the inlet and the substrate. These trajectories are size-dependent, so that a polydispersity will automatically lead to unequal distribution of particle sizes over the substrate. Using narrowly distributed particles avoids size inhomogeneity over the films. Still, methods have to be found to obtain films with uniform thickness, which will be discussed later in this chapter.

The multi-step, size-selection based synthesis method including inflight annealing is not simple. Many, mostly technological obstacles have to be overcome. We will list here some of the most encountered ones.

(1) *Choice of the most suitable synthesis technique.* As the synthesis and deposition is usually carried out over an extended period of time without a person continuously

attending the experiments, the use of chemical precursors is better avoided. Instead, physical synthesis has proven to be a reliable source of nanoparticles over long periods. The advantages of physical processes applying inert carrier gas are: (I) absence of liquid by-products, (II) more economic particle-from-medium separation than in liquid medium, (III) avoidance of surface-adhering and -contaminating surfactants and possibility of using high-purity inert gases, (IV) possibility of high-temperature annealing of the product in-flight, so that high-temperature phases or better crystalline products can be obtained, (V) easier scale-up of supporting unit operations such as transport and collection, (VI) no limitations for obtaining a desired particle size, (VII) possibility of size-selection methods directly in the gas phase. We apply mainly direct thermal evaporation in a hot-wall furnace for substances having sufficient vapour pressure at temperatures below $1100\,^{\circ}$C. For low-vapour pressure metals, methods resulting in a high-temperature plasma can be used, such as spark discharge or mini-arc evaporation (Fig. 4.1). For compounds or alloys which do not congruently evaporate, the more costly laser ablation technique has to be used.

(2) *Reduction of particle concentration.* The size selection procedure based on the electrical mobility limits the maximal attainable particle concentration of monodisperse particles, depending on material and size between 10^4 and $10^6\,\mathrm{cm}^{-3}$. The only way to obtain a substantial larger amount of particles is to increase the volumetric flow of the gases, in commercial DMAs limited to $\sim$1 slm (l/min at standard conditions). This necessitates the construction of new DMAs. A new cylindrical DMA has been designed and investigated [6], operating up to 100 slm, thereby principally allowing a 100 times larger production rate. Further scale-up is in progress (transferproject SFB 445 T01).

(3) *No commercial equipment existing.* Commercial DMAs are not designed to operate with high-purity gases and do not possess vacuum-tight fittings. This causes two problems: it is not possible to evacuate the setup in order to eliminate gaseous impurities and moisture (a standard procedure in inert gas evaporation), and during the synthesis oxygen will leak in causing oxidation of the nanoparticles. As an example, the size selection of FePt nanoparticles with commercial DMAs leads to a large fraction of FeOx in the nanoparticles. A vacuum-tight DMA has been constructed in our institute and allows the size selection of oxidation-sensitive metals and semiconductors. Also other components such as radioactive neutralizers and ceramic tube fittings are fabricated in-house having vacuum (KF) connections.

(4) *Consumption of inert gas.* DMAs require the use of a an inert sheath gas, in practise with volume flow rates approximately 10 times larger than that of the nanoaerosol when geometric standard deviations below 1.10 are sought for. In order to limit the consumption of inert carrier gas, a sheath gas recirculation system has been designed (sketch is shown in Fig. 4.2) and is now used in virtually all size selection setups in our laboratory. It allows sheath gas flow rates up to 20 slm and is constructed from vacuum-tight components.

(5) *Low efficiency of particle charging and multiple charges.* The use of bipolar chargers such as Kr-85 sources leads to the attainment of bipolar charge

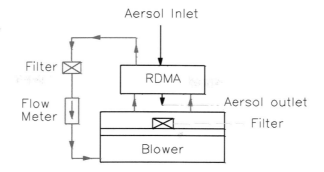

Fig. 4.2 Schematic diagram of the size-selection assembly, showing a radial differential mobility analyzer (RDMA) and a recirculation assembly for the sheath gas

equilibrium, in which smaller particles have smaller charging probabilities. This limits the number concentration of size-selected nanoparticles, because only charged ones can be selected. Several approaches can be followed to increase the charging probability. In some cases, there is no need for a charger, as the nanoaerosol can leave the synthesis region already charged under some conditions. As an example, the FePt particles coming from the laser ablation reactor are already charged and do not require charging. The charging efficiency can be improved by using unipolar chargers based on corona dischargers or UV photoionizers [7].

(6) *Particle losses*. These are caused by diffusional losses to the wall (mainly dependent on the total tube length and particle size), thermophoretic losses at the outlet of the hot-wall furnace used for in-flight annealing and electrostatic losses of the charged particles.

(7) *Finding the right annealing temperature*. Several problems can occur when performing in-flight annealing and sintering. Ideally, the (partial) coalescence caused by the annealing leads to the attainment of a mobility-based diameter which does not change upon further temperature increase, indicating that a quasi-spherical or crystal-like form has been attained. But often evaporation occurs simultaneously, so that the mobility diameter is continuously decreasing upon temperature increase [14]. Another problem is charge loss during the high-temperature treatment [12].

In the following chapter, we will cover three topics illustrating these challenges, and also demonstrating some solutions.

4.2 Synthesizing Monodisperse Nanoparticles by Means of Spark Discharge

4.2.1 Spark Discharge

The spark discharge technique has been utilized to synthesize metallic (Au) and semiconductor (Ge) nanoparticles. This technique was pioneered by Schmidt-Ott

Fig. 4.3 Schematic diagram of the spark generator

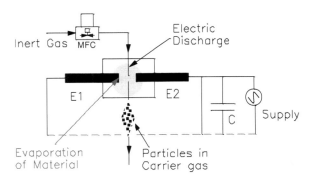

[16], the configuration used in our work is based on his design. Two electrodes are placed about 1–2 mm apart and are enclosed inside a vacuum-tight chamber, evacuable down to 10^{-5} mbar. Evacuation prior to synthesis as well as the use of purified carrier gas minimizes oxidation of primary particles during their formation. One of the electrodes (E2) is connected to the higher voltage power supply via a capacitor C (capacitance 26 nF), as schematically shown in Fig. 4.3. Another electrode (E1) is connected to earth and also connected to the micrometer in order to adjust the distance between the electrodes for successive measurements. When the high voltage supply is switched on, a capacitor is charged and upon reaching the breakdown voltage discharges, resulting in the ionization of the carrier gas between the electrodes. The ionized species i.e. electrons and positive ions thus generated move towards the opposite electrodes and evaporate the electrode materials. Due to very short duration of the pulsed discharge, evaporated species cool down rapidly which creates supersaturation conditions leading to homogeneous nucleation and condensation and thereby formation of tiny (primary) particles. These primary particles grow in size due to coagulation or coalescence before being carried away from the spark chamber by carrier gas. The repetition frequency (f_s) of the discharge can be changed by the charging current (I_c) of the capacitor and breakdown voltage (V_{bd}) as

$$f_s = \frac{I_c}{CV_{bd}} \qquad (4.1)$$

The energy E_d transferred from the capacitor at discharge,

$$E_d = \frac{CV_{bd}^2}{2} \qquad (4.2)$$

The power P_{SG} delivered during discharge is

$$P_{SG} = f_s E_d \equiv \frac{I_c V_{bd}}{2} \qquad (4.3)$$

4 Synthesis and Film Formation of Monodisperse Nanoparticles

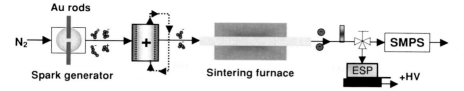

Fig. 4.4 Schematic representation of synthesis setup for preparing monodisperse Au nanoparticle

Thus, as long as V_{bd} stays reasonably constant, the delivered power and hence the particle production rate increases with charging current or frequency [19].

4.2.2 Synthesis of Monodisperse Au Nanoparticles

In order to obtain the highest possible concentration of monodisperse spherical gold nanoparticles, a study was made of both the optimal experimental setup as well as the process parameters. Figure 4.4 represents schematically different components of Au synthesis setup, which consists of a spark generator having two Au electrodes, a radial DMA, a furnace for sintering and an electrostatic precipitator (ESP). A neutralizer for particle charging is not included in the setup, as particles leaving the generator are to a great extent already charged. A commercial scanning mobility particle sizer (SMPS, TSI 3088) allows online measurement of the mobility-based particle size distribution.

The effect of spark frequency and carrier gas (N_2) i.e. aerosol flow rate (Q_a) on the average particle size and the size distribution was studied. By maintaining a constant Q_a of 1.0 slm, the size distribution of Au agglomerates is monitored as a function of charging current (I_c), as shown in Fig. 4.5a. As evident from equation (1), the charging current is related to the spark frequency which was therefore also measured. With the increase in the charging current, spark frequency increases, as shown in Fig. 4.5b. For a fixed charging current value, there is a range of spark frequencies (shown by the error bar) rather than a fixed value, hence points towards the instability of discharging. Capacitor charging voltage slightly decreases with the increase in charging current, as clear from Fig. 4.5b. Reversal in the polarity of voltage on discharging has also been observed, as shown by monitored waveform in Fig. 4.5c. The size distribution of the Au agglomerates shifts to higher mobility diameter (D_m) with increase in the charging current, which corresponds to increase in the spark frequency from 8 to 115 Hz. With the increase in the spark frequency, the sputtered mass per unit volume of the process gas increases, resulting in higher supersaturation which leads to an increase in the primary particle number concentration. This increase in primary number concentration results in higher coagulation between them and consequently leads to shift in the size distribution towards larger D_m values. Upon increasing the charging current, the total particle number concentration (N) increases from

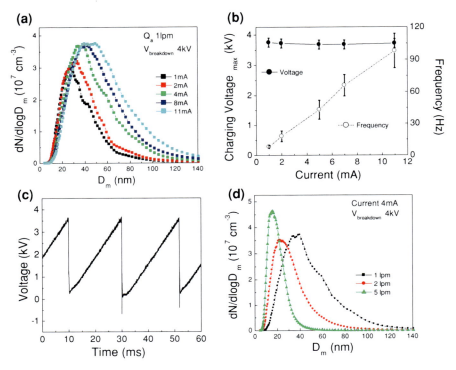

Fig. 4.5 a Size distribution of generated Au agglomerates at different charging current values, while maintaining Q_a and $V_{breakdown}$ at 1 slm and 4 kV, respectively. **b** Charging voltages of capacitor and spark frequencies as function of charging current. **c** Voltage reversal during some cycles of discharging as monitored by an oscilloscope. **d** Size distribution at different Q_a, keeping capacitor charging current and $V_{breakdown}$ fixed at 4 mA and 4 kV, respectively

1.39×10^7 to $2.06 \times 10^7 \, cm^{-3}$, while the geometric standard deviation (σ_g) changes from 1.55 to 1.62.

On increasing Q_a from 1 to 5 slm, the particle size distribution shifts to the lower mobility diameter values (Fig. 4.5d). Geometric mean mobility diameter (D_{gm}) values reduce from 36 to 16 nm and the value of σ_g decreases from 1.59 to 1.44, on increasing the Q_a from 1 to 5 slm. In our case, with the increase in Q_a, particle number concentration increases, which is contrary to reported results [9, 10]. Apparently, in our case, increase in Q_a acts like a virtual dilution flow, which reduces the residence time of generated particles inside the chamber. The increased Q_a also lessens the probability of the generated particles to deposit on the chamber walls before being carried out of the chamber. Therefore, reduction in residence time enhances the particle number concentration with the increase in Q_a [5]. Coagulation depends linearly on the square of the particle number concentration and on residence time. The particle number concentration increases only a factor of 1.05, whereas residence time decreases by a factor of 5, with the increase in Q_a from 1 to 5 slm. Therefore, the

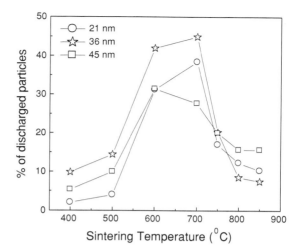

Fig. 4.6 Percentage of discharged particles as a function of sintering temperature for Au nanoparticles having three different mobility diameters

effect of reduction in residence time is more pronounced than the increase in particle number concentration on the coagulation rate. As a result of reduction in coagulation rate, particle size distribution shifts to lower D_m values with the increase in Q_a from 1 to 5 slm

The synthesis setup shown in Fig. 4.4 has one drawback, however. In order to obtain spherical particles, Au agglomerates have to be annealed in-flight. This leads to a (partial) loss of charges on the particles. In Fig. 4.6, it can be seen that particles lose charges upon increasing annealing temperature. At still higher temperatures, the particles start to get re-charged but the well-defined charge state is lost. It is important to know that for creating pair particles, the aerosol has to consist purely of particles having a single charge. Only 0.5 % discharging of particles has been observed on performing sintering prior to the size-selection step. Therefore, for further study, sintering is performed before the size-selection. As this can lead to particles getting negatively charged, applying a positive polarity in the DMA leads to the highest number concentration of size-selected particles. A sintering temperature of 800 °C is sufficiently high enough to convert Au agglomerates into spherical and crystallized nanoparticles. Therefore, for further study, sintering temperature was kept constant at 800 °C.

Several different size-selected samples have been generated, using an aerosol to sheath gas ratio of 7.5 or 10. In Fig. 4.7, a transmission electron microscopy (TEM) image is shown as well as the size distributions based on the mobility measurements and the TEM micrographs analysis. The presence of double charged particles is clearly visible in the mobility-based distribution. Particles having larger selected sizes show increasingly a deviation from the spherical shape and a more pronounced crystal shape, as clear from high-resolution TEM (HRTEM) images shown in Fig. 4.8. The larger particles also show more multiple twinning; only the smallest ones are sometimes perfectly monocrystalline.

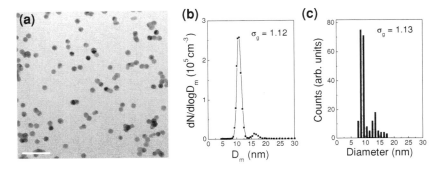

Fig. 4.7 **a** Bright field TEM images of Au nanoparticles of geometric mean diameter of 9.87 nm, **b** size distribution by SMPS measurement and **c** size distribution calculated by TEM images

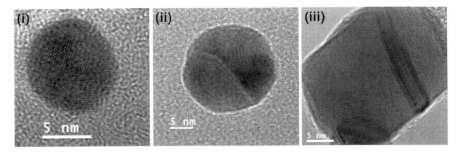

Fig. 4.8 Typical HRTEM images of Au nanoparticles of geometric mean diameters of (i) 9.87, (ii) 19.9 and (iii) 32.2 nm

4.2.3 Synthesis of Ge Nanoparticles

The synthesis of semiconducting nanoparticles by means of spark discharge is more challenging than that of metallic particles due to the lower conductivity of the material and to the increased tendency of 'splashing', a phenomenon also known in laser ablation which leads under some conditions to the presence of particles in the range 0.1–1 μm. To prepare Ge particles, two Ge rods having a diameter of ∼6.35 mm are used. Figure 4.9 shows the size distribution of Ge particles at varying charging currents and breakdown voltages, while keeping Q_a fixed at 1 slm.

With a breakdown voltage of 4 kV (Fig. 4.9a), it is observed that the value of D_{gm} decreases from 19.9 to 10.1 nm, on increasing the charging current from 2 to 13 mA, although the sparking frequency increases from 7 to 170 Hz, as shown in Fig. 4.9b. This observation is in contrast to the case of Au, as shown in Fig. 4.5a. In order to understand this trend the charging voltage of the capacitor is also measured. It was found that the discharging voltage decreased with an increase of the charging current, in contrast to the Au system where it remains constant. Per discharge therefore less Ge is removed from the electrodes. In order to shift the size

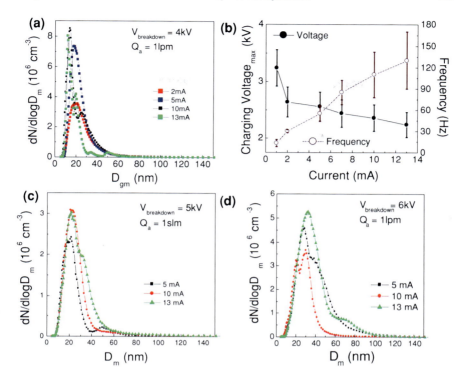

Fig. 4.9 Size distribution of Ge agglomerates generated with different charging current values at different breakdown voltages **a** 4 kV, **c** 5 kV and **d** 6 kV, while maintaining Q_a fixed at 1 slm. **b** Charging voltage (charging voltage maximum = discharge voltage) and spark frequency measurements during the sparking at breakdown voltage of 4 kV

distribution to larger sizes, the distance between the Ge electrodes is increased, which leads to an increase in the breakdown voltage. The value of D_{gm} increases from 10 to 21 nm and then to 30 nm on increasing the breakdown voltage from 4 to 5 and then to 6 kV, at a charging current of 13 mA, as shown in Fig. 4.9c, d. The value of σ_g is always in the range of 1.32–1.42. Therefore, for the further studies, Ge agglomerates are generated with a breakdown voltage of 6 kV and a charging current of 13 mA or a frequency leading to discharge frequencies of 90–170 Hz.

For the size-selected Ge particles, the same phenomenon of particle dis- and re-charging was observed, caused by the thermal annealing, starting at 500 °C. Thus, the sintering was also performed prior to size-selection. A typical TEM image of the nanoparticles and corresponding size distribution as measured and calculated by SMPS and TEM, respectively, are shown in Fig. 4.10.

HRTEM images and corresponding FFT patterns revealed that some nanoparticles are monocrystalline while others contained multiple domains. The FFT patterns show reflections corresponding to the (111), (200), (220) and (311) planes with

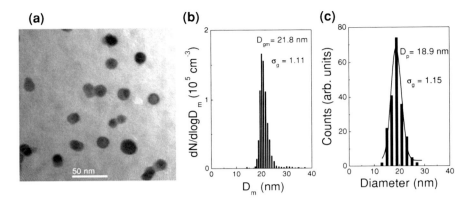

Fig. 4.10 a TEM image of Ge nanoparticles, obtained after sintering at 900 °C, **b** size distribution as monitored by SMPS and **c** size distribution obtained by TEM images analysis

interplanar spacings 0.326, 0.279, 0.203 and 0.174 ± 0.003 nm, respectively, which are analogous to that observed for cubic structure of Ge in bulk [18].

4.3 Formation of Alloy and Pair Nanoparticles

4.3.1 Motivation and Synthesis Approach

Nanoscale engineered hybrid systems such as, bimetallic nanoparticles, metal-semiconductor nanoparticles, bisemiconducting nanoparticles and metal nanoparticles imbedded in thin films are attracting interest as they offer a wide scope of applications [9]. The interaction of two types of atoms induces new physical, chemical and mechanical properties. Additionally, the composite nanoscale systems can have phase-diagrams different from the bulk [19]. Moreover, stability of bulk metastable phases at nanoscale has also been reported [8, 19]. Bimetallic nanoclusters/nanoparticles have demonstrated to be most attractive for catalytic applications with several important advantages over the monometallic one. In particular, Au is reported to have excellent catalytic and optical properties when combined with Pt, Pd, Ag, Cu and Zn [8, 17]. Optical coupling of semiconductor nanoparticles with metal nanoparticles can be a viable route to modify the spectral features of semiconductor nanoparticles or vice versa. For example, Au nanoparticles exhibit a peculiar absorption peak at visible wavelengths due to surface plasmon resonance (SPR), which in turn depends on the size, shape and refractive index of the external medium. Therefore, SPR of Au nanoparticles can be utilized to enhance emission of semiconductors nanoparticles. The ability of noble metals to modify the optical properties of Si has already been demonstrated and a remarkable enhancement in the radiative emission rate has been reported [2]. Furthermore, semiconductor nanowhiskers normally require seed particles, often Au nanoparticles, to grow via the vapor-liquid-solid mechanism. Alloy formation at the initial stage of growth takes place between seed (Au) and species containing the desired semiconductor material, therefore, the composition of the seed particles can be utilized to induce modification/doping in subsequently growing one-dimensional nanostructures. Therefore, the motivation of the present work is to produce alloy nanoparticles of a desired composition that can be further used to grow nanowhiskers of III–V. Si, Sn and Ge provide n-type doping to GaAs, whereas, Zn, Mn and In induce p-type doping. For our study, we have focused on the system Au-Ge.

To prepare alloy nanoparticles, two different approaches have been utilized. In the first approach, two spark generators, one equipped with Au electrodes and another with Ge electrodes have been utilized to produce particles of Au and Ge, respectively. Au particles of one polarity and Ge particles of opposite polarity are selected with the separate DMAs. Subsequently, both Au and Ge size-selected particles are subjected to sintering during their flight in separate furnaces and then allowed to pass through a mixing tube. In the mixing tube, particles of Au and Ge having opposite charges form Au-Ge pair particles. After that, the formed Au-Ge pair particles pass through a sintering furnace in order to obtain Au-Ge alloy nanoparticles. The formed alloy nanoparticles can be collected on substrates either via low pressure impactor or electrostatic precipitation. In order to use this approach to produce pair and subsequently alloy nanoparticles, it is imperative that particles of each type must carry relatively opposite charges. In addition to this, to prepare well-distinguishable pair nanoparticles, particles must be spherical in shape. This approach of preparing alloy nanoparticles is advantageous as the composition of the finally prepared alloy nanoparticles can be controlled by selecting the Au and Ge particles of desired sizes, by performing size-selection step.

In the second approach, one electrode made of Au and another made of Ge has been placed inside the spark chamber. In this case, a mixture of Au and Ge atoms is produced as a result of co-discharging, which might lead to AuGe alloy nanoparticles after size-selection and sintering steps. In this approach, it is not possible to control the composition of the synthesized nanoparticles as the evaporation rates of both the electrodes are not identical during codischarging and the particle formation itself is not controlled. Cathode material evaporates faster compared to anode material. Therefore, depending upon the selection of Au either as anode or as cathode, the evaporated amount of Au in the initial mixture can be changed in comparison to Ge. However, it is not possible to control the final compositions of the obtained composite/alloy nanoparticles.

4.3.2 Synthesis of Au-Ge Pair Nanoparticles

For producing AuGe alloy nanoparticles of controlled composition, the Au and Ge nanoparticle synthesis setups are coupled together as shown in Fig. 4.11. The

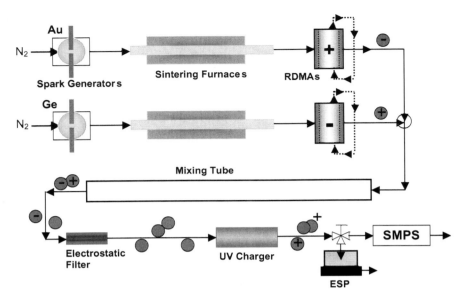

Fig. 4.11 Schematic diagram of Au-Ge pair nanoparticle synthesis setup

generated aerosols of Au and Ge nanoparticles of desired sizes but carrying opposite charges are allowed to pass through a long aggregation tube in order to produce pair particles. During the flow through the long tube, charged Au or Ge nanoparticles may loose their charges, which is problematic because they will also be able to pass through the electrostatic filter (EF) applied to separate the particle pairs from the noncollided Au or Ge particles. Therefore, discharging loss is first monitored by allowing only an Au aerosol of known particle size and concentration to pass through the long tube. An EF is placed after the long tube to separate out charged Au nanoparticles, thus allowing to determine the concentration of discharged particles. The percentage of particles getting discharged, while passing through the long tube, increases with the increase in the residence time of the nanoparticles, as depicted in Fig. 4.12. The mechanism of this spontaneous discharging is not clear, but might be due to spontaneous ion formation such as also occurring in the atmosphere (due to cosmic radiation).

Pair particle formation has been monitored by online SMPS, prior to depositing them on a substrate. Au and Ge nanoparticles of D_{gm} values of 21.7 and 20.9 nm, respectively, are selected as shown in Fig. 4.13a and b. The Au and Ge nanoparticles have undergone sintering at temperatures of 800 and 400 or 800 °C, respectively, before size-selection. As is clear from the measured size distributions, besides singly charged nanoparticles, both Au and Ge aerosols also contain nanoparticles having double charges. Subsequently, both Au and Ge aerosols, each with flow rate of 1 slm, are then combined in the mixing tube of diameter and length of 7.0 and 400 cm, respectively, leading to a residence time of 7.7 min. The mobility

4 Synthesis and Film Formation of Monodisperse Nanoparticles

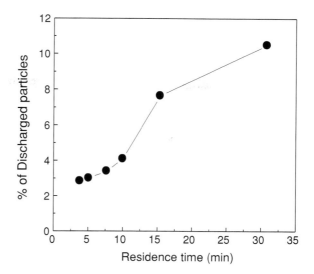

Fig. 4.12 Percentage of discharged particles having a mobility diameter of 28 nm, as a function of residence time. The mixing tube has a diameter of 7.0 cm and a length of 400 cm

diameter of a pair (D) consisting of identical mobility diameter (d) particles is given by $D = 1.32 \times d$. Therefore, the mobility diameter of a nanoparticle pair, formed by Au and Ge nanoparticles of 21 nm mobility diameters, must be 27.7 nm. Similarly, the mobility diameter of pair particles should be ∼40 nm when formed by doubly charged Au and Ge nanoparticles each having a mobility diameter of 31 nm. The size distribution of the mixed aerosol containing both Au and Ge nanoparticles is monitored after the mixing tube. Figure 4.13c shows the size distribution after mixing of Au and Ge aerosols. One maximum at 22.5 nm coincides with the selected sizes of Au and Ge nanoparticles hence, corresponding to the original Au or Ge nanoparticles. The other maximum at 28.9 nm, which is quite close to the value of 27.7 nm, if 21 nm sized Au and Ge nanoparticles form pairs, and thereby indicates formation of Au-Ge pair nanoparticles. Much more single Au or Ge particles than particle pairs are detected in that case. On applying 2 kV to the EF after the mixing tube in order to remove singly charged Au and Ge particles, the curve shows a sharp decrease in the presence of single particles around 22 nm. Clearly, most of the Au and Ge particles are now removed, however the Au and Ge nanoparticles which spontaneously discharge during their residence in the mixing tube cannot be removed. As these particles would strongly disturb the interpretation of the effect of seed particles on the VLS mechanism, efforts are now carried out to size separate the pairs from the single particles.

The SEM micrographs of the mixture of pair particles and single particles are shown in Fig. 4.14. In the SEM micrograph, sintering of Au and Ge particles is carried out at 800 and 400 °C, respectively. An energy dispersive x-ray spectroscopic (EDS) analysis by line scan on the Au-Ge pair particle confirms the presence of both Au and Ge in the pair.

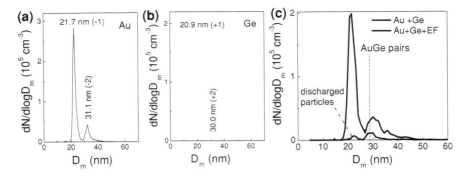

Fig. 4.13 Tandem DMA measurement showing size distributions of **a** selected Au nanoparticles, **b** selected Ge nanoparticles and **c** pair formation. Au and Ge nanoparticles are sintered at temperatures 800 and 400/800 °C, respectively, before size-selection showing pair formation. Au+Ge indicates, measurement after mixing aerosols of Au and Ge inside the mixing tube, Au+Ge+EF indicates that electrostatic filtering is applied to Au+Ge

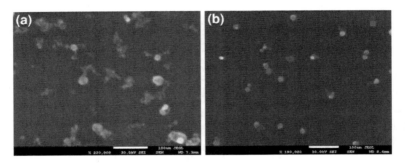

Fig. 4.14 SEM images showing formation of pair nanoparticles in which Ge nanoparticles are subjected sintering at **a** 400 °C and **b** 800 °C. The sintering of Au nanoparticles is performed at 800 °C

4.3.3 Synthesis of AuGe Alloyed Nanoparticles by Co-Sparking

Co-sparking between the electrodes of Au and Ge is performed in order to investigate whether the complicated bipolar mixing procedure can be avoided. With a distance between the Au and Ge electrodes of ∼1 mm, breakdown is observed to occur at ∼4.1 kV and this value of breakdown voltage changes negligibly with increasing flow rates from 1 to 10 slm. A study was made of the effect of variation of the gas flow rate as well as charging current using the SMPS as shown in Fig. 4.15.

It can be seen that at a flow rate of 1 slm, an insufficient number concentration is generated. At higher flow rates, the number concentration seems to be fairly independent of the charging current whereas the geometric mean diameter slightly decreases with the flow rate. For the size-selected samples, a flow rate of 2 slm was

4 Synthesis and Film Formation of Monodisperse Nanoparticles

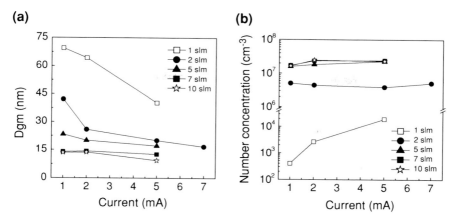

Fig. 4.15 Variations in **a** D_{gm} and **b** number concentration as a function of charging current at different carrier gas flow rate

Table 4.1 Atomic percentage of Au and Ge in Au-Ge mixture generated at different charging current values. The corresponding spark frequencies with the error range are also mentioned

Current (mA)	Most probable frequency (±error)	Atomic (%) Au	Atomic (%) Ge
1	8 (±4)	2	98
2	21 (±8)	3	97
5	55 (±10)	17	83
7	74 (±10)	30	70

subsequently used. By means of EDS, the relative atomic percent (at. %) of Au and Ge was investigated as function of charging current (Table 4.1). Hereby, occasional larger particles as a result of splashing were excluded from the analysis, as they play no role in the size-selected samples.

The atomic percent of Au increases from 3.12 to 30.09 on increasing the spark frequency (most probable) from 8 to 74 Hz, i.e. increasing charging current from 1 to 7 mA. Noteworthy, in the present setup, Au is acting as an anode, therefore, will sputter less than the cathode i.e. Ge, which is also clearly reflected by the EDS measurements. Besides, there is a reasonably increase in at. % of Au, at higher charging current/spark frequency i.e. at ≥5 mA/74 Hz. During the monitoring of waveform at charging current 5 mA, reversing of voltage has been observed during some cycles of discharging, which leads to more evaporation of Au and hence larger atomic concentration. Therefore, to prepare well-defined AuGe nanoparticles, higher charging current has been utilized.

For the size-selected AuGe nanoparticles, in-flight annealing was performed at 900 °C. Samples were produced at different charge currents as well as different selected sizes. From the TEM micrographs, in Fig. 4.16, it was found that the particles

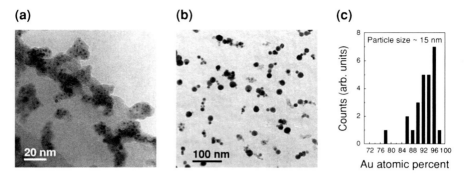

Fig. 4.16 TEM images of the samples prepared with **a** 55 Hz and **b** 74 Hz, most probable frequency and with sintering at 900 °C. **c** Histogram of Au concentration in around 20 nanoparticles of 15 nm size, as measured by EDS

consisted mostly of multiple crystalline Au crystallites containing some Ge, imbedded in a Ge amorphous matrix. The size of the Au inclusions was found to be a function of the charging current. It can be concluded that the samples are much more inhomogeneous than in the case of bipolar mixing. This also becomes clear from the histogram showing the Au atomic compositions of some 20 individual particles.

4.4 Film Formation by Electrostatic Means

A thin nanoparticle layer is often necessary for the characterisation of properties, such as particle size, shape, chemical composition and crystallinity. A growing group of applications require the use of a thin, high-quality film composed of monodisperse nanoparticles, as the specific size-dependent properties will then be the most prominent. Depending on the application or characterisation technique, one or more of the following film properties have to be controlled: (1) diameter of the deposit, (2) film thickness, (3) homogeneity of the deposit and (4) porosity. Furthermore, process parameters such as gas flow rate and deposition efficiency play a role. There are three main methods for collecting nanoparticles out of the gas phase on a substrate, low-pressure impaction [4], thermophoretic deposition [13] and electrostatic precipitation [3]. Here, we concentrate on electrostatic precipitation due to its high deposition efficiency when the nanoparticles are already charged, the low pressure loss and the potential to create deposits having large diameters. From an application point of view, control over the diameter of the deposit (spot size) is important.

In this work, an ESP was developed which does allow to produce a homogeneous deposit of nanoparticles. Its performance was investigated using size-selected nanoparticles in the size range 10–70 nm. The design shown in Fig. 4.17, demonstrated the best performance. It allows the distance between the nozzle and the surface of the central electrode z to be varied. Further parameters which can be varied

4 Synthesis and Film Formation of Monodisperse Nanoparticles

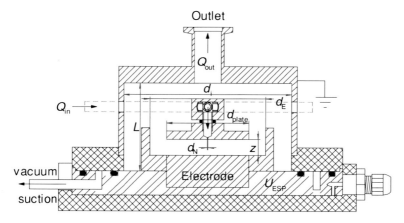

Fig. 4.17 Final version of the electrostatic precipitator for nanoparticles deposition

are the nozzle diameter d_N, the electrode diameter d_E, the sample flow rate Q_{in} and the electrode voltage U_{ESP}. Differences to previous versions are: (1) instead of a tube-like aerosol inlet, a plate with an orifice is used, leading to a more uniform electric field and a more suitable gas velocity profile without a recirculation zone, (2) the aerosol does not flow in the vicinity of the isolator, avoiding deposition of charges on the isolator which causes remnant fields and bad reproducibility, (3) the gas outlet is placed central, as to obtain the most symmetric outflow profile as possible which is necessary to obtain perfectly spherical deposits and (4) attachment of housing to base by means of vacuum suction.

The performance of the ESP was investigated experimentally with the help of monodisperse nanoparticles in the size range 10–70 nm, generated by means of evaporating PbS powder from a tube furnace at 680 °C, followed by a radioactive charger (^{85}Kr), a sintering furnace operating at 400 °C and a DMA (Model 3085, TSI, Shoreview, MN, USA). The aerosol flow rate Q_{in} was 1 slm, further $z = 10$ mm; $d_N = 1$ mm. The electrostatic deposition efficiency is defined as the ratio between the number concentration at the outlet of the ESP with and without applying the high voltage. Figure 4.18a shows that the efficiency increases with the applied voltage, leading to larger electrostatic forces, and decreases with increasing particles size, which is a result of the electric mobility decreasing with size. Images of the deposit were taken with a digital camera (Canon D90), allowing to determine the deposition diameter. Figure 4.18b shows that the spot size is a strong function of the particle diameter, and to a lesser extent of the applied voltage. It can be seen that the variability of the spot diameter is limited when the applied voltage is used as control parameter. The spot size in case of the smallest nanoparticles is mainly controlled by the inlet diameter of the nozzle, whereas for the larger particles the spot diameter is much larger than the nozzle diameter. Notice that the difference in deposition behaviour will lead to preferential sampling at different radial positions when depositing polydisperse aerosols.

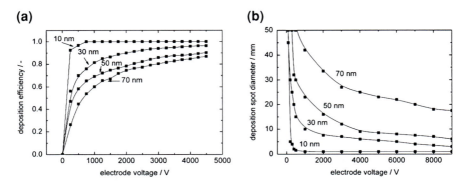

Fig. 4.18 a The deposition efficiency of ESP for nanoparticles with diameters d_P of 10, 30, 50 and 70 nm. **b** The measured deposition spot diameters using PbS nanoparticles with diameters d_P of 10, 30, 50 and 70 nm as function of applied electrode voltage

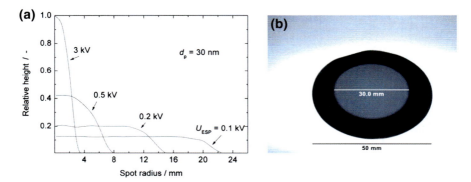

Fig. 4.19 a The relative height of the deposition spot formed for PbS nanoparticles having a diameter d_P of 30 nm as a function of radial position and electrode voltage U_{ESP}. 2D scans were executed with an Ambios XP 200 contact profilometer being started at the center of the deposition spot and **b** a digital image showing a Si wafer with a diameter of 50 mm being covered by PbS nanoparticles having an equivalent particle diameter d_P of 50 nm, by applying an electrode voltage of 500 V

For the determination of the layer thickness of the deposition spot, a contact profilometer was used (XP 200, Ambios), which measures small surface variations due to vertical displacement of a stylus as a function of position. Vertical features ranging in height from 10 nm to 65 μm can be traced by this purely mechanical method. As the stylus can deplace particles and create particle-free tracks, the measurement was started in the center of the deposit using a very small force (30 μg), avoiding this effect. In Fig. 4.19a, the height profile of the deposition spot is shown for 30 nm diameter nanoparticles. The film thickness at the center was 200 nm at $U_{ESP} = 0.1$ kV. It can be seen that for the higher voltages a Gaussian deposition profile is found, while for the smaller voltages the deposition profile is constant over a large portion

4 Synthesis and Film Formation of Monodisperse Nanoparticles

of the film, which is clearly the most desired case. In Fig. 4.19b, the homogeneity as well as the perfect sphericity can clearly be seen.

References

1. X. Bai, T.L. Wang, N. Ding, J.H. Li, B.X. Liu, Nonequilibrium alloy formation in the immisicble Cu-Mo system studied by thermodynamic calculation and ion beam mixing. J. Appl. Phys. **108**, 073534-1–073534-5 (2010)
2. M. Bassu, M.L. Strambini, G. Barillaro, F. Fuso, Light emission from silicon/gold nanoparticle systems. Appl. Phys. Lett. **97**, 143113-1–143113-3 (2010)
3. J. Dixkens, H. Fissan, Development of an electrostatic precipitator for off-line particle analysis. Aerosol Sci. Technol. **30**, 438–453 (1999)
4. J. Fernandez de la Mora, S.V. Hering, N. Rao, P.H. McMurry, Hypersonic impaction of ultrafine particles, J. Aerosol Sci. **21**(2), 169–187 (1990)
5. C. Helsper, W. Mölter, F. Löffler, C. Wadenpohl, S. Kaufmann, Investigation of new aerosol generator for the production of carbon aggregate particles. Atm. Environ. **27**(8), 1271–1275 (1993)
6. E. Hontañón, F.E. Kruis, A Differential Mobility Analyzer (DMA) for size selection of nanoparticles at high flow rates. Aerosol Sci. Technol. **43**(1), 25–37 (2009)
7. E. Hontañón, F.E. Kruis, Single charging of nanoparticles by UV photoionization at high flow rates. Aerosol Sci. Technol. **42**(4), 310–323 (2008)
8. E. Juárez-Ruiz, U. Pal, J.A. Lombardero-Chartuni, A. Medina, J.A. Ascencio, Chemical synthesis and structural characterization of small AuZn nanoparticles. Appl. Phys. A **86**, 441–446 (2007)
9. S. Kala, B.R. Mehta, F.E. Kruis, A dual-deposition setup for fabricating nanoparticle-thin film hybrid structures. Rev. Sci. Instrum. **79**(1), 013902 (2008)
10. S. Kala, B.R. Mehta, F.E. Kruis, V.N. Singh, Synthesis and oxidation stability of monosized and monocrystalline Pr nanoparticles. J. Mater. Res. **24**(7), 2276–2285 (2009)
11. F.E. Kruis, K. Nielsch, H. Fissan, B. Rellinghaus, E.F. Wassermann, Preparation of size-classified PbS nanoparticles in the gas phase. Appl. Phys. Lett. **73**, 547-1–547-3) (1998)
12. M.H. Magnusson, K. Deppert, J. Malm, J. Bovin, L. Samuelson, Gold nanoparticles: Production, reshaping, and thermal charging. J. Nanopart. Res. **1**(2), 243–251 (1999)
13. A.D. Maynard, The development of a new thermophoretic precipitator for scanning transmission electron microscope analysis of ultrafine aerosol particles. Aerosol Sci. Technol. **23**, 521–533 (1995)
14. K.K. Nanda, A. Maisels, F.E. Kruis, Surface tension and sintering of free gold nanoparticles. J. Phys. Chem. C **112**(35), 13488–13491 (2008)
15. S. Schwyn, E. Garwin, A. Schmidt-ott, Aerosol generation by spark discharge. J. Aerosol. Sci. **19**(5), 639–642 (1988)
16. S. Senapati, A. Ahmad, M.I. Khan, M. Sastry, R. Kumar, Extracellular biosynthesis of bimetallic Au-Ag alloy nanoparticle. Small **1**, 517 (2005)
17. B.R. Taylor, S.K. Kauzlarich, G.R. Delgado, H.W.H. Lee, Solution synthesis and characterization of quantum confined Ge nanoparticles. Chem. Mater. **11**, 2493–2500 (1999)
18. N.S. Tabrizi, M. Ullmann, V.A. Vons, U. Lafont, A. Schmidt-ott, Generation of nanoparticles by spark discharge. J. Nanopart. Res. **11**, 315–332 (2009)
19. C.X. Wang, G.W. Yang, Thermodynamics of metastable phase nucleation at the nanoscale. Mater. Sci. Eng. R **49**, 157–202 (2005)

Part II
Structure and Dynamics

Chapter 5
Diffusion Enhancement in FePt Nanoparticles for L1$_0$ Stability

Mehmet Acet, M. Spasova and A. Elsukova

Abstract FePt has a high magnetic anisotropy in the thermodynamically stable L1$_0$ phase, so that nanoparticles of this material could remain ferromagnetic at small sizes and be used for magnetic storage. However, the stabilization of the L1$_0$ phase is not straight forward at small particle sizes, and instead, twinned icosahedral structures are formed which are thought to be metastable. In order to enhance the formation of the L1$_0$ phase, the lattice has to be agitated so that icosahedral structures will destabilize. In addition to thermal annealing, we introduce agitation methods related to oxygen reactive sputtering, nitriding-denitriding, and segregation in FePt nanoparticles prepared in the gas phase. We discuss principally the structural properties of the particles obtained prior to and after employing the various agitation techniques.

5.1 Introduction

The advancement in high-density magnetic storage media technology relies much on the progress in developing materials that can sustain memory at small-as-possible dimensions. The miniaturization of the individual magnetic 'bits' is, therefore, one of the challenges that aims to provide a high-as-possible storage density; next to favorable magnetic properties such as substantial magnetization and magnetic anisotropy. Currently, magnetic storage densities close to the Tb/in^2 level are being aimed at; the attainability being limited by superparamagnetism. Miniature magnets of nanometer size are therefore by nature attractive materials that could contribute to magnetic storage media technology.

As the particle size decreases and the superparamagnetic limit is approached, the magnetic anisotropy energy can overcome the effect of the thermal energy and fix the direction of the magnetization to provide the 'bit' information. If the thermal

---

M. Acet (✉) · M. Spasova · A. Elsukova
Faculty of Physics and CENIDE, University of Duisburg-Essen,
Lotharstraße 1, 47057 Duisburg, Germany
e-mail: mehmet.acet@uni-due.de

A. Lorke et al. (eds.), *Nanoparticles from the Gas Phase*, NanoScience and Technology,
DOI: 10.1007/978-3-642-28546-2_5, © Springer-Verlag Berlin Heidelberg 2012

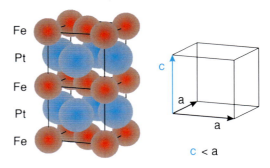

Fig. 5.1 The L1$_0$ structure of FePt. Fe and Pt layers alternate along the [001] direction. The structure has a $(c/a) < 1$ tetragonal distortion

energy is too high, magnetization orientation can no longer be sustained, and the storage information is lost. In order to overcome the adversity of the thermal effect, nanoparticles with high uniaxial magnetic anisotropy are required [1]. FePt and CoPt compounds in the L1$_0$ crystallographic phase (Fig. 5.1) are known to possess a large magnetic anisotropy energy density, $K_u \sim 7 \times 10^{-6}\,\mathrm{Jm}^{-3}$ [2–4]. The magnetic moment of FePt being somewhat larger than that of CoPt is a further advantageous factor in applications. For these reasons, we have chosen FePt as the prototype system for studying the structural and magnetic properties of high magnetic anisotropy nanoparticles.

Nanostructured FePt can be fabricated as finely grained thin-films [4–7] or as nanoparticles. Here, we focus on the latter. Nanoparticles can be prepared by chemical processes, where the product is usually a suspension in liquid, or by gas-phase preparation techniques, where the particles form in the vapor state prior to depositing onto a sample carrier. Chemical processes yield particles that are usually coated with organic molecules. These particles usually have a face-centered-cubic (fcc) crystal structure, and post-deposition annealing is required to stabilize the L1$_0$ phase. However, post-deposition annealing can lead to particle agglomeration and therefore to unwanted size-growth [8–14]. Additionally, the organic molecules are also affected by post-deposition annealing so that impurities, mainly carbon, also constitute a part of the sample.

We use a gas-phase preparation technique that combines rapid cooling of high-pressure-sputtered ($\sim$1 mbar) FePt to form the particles and their subsequent in situ annealing in a single apparatus as discussed in Sect. 5.3. Gas-phase preparation has the advantage of having particles that are free of any organic material. Because the particles are annealed in the gas phase itself, post-deposition annealing is not required to form the L1$_0$ structure [15–19]. The drawback of the gas-phase method is that the yield is relatively lower than that in chemical methods. Both methods are capable of producing self-organized arrays on substrates [20, 21]. For self-organization of gas-phase prepared particles the substrate surfaces have to be coated with organic material prior to deposition [22, 23].

Although L1$_0$ is thermodynamically the stable structure of FePt at room temperature, this structure does not readily occur in nanoparticles unlike the case in

Fig. 5.2 Nanoparticle generator

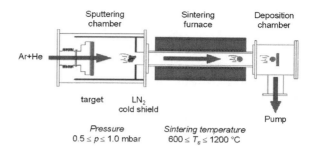

bulk form. This is true for nanoparticles prepared both by wet chemical and gas-phase methods. The fcc-L1$_0$ order/disorder transformation requires the diffusion of the constituent atoms and relies on the availability of vacancies over which atoms can exchange positions and acquire their equilibrium positions. It is thought that if vacancies are not present, or at least insufficient, diffusion cannot take place, and the system remains in the metastable fcc state.

To sort out this issue, we have undertaken a series of studies where we aim to enhance the diffusion in fcc FePt gas-phase prepared nanoparticles. This we do essentially by reactive sputtering, where we introduce oxygen or nitrogen along with argon in the sputtering process. Oxygen is expected to act as a surfactant, and nitrogen is expected to be trapped interstitially during the initial coagulation into nanoparticles. In both cases, subsequent in situ annealing releases the oxygen surfactant or the nitrogen interstitial; in the process in which diffusion is expected to be enhanced. We additionally investigate the effect of enhanced diffusion stimulated by substitutional atoms in FePt. We introduce Cu into the FePt matrix, and study the structure of annealed and non-annealed particles. We report on the magnetic properties of FePt nanoparticles prepared under these various conditions. The studies are undertaken using transmission electron microscopy (TEM) and magnetization.

5.2 Gas-Phase Preparation of FePt Nanoparticles

Although sputtering and in-flight annealing serves as a "clean" method to prepare FePt nanoparticles, the L1$_0$ state is not readily stabilized. Instead, random solid solutions of Fe and Pt are formed with multiply twinned icosahedral or polycrystalline fcc structures due to lack of diffusion in particles of several nanometers. Therefore, diffusion has to be (and can be) promoted by employing various techniques. However, before introducing and discussing these techniques, we review briefly the properties of gas-phase prepared FePt nanoparticles.

FePt nanoparticles are produced by in-flight annealing of primary particles obtained by high-pressure (0.5–1 mbar) Ar-sputtering of FePt alloy targets. The nanoparticle-generator setup is shown schematically in Fig. 5.2. The sputtering

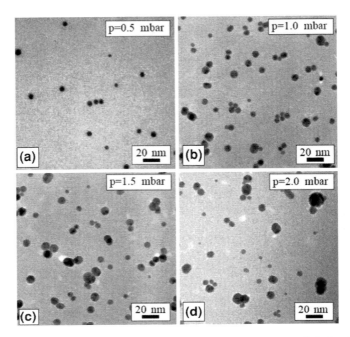

Fig. 5.3 Overview of particles prepared at $T_s = 1{,}273$ K and at pressures **a** 0.5 mbar, **b** 1.0 mbar, **c** 1.5 mbar, and **d** 2.0 mbar (after [15])

chamber is surrounded by a cold-shield cooled with liquid nitrogen so that the sputter product is rapidly cooled in the presence of He exchange gas introduced along with Ar. This causes the sputter-product to coagulate and to form primary nanoparticles. The primary particles are carried further through the furnace with the gas-flow and are annealed in situ at elevated temperatures prior to depositing onto a substrate. The residence time of the particles within the sintering furnace is about 0.05–0.10 s and depends on the gas-flow. The particle sizes can be controlled between about 5 and 20 nm depending on the chosen parameters such as, pressure, gas-flow, and sintering temperature [15].

An example of the pressure dependence of the particle size is shown in the TEM images in Fig. 5.3a–d, where the chamber pressure p ranges from 0.5 to 2 mbar with the sputtering time constant at 20 min. The coverage is low for $p = 0.5$ mbar and becomes higher above 1.0 mbar. The size distribution also appears to broaden with increasing pressure. Indeed, the size distributions given in Fig. 5.4a, b for particles prepared under 0.5 and 1.5 mbar, respectively, show that for $p = 0.5$ mbar, the mean particle diameter is about 6 nm with a geometric standard deviation $\sigma = 1.1$, whereas for particles prepared at $p = 1.5$ mbar, the mean particle diameter is about 8 nm with $\sigma = 1.3$. The broadening of the distribution with increasing pressure is related to an increased particle concentration because of increased sputter yield. This results in

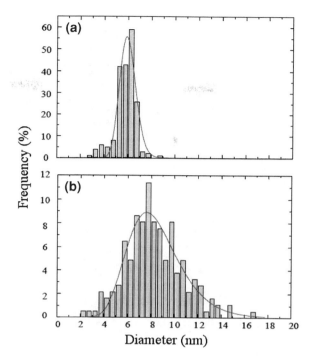

Fig. 5.4 Size distributions of particles prepared at $T_s = 1{,}273\,\text{K}$ under **a** 0.5 mbar and **b** 1.5 mbar. The data are fitted by log-normal distribution functions

an increase in the probability for interparticle collisions to occur, thereby enhancing the agglomeration.

The particle size is an important parameter that influences the morphology and the crystal structure. Figure 5.5a shows a HRTEM image of a twinned icosahedral particle prepared under $p = 1.0\,\text{mbar}$ and $T_s = 600\,°\text{C}$. The twin-boundary is indicated by the heavy white line in the figure. Figure 5.5b shows the power spectrum of the particle pertaining to the twinned structure. The particles prepared at a higher sintering temperature, $T_s = 1{,}000\,°\text{C}$, show more crystalline nature, and the number of multiply twinned particles decrease. Some of these particles also show $L1_0$ ordering as seen in a typical HRTEM image in Fig. 5.6a. The particle is oriented close to the [1 0 0] zone axis, and the ordering is observed by the contrast of the lattice planes along the [1 0 0] direction. The corresponding power spectrum is shown in Fig. 5.6b, where {0 0 1} superstructure reflections are indicative of the $L1_0$ structure. The power spectrum gives a c/a ratio of 0.975 in good agreement with earlier studies on bulk specimens (0.96–0.98) [24, 25], providing further evidence for the formation of the $L1_0$ phase.

The morphology of the particles in relation to preparation parameters are summarized in Fig. 5.7. In the given sintering temperature ranges only icosahedral particles are formed at low sputtering pressures, whereas polycrystalline and $L1_0$ form at higher pressures; the $L1_0$ formation requiring also higher sintering temperatures.

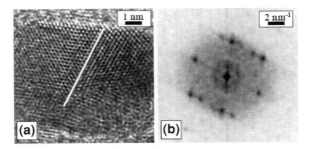

Fig. 5.5 HRTEM image of an FePt nanoparticle prepared under 1.0 mbar and $T_s = 873$ K. **a** Twinned particle. The twin boundary is shown with the heavy white line. **b** Power spectrum of the image in (**a**)

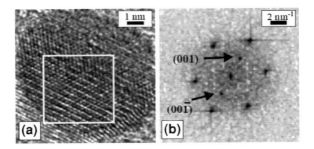

Fig. 5.6 HRTEM image of an FePt nanoparticle prepared under 1.0 mbar and $T_s = 1{,}273$ K. **a** HRTEM micrograph of a single crystal region of an FePt nanoparticle observed along the [1 1 0] zone axis. **b** The power spectrum of the marked area in (**a**)

Fig. 5.7 The $L1_0$ structure of FePt. Fe and Pt layers alternate along the [001] direction. The structure has a $(c/a) < 1$ tetragonal distortion

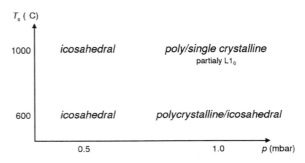

Certainly, higher sputtering pressures lead to larger particle size, as evidenced in Fig. 5.4. Larger particle size facilitates $L1_0$ stabilization, mainly because the probability of introducing lattice imperfections—especially vacancies—increases. This provides the necessary diffusion to form the $L1_0$ phase that readily stabilizes in bulk samples.

5.3 Diffusion Enhancement in FePt

FePt particles with twinned structures have low magnetic anisotropy and are unsuitable for magnetic data storage. In order to produce ferromagnetic nanoparticles, it is necessary to avoid their formation. We now discuss various methods that are expected to enhance atomic diffusion in FePt nanoparticles so that the high-anisotropy $L1_0$ phase can be formed. The methods are based on (1) using oxygen as surfactant in the sample preparation process, (2) introducing nitrogen as 'mobility enhancer', and (3) taking advantage of the segregation of a third element introduced into FePt to enhance diffusion.

5.3.1 Oxygen Mediated Destabilization of Twinned Structures

It is known that in the gas-phase based preparation of transition metal nanoparticles (e.g., Cu or Ni), the formation of twinned structures is suppressed in the presence of oxygen, and single crystalline particles are favored [26–28]. We adopt this method for gas-phase prepared FePt nanoparticles and show the influence of reactive sputtering with oxygen on the structural properties.

Particles obtained under various oxygen concentrations and sputtering pressure $p = 1.0$ mbar are shown in Fig. 5.8a–c. With increasing concentrations of oxygen, the mean particle size decreases, and the particle size distributions broaden. At an oxygen concentration of 1.7 % (Fig. 5.8d), the particles agglomerate, and by further increasing the oxygen content to 3.4 % (Fig. 5.8e), the size of the particles making up the agglomerate decrease to 1–2 nm. Normally, sintering at elevated temperatures should lead to an increase in particle size and to the formation of particles with spherical morphology. However, it appears that oxygen modifies the growth characteristics in the nucleation zone. At the early stage of growth, the particle sizes are smallest and the influence of the particle's surface is greatest. There are two main effects of oxygen on free surfaces of FCC metals: (1) Diffusion of oxygen on a metal surface is slower than the surface self-diffusion of the metal so that it leads to an increase in the activation energy for surface diffusion (reduced surface diffusivity) as in Cu and Pt [29, 30]. (2) Oxygen can also reduce the surface free energy of metals (e.g., Cu, γ-Fe, Ni [27, 30, 31]). If the surface diffusivity is reduced, the critical size for nucleation decreases leading to smaller primary particles. At the same time, the nucleation rate and, thereby, the particle number concentration increases [32]. Without oxygen, the particle number concentration is just below the critical limit for interparticle collisions [15]. The higher nucleation rate leads to the onset of particle–particle collisions and, as a result, agglomeration sets in. However, effect (2) gains weight by further increasing the oxygen concentration to 6.6 % leading again to nonagglomerated individual particles (Fig. 5.8f). The size of the particles increase again as also shown in Fig. 5.9. When the second effect gains dominance, oxygen reduces the surface free energy giving once again rise to the

Fig. 5.8 TEM images of FePt nanoparticles prepared at different oxygen concentrations. The particle sizes and the geometrical standard deviation are indicated in the panels.
a No oxygen, **b** 0.5%, **c** 0.9%, **d** 1.7%, **e** 3.4%, and **f** 6.6%. The particles are sintered at $T_s = 873$ K

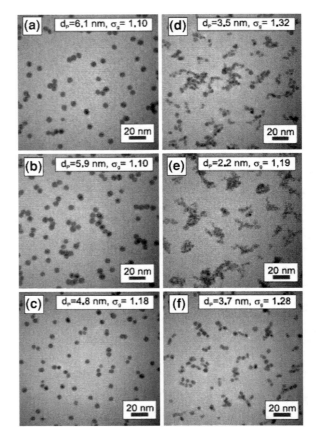

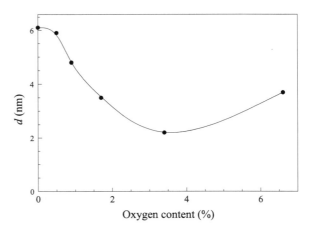

Fig. 5.9 Oxygen concentration dependence of the particle size

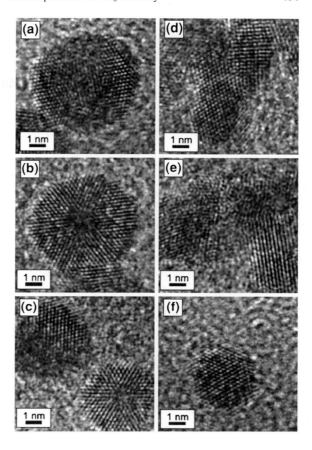

Fig. 5.10 HRTEM micrographs of particles as prepared at different oxygen concentrations. **a** No oxygen, **b** 0.5%, **c** 0.9%, **d** 1.7%, **e** 3.4%, and **f** 6.6%. The particles are sintered at $T_s = 873$ K

formation of larger particles. Since the gas pressure is kept constant and, thereby, the collision time is not affected, the coalescence properties in the presence of a critical amount of oxygen have to be modified. A similar effect has also been observed for Cu nanoparticles [27].

The crystal structure of the FePt nanoparticles prepared under various oxygen concentrations are shown in Fig. 5.10. Particles prepared in the absence of oxygen have icosahedral structures [33]. Figure 5.10a shows a typical icosahedral particle along its 2-fold axis [34]. At low oxygen concentrations there is no major modification of the morphology as in Fig. 5.10b where the icosahedral particle is oriented with its 3-fold symmetry axis parallel to the viewing direction. When the oxygen content is further increased, single crystal fcc particles begin to form as seen in Fig. 5.10c. In Fig. 5.10d, we show a HRTEM image of agglomerated particles prepared under 1.7% oxygen, where the crystalline nature of the polycrystalline particle becomes more apparent. At 3.4% oxygen (Fig. 5.10e), the crystallinity once again decreases, and the minimum particle size is reached (cf. Fig. 5.8e). At higher oxygen concentrations,

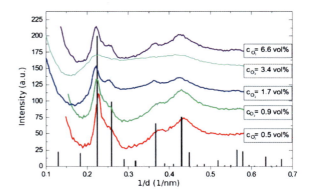

Fig. 5.11 Electron diffraction of FePt nanoparticles prepared under different oxygen concentrations. The *vertical lines* indicate the position of Bragg peaks for bulk FePt. The *gray lines* show reflection positions for iron oxides

the particle size increases again and mainly cuboctahedral single crystal fcc FePt nanoparticles are observed (Fig. 5.10f).

Figure 5.11 shows electron diffraction patterns obtained from FePt nanoparticles shown in Figs. 5.8 and 5.10. The reflection positions agree with those for bulk FePt (black bars). The peaks broaden with increasing oxygen concentration indicating decreasing crystallite size. At 3.4 %, the peaks are broad in agreement with the small size and the low degree of crystallinity. At 6.6 % oxygen the crystallinity is recovered.

These results show that reactive sputtering with oxygen enhances the diffusion so that predominantly crystalline nanoparticles are stabilized. However, with this method, the formation of the $L1_0$ phase is not observed with the preparation conditions employed.

5.3.2 $L1_0$ Stabilization by Interstitial Nitriding–Denitriding

If nascent nitrogen can be introduced into FePt nanoparticles during particle coagulation in the condensation chamber (nitriding), it can then be allowed to effuse by in-flight annealing (denitriding), and while effusing, nitrogen can agitate the lattice and enhance the atomic mobility of Fe and Pt. This can favor the formation of the $L1_0$ phase. In an FCC lattice, nitrogen takes up octahedral interstitial positions in Fe alloys and effuses out of the lattice above 700 °C [35]. The method was employed in FePt thin films grown by reactive sputtering with nitrogen that gave rise to enhanced coercivity due to $L1_0$ formation when the films were annealed [36, 37]. This process can also take place in gas-phase prepared FePt nanoparticles as discussed below.

We show in Fig. 5.12a, a TEM image of FePt nanoparticles prepared in the absence of nitrogen at a sputter pressure of 0.5 mbar and an annealing temperature of 1,000 °C. The particles have a mean diameter of 6 nm. The upper part of Fig. 5.12b shows a high resolution image of a typical particle with multiply twinned icosahedral morphology. The image shows contrast related to an icosahedron along its 3-fold symmetry axis. The power spectrum of the image, shown in the lower part of Fig. 5.12b, exhibits the

5 Diffusion Enhancement in FePt Nanoparticles for L1$_0$ Stability   133

Fig. 5.12 Particles prepared in the absence of nitrogen. **a** Low magnification image. **b** *Upper panel* is a high resolution image showing the multiply twinned morphology of the particles. The *lower panel* is the power spectrum exhibiting the characteristic elongated {111} reflections

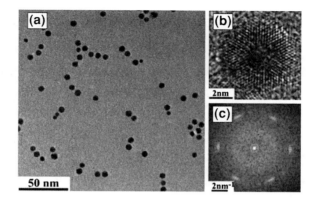

characteristic pattern with elongated {111} reflections. These particles are stable up to an annealing temperature of 1,200 °C and no L1$_0$ ordering occurs. About 87 % of the particles prepared under such conditions are multiply twinned and the remaining are polycrystalline FCC.

Figure 5.13a shows a TEM image of particles prepared by nitrogen reactive sputtering. The remaining conditions are the same as for preparing particles without nitrogen. The particles again have a mean diameter of 6.0 nm. The HRTEM image in the upper part of Fig. 5.13b shows a single-crystalline FCC FePt nanoparticle where the plane of the figure is perpendicular to the [110] direction. The particle is bounded by {100} and {111} facets and has a cuboctahedral morphology [38, 39]. The power spectrum of the image in the lower part of Fig. 5.13b is indicative of the FCC structure. In the presence of nitrogen, the amount of icosahedral nanoparticles reduces substantially from 87 to 6 %. Single-crystalline particles amount to 70 %, while the remaining 24 % are polycrystalline. The image in the upper part of Fig. 5.13c exhibits bright and dark contrasts of the lattice planes which is due to the ordering in the L1$_0$ structure. The power spectrum in the lower panel shows {001} superstructure reflections confirming the presence of the L1$_0$ structure.

The layered contrast in Fig. 5.13c can be observed only when the particles are favorably aligned with respect to the electron beam meaning that not all particles in the L1$_0$ state can be observed. Using a statistical method [40], we estimate that about 70 % of all single crystalline and polycrystalline FePt nanoparticles have an L1$_0$ structure.

The morphology diagram of the of FePt nanoparticles prepared by reactive sputtering under various nitrogen concentrations is summarized in Fig. 5.14. Sputtering in the presence of nitrogen leads to considerable changes in structure and morphology of the particles, and under optimum conditions, the L1$_0$ phase can be stabilized.

Electron energy loss spectroscopy (EELS) and X-ray absorption spectroscopy (XAS) studies on primary and sintered particles provide information about the incorporation of nitrogen in the lattice structure of FePt. Figure 5.15 shows EELS spectra of primary and sintered particles in the energy range from 380 to 500 eV with the

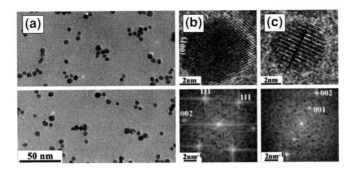

Fig. 5.13 Particles prepared in the presence of nitrogen. **a** Low magnification image. **b** The *upper* high resolution image shows an FCC cuboctahedral particle with (111) and (100) facets. The *lower panel* shows the power spectrum related to the FCC structure. **c** L1$_0$ particle (*upper panel*) with alternating *bright* and *dark layers* along the c-axis. The *lower panel* shows the power spectrum related to the L10 structure

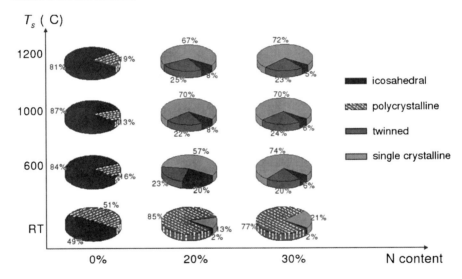

Fig. 5.14 The morphology diagram of FePt nanoparticles prepared by reactive nitrogen sputtering

background subtracted. Smoothed data by adjacent averaging is shown with the heavy lines. The spectrum for primary particles shows a peak at the nitrogen ionization edge at about 400 eV confirming that nitrogen is introduced into the FePt structure [41]. The spectrum for the sintered particles show no feature at this energy indicating that nitrogen is expelled from the particles on annealing.

XAS studies give a clearer indication and further confirm the presence of nitrogen in the structure. Figure 5.16 shows the normalized XAS in the energy range 395–410 eV. The spectrum of the primary nanoparticles (filled circles) shows two

5 Diffusion Enhancement in FePt Nanoparticles for L1$_0$ Stability

Fig. 5.15 Background treated EELS data on primary and sintered particles. For better visualization, the data are smoothed by 20-point adjacent averaging and shown with the *heavy line*

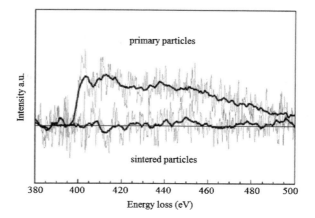

Fig. 5.16 XAS spectrum of the primary and sintered nanoparticles

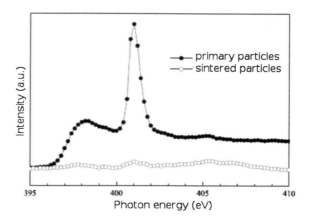

pronounced absorption features with maxima at about 398 and 401 eV. The first corresponds to the the binding energy range of various iron nitrides measured using X-ray photoelectron spectroscopy [42, 43]. The second peak in the XAS spectrum centered around 401 eV is due to the presence of molecular nitrogen, which is either trapped or adsorbed on the surface of the particle. The spectrum of the sintered nanoparticles is shown with open circles. The spectrum shows no substantial feature, indicating that nitrogen has been expelled out of the particles during sintering.

5.3.3 L1$_0$ Stabilization by the Mobility Enhancement of a Substitutional Element

A further method that can be used to stabilize the L1$_0$ structure of FePt is to add a third element to the compound that would segregate on annealing (at least partially) such

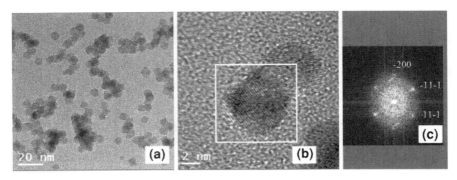

Fig. 5.17 Structural characterization of primary Fe–Pt–Cu nanoparticles. **a** Low magnification, **b** high resolution (beam along [011]), **c** power spectrum of area shown in (**b**)

as Cu. Sputtering an Fe–Pt–Cu target under conditions similar to those for preparing FePt nanoparticles would first yield a nanoparticle as a metastable solid solution of Fe–Pt–Cu due to rapid condensation in the nucleation chamber. When the particle is annealed in flight, Cu is expected to partially segregate when the Cu concentration is higher than its solubility limit in Fe–Pt. On doing so, the mobility of the atoms is expected to increase, and the $L1_0$ can be expected to form. Studies on thin films of Fe–Pt–Cu have shown that post-preparation annealing of the films have indeed led to the stabilization of the $L1_0$ structure with which the particles also exhibited a substantial coercive field.

Fe–Pt–Cu was sputtered under $p = 0.5$ mbar and obtained as primary particles and as particles in-flight-annealed at 1,000 °C. The composition of both primary and annealed particles is determined as $(Fe_{50}Pt_{50})_{75}Cu_{25}$. Figure 5.17 shows structural characterization of the particles using TEM. Figure 5.17a shows that the particles are fairly monodisperse, and the size is roughly 6 nm. A representative high resolution image in Fig. 5.17b shows that the particles have a high degree of crystallinity, and no icosahedral particles are observed. The power spectrum of the selected area shown in Fig. 5.17c indicates that the particles have an FCC structure.

The structural characterization of the annealed particles in Fig. 5.18a show that in-flight annealing of the particles definitely alter the structure. The size of the particles are nearly the same, as seen in Fig. 5.18b, however, the structure is now $L1_0$. This is also confirmed by the power spectrum given in Fig. 5.18c, where {0 0 1} superstructure reflections are observed.

At 25 % Cu concentration, Fe–Pt–Cu can only form a metastable random solid solution when rapidly cooled, as would be the case when the particles rapidly coagulate in the plasma chamber. On annealing, partial segregation would have to occur that could lead to a compositional gradient leaving behind Cu-rich areas. These properties have to be further studied to give the true compositional range of the occurrence of the $L1_0$ phase in Fe–Pt–Cu. Nevertheless, the $L1_0$ phase definitely forms in these alloys.

5 Diffusion Enhancement in FePt Nanoparticles for L1₀ Stability

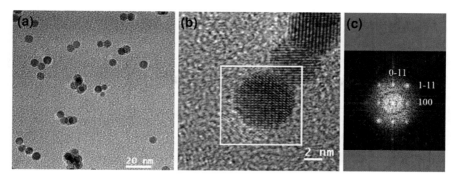

Fig. 5.18 Structural characterization of Fe–Pt–Cu nanoparticles in-flight annealed at 1,000 °C. **a** Low magnification, **b** high resolution (beam along [011]), **c** power spectrum of area shown in (**b**)

5.4 Conclusion

The search for high-magnetic-anisotropy nanoparticles remains a topic of current research with the aim of producing magnetically stable nanoparticles for high-density magnetic storage media. FePt in nanoparticle form is certainly a candidate, but much remains to be learned about the L1₀ phase properties of these particles. In particular, how it will be possible to stabilize this phase routinely at small sizes, and how the monodispersity can be better controlled are basic questions related to the structural aspects of the problem, with which we have mainly dealt with here. The aspect of magnetic properties and the problems related to it is another topic of its own.

FePt does not readily stabilize in the L1₀ phase and icosahedral structures are favored especially at small sizes. However, since the icosahedral structure is normally not the thermodynamically stable sate, inducing somehow lattice agitation can cause the structure to destabilize and form the L1₀ structure. In this work, we have shown that reactive sputtering with oxygen, nitriding–denitriding, and inducing segregation are all methods that provide lattice agitation that destabilize the icosahedral structure. It is now necessary to pursue these studies by optimizing parameters and introducing new lattice agitation techniques to aid the formation of the L1₀ phase in FePt.

Acknowledgments We thank our collegues with whom we have worked over the years, B. Rellinghaus, S. Stappert, O. Dmitrieva, G.Ünlü, G. Dumpich, and J. Kästner, and who have carried out much of the works presented in this article. We gratefully acknowledge the Deutsche Forschungsgemeinschaft (SFB445) for financial support.

References

1. D. Weller, A. Moser, L. Folks, M.E. Best, W. Lee, M.F. Toney, M. Schwickert, J.U. Thiele, M. Doerner, IEEE. Trans. Mag. **36**, 10 (2000)
2. D. Weller, A. Moser, IEEE Trans. Magn. **35**, 4423 (1999)
3. O. Gutfleisch, J. Lyubina, K.-H. Müller, L. Schultz, Adv. Eng. Mater. **7**, 208 (2005)

4. D.J. Sellmyer, Y. Yu, M. Yan, Y. Sui, J. Zhou, R. Skomski, JMMM **303**, 302 (2006)
5. D.J. Sellmyer, M. Yan, Y. Yu, R. Skomski, IEEE Trans. Magn. **41**, 560 (2005)
6. K. Sato, B. Bian, Y. Hirotsu, J. Appl. Phys. **91**, 8516 (2002)
7. K. Sato, T. Kajiwara, M. Fujiyoshi, M. Ishimaru, Y. Hirotsu, T. Shinohara, J. Appl. Phys. **93**, 7414 (2003)
8. S. Sun, C.B. Murray, D. Weller, L. Folks, A. Moser, Science **287**, 1989 (2000)
9. Z.R. Dai, S. Sun, Z.L. Wang, Nano Lett. **1**, 443 (2001)
10. Y.K. Takahashi, M. Ohnuma, K. Hono, Jpn. J. Appl. Phys. **40**, 1367 (2001)
11. J.W. Harrel, S. Wang, D.E. Nikles, M. Chen, Appl. Phys. Lett. **79**, 4393 (2001)
12. D.H. Ping, M. Ohnuma, K. Hono, M. Watanabe, T. Iwasa, T. Msumoto, J. Appl. Phys. **90**, 4708 (2001)
13. T. Thomson, B.D. Terris, M.F. Toney, S. Raoux, J.E.E. Baglin, S.L. Lee, S. Sun, J. Appl. Phys. **95**, 6738 (2004)
14. A. Ethirajan, U. Wiedwald, H.-G. Boyen, B. Kern, L. Han, A. Klimmer, F. Weigl, G. Kästle, P. Ziemann, K. Fauth, J. Cai, J. Behm, P. Oelhafen, P. Walther, J. Biskupek, U. Kaiser, Adv. Mater. **19**, 406 (2007)
15. S. Stappert, B. Rellinghaus, M. Acet, E.F. Wassermann, J. Cryst. Growth **252**, 440 (2003)
16. Y.H. Huang, Y. Zhang, G.C. Hadjipanayis, D. Weller, J. Appl. Phys. **93**, 7172 (2003)
17. J. Qiu, J.H. Judy, D. Weller, J. Wang, J. Appl. Phys. **97**, 10J319 (2005)
18. S. Stoyanov, Y. Huang, Y. Zhang, V. Skumryev, G.C. Hadjipanayis, D. Weller, J. Appl. Phys. **93**, 7190 (2003)
19. B. Rellinghaus, E. Mohn, L. Schultz, T. Gemming, M. Acet, A. Kowalik, B.F. Kock, IEEE Trans. Magn. **42**, 3048 (2006)
20. S. Sun, S. Anders, T. Thomson, J.E.E. Baglin, M.F. Toney, H. Hamann, C.B. Murray, B.D. Terris, J. Phys. Chem. B **107**, 5419 (2003)
21. C. Verdes, R.W. Chantrell, A. Satoh, J.W. Harrell, D. Nikles, JMMM **304**, 27 (2006)
22. U. Queitsch, E. Mohn, F. Schäffel, L. Schultz, B. Rellinghaus, A. Blüher, M. Mertig. Appl. Phys. Lett. **90**, 113114 (2007)
23. A. Terheiden, O. Dmitrieva, M. Acet, Ch. Mayer, Chem. Phys. Lett. **431**, 113 (2006)
24. J.-U. Thiele, L. Folks, M.F. Toney, D. Weller, J. Appl. Phys. **84**, 5686 (1998)
25. M.R. Visokey, R. Sinclair, Appl. Phys. Lett. **66**, 1692 (1995)
26. J. Urban, H. Sack-Kohngehl, K. Weiss, High Temp. Mat. Sci. **36**, 155 (1996)
27. D.L. Olynick, J.M. Gibson, R.S. Averback, Phil. Mag. A **77**, 1205 (1998)
28. S. Stappert, Diploma thesis, Duisburg, 2000
29. M. Yata, H. Rouch, K. Nakamura, Phys. Rev. B **56**, 10579 (1997)
30. H. Landolt, R. Brnstein, *Numerical Data and Functional Relationships in Science and Technology: New Series III*, vol. 42A (Springer, Berlin, 1992), pp. 50–52, 478–479
31. L. Li, A. Kida, M. Ohnishi, M. Matsui, Surf. Sci. **493**, 120 (2001)
32. R.C. Flagan, M.M. Lunden, Mater. Sci. Eng. A **204**, 113 (1995)
33. S. Stappert, B. Rellinghaus, M. Acet, E.F. Wassermann, Mat. Res. Soc. Proc. **704**, 73 (2002)
34. J. Urban, Cryst. Res. Technol. **33**, 1009 (1998)
35. M. Acet, B. Gehrmann, E.F. Wassermann, H. Bach, W. Pepperhoff, J. Magn. Magn. Mater. **232**, 221 (2001)
36. H.Y. Wang, W.H. Mao, X.K. Ma, H.Y. Zhang, Y.B. Chen, Y.J. He, E.Y. Jiang, J. of Appl. Phys. **95**, 2564 (2004)
37. H.H. Hsiao, R.N. Panda, J.C. Shih, T.S. Chin, J. Appl. Phys. **91**, 3145 (2002)
38. M.J. Yacaman, J.A. Ascencio, H.B. Liu, J. Gardea-Torresdey, J. Vac. Sci. Technol. B. **19**, 1091 (2001)
39. Z.R. Dai, S. Sun, Z.L. Wang, Surf. Sci. **505**, 325 (2002)
40. O. Dmitrieva, Ph.D Thesis, Duisburg, 2007
41. O.L. Krivanek, C.C. Ahn, EELS Atlas (Gatan Inc., Pleasanton, 1983)
42. X. Wang, W.T. Zheng, H.W. Tian, S.S. Yu, W. Xu, S.H. Meng, X.D. He, J.C. Han, C.Q. Sun, B.K. Tay, Appl. Surf. Sci. **220**, 30 (2003)
43. P.C.J. Graat, M.A.J. Somers, E.J. Mittemeijer, Appl. Surf. Sci. **136**, 238 (1998)

Chapter 6
Simulation of Cluster Sintering, Dipolar Chain Formation, and Ferroelectric Nanoparticulate Systems

Anna Grünebohm, Alfred Hucht, Ralf Meyer, Denis Comtesse and Peter Entel

Abstract Magnetic and ferroelectric nanoparticles are subjects of increasing basic research for future technologies. In this work Fe and Ni clusters and near-ferroelectric TiO_2 clusters have been chosen as representatives in order to discuss fundamental issues such as sintering of magnetic and near-ferroelectric clusters as well as configurations resulting from cluster agglomeration due to magnetic dipolar interactions.

6.1 General Introduction

The formation, structure, and properties of nanoparticles grown in the gas or liquid phase belongs to an active field of basic research, which is of interest for future applications in different areas, ranging from computer technology to catalytic and biomedical applications (e.g., see [1, 2]). Besides this, magnetic nanoparticles show a variety of additional properties when compared to non-magnetic particles (for an overview, see [3]), which can often be attributed to the anisotropic and long-range dipolar interaction between the particles. For example, in recent experiments on magnetic nanoparticles with high mobility the formation of chains, rings, and network-like structures has been observed [4–6]. Using a classical model, we discuss the influence of particle size and temperature on the chain formation in Sect. 6.2. When nanoparticles are produced in the gas phase, a certain degree of agglomeration of the nanoparticles into larger units is difficult to avoid. In addition to a change in the size of the particles, the agglomeration leads to irregularly shaped entities with lattice defects due to the misalignment of the crystal lattices of the agglomerated particles. These structural shortcomings can lead to substantial variations of the secondary

---

A. Grünebohm (✉) · A. Hucht · D. Comtesse · P. Entel
University of Duisburg-Essen and CENIDE, Lotharstraße 1, 47057 Duisburg, Germany
e-mail: anna@thp.uni-due.de

R. Meyer
Laurentian University, 935 Ramsey Lake Road, Sudbury, ON P3E 2C6, Canada

A. Lorke et al. (eds.), *Nanoparticles from the Gas Phase*, NanoScience and Technology, DOI: 10.1007/978-3-642-28546-2_6, © Springer-Verlag Berlin Heidelberg 2012

particles' physical properties and make it more difficult to control the quality of the product.

In order to reverse the negative effects of agglomeration and to obtain particles with regular shapes and well-controlled properties, gas phase-produced nanoparticles are typically sintered. During the sintering, the particles are heated to temperatures of several hundred Kelvin above the room temperature. At these elevated temperatures, the tendency of the particles to minimize their excess energy caused by the surface and lattice defects leads to the removal of most of the structural defects and the formation of nearly spherical particles. Although the general principle of the nanoparticle sintering is easy to understand, there are many open questions concerning the details of the process. In Sect. 6.3 we focus on the following issues: by which mechanisms can the crystal lattices of agglomerated nanoparticles be aligned? What are the remaining structural defects after the sintering? What is the time scale of the dominant diffusion process driving the sintering process? And, how do agglomeration or sintering processes interplay with structural phase transitions? In order to address these questions we have performed molecular-dynamics (MD) simulations of the agglomeration and sintering of two nanoparticles.

Nanocomposites which consist of several agglomerated or sintered nanoparticles possess advantages over conventional materials such as exceptional mechanical properties, and the use of mixed composite systems allows for multifunctional devices on small length scales. In this context, especially the mixture of ferroelectric or paraelectric materials with a high polarization response together with magnetic materials may serve as multicomponent, multiferroic, or magnetoelectrical coupled materials. In such multicomponent systems the ferroelectric and ferromagnetic phases are coupled through strain effects and hybridization at the interfaces. Thus, they can overcome the failure of single-phase materials, which do not show a considerable magnetoelectric effect at room temperature due to the different origins of the ferroic properties [7]. While ferromagnetism is mainly moderated through highly localized d- and f-electrons, the ferroelectricity is mainly moderated by cation off centering which is based on empty d-shells. Recently, it has been shown that ferroelectricity can appear in dimensions of several Å, see review [7], and thus the formation of multistructures at the nanorange is indeed possible. As a first step into the direction of multiferroic nanocomposites, we study the interaction and agglomeration of oxidic nanoparticles with a large polarizability and their possible ferroelectric behavior in Sect. 6.4. As the high polarizability of such oxidic systems makes the use of model potentials questionable, we perform density functional theory simulations instead. Here, we have to restrict ourselves to the simulation of the particle interaction and first agglomeration steps at $T = 0\,\mathrm{K}$. In return we are able to describe the atomic arrangement and electronic structure on an ab inito level.

6.1.1 Methods

In order to study the structural properties of pairs of nanoparticles during agglomeration and sintering, we have employed the method of MD simulations in Sects. 6.2

and 6.3. For an overview of this method, see [8]. In a MD simulation the classical equations of motion of an ensemble of particles are integrated numerically. Since the method can be used on an atomistic level, the evolution of structural defects can be studied in great detail and with precise time information. In addition to this, there are no restrictions on the positions of the atoms so that arbitrary defect configurations can be studied with MD simulations.

In order to integrate the equations of motions of the atoms, a MD simulation needs to employ a model for the forces acting upon the atoms. Although density functional theory methods (DFT) can in principle be used for this task [9], their use is restricted to several hundreds of atoms and tiny periods of a MD simulation even on supercomputers.

We used instead semi-empirical many-body model potentials of the embedded-atom method [10, 11] (Sect. 6.3.2) and tight-binding second-moment [12] (Sect. 6.3.1) type which are computationally much less demanding and therefore allow for longer simulations of larger systems. Even with such model potentials the sintering simulations remained challenging since the diffusion processes that play an important role in the sintering process require long simulations in the nanosecond range.

Attention had to be paid to the control of the temperature of the system over the course of the simulation. During the simulation the reduction of surface area and removal of defects leads to a reduction of the potential energy. The excess energy is initially converted into kinetic energy and thus increases the temperature of the system. In a real situation, the excess energy is eventually removed from the particle through collisions with atoms of the surrounding gas. Since there is no surrounding gas in our simulations we employed the Nosé-Hoover thermostat method [13] (Sect. 6.3.1), respectively, the Andersen thermostat [14] (Sects. 6.2, 6.3.2) to keep the temperature in our system constant.

In Sect. 6.2 we are focusing on the large dipolar interaction between different magnetic particles. Because of this, it is not necessary to calculate the particle interactions on an atomistic level within our MD simulation. Instead, we make use of a simple particle–particle interaction potential, which includes magnetic interactions explicitly, for details, see Sect. 6.2.

In Sect. 6.4 we use DFT to study the oxidic and highly polar TiO_2-system. In contrast to metallic systems, simple model potentials cannot describe the high polarizability and thus the large dependency of the electronic charges on the local atomic arrangement. These properties are especially important for applications and have a large influence on sintering and agglomeration, see [15]. In the past, nanoparticles and sinter processes of TiO_2 nanoparticles have been investigated by means of various MD calculations, in part by employing highly sophisticated model potentials and corresponding variable-charge MD.

Until now, it is not clear whether all system properties are described properly, as these highly optimized potentials have a low transferability and although structural properties are reproduced for the different TiO_2 morphologies anatase and rutile, important properties such as the Born effective charges, which are very important for possible ferroelectric applications are drastically underestimated [16]. As DFT simulations are computationally very demanding we are not able to perform finite temperature MD simulations. Instead, we stick to $T = 0\,\mathrm{K}$ and model the basic features of the nanoparticle interaction through pressure-induced agglomeration by a gradual shrinking of the particle–particle distances with successive relaxation of the atomic positions with the conjugate gradient method.

6.2 Molecular-Dynamics Simulations of the Dipolar-Induced Formation of Magnetic Nanochains and Nanorings

In this chapter, we present results for a diluted gas of spherical magnetic nanoparticles. The simulations are performed at fixed volume V, particle number N, temperature T, and magnetic field $\mathbf{B} = 0$; for a discussion of finite magnetic fields, see [17].

We use the simple interaction potential $\mathscr{U}_{ij} = \mathscr{U}_{ij}^{\mathrm{d}} + \mathscr{U}_{ij}^{\mathrm{h}}$ of two particles i and j at distance $\mathbf{r}_{ij} = \mathbf{r}_j - \mathbf{r}_i$, where $\mathscr{U}_{ij}^{\mathrm{d}}$ is the long-range anisotropic dipolar interaction,

$$\mathscr{U}_{ij}^{\mathrm{d}} = \frac{\mu_0 \mu^2}{4\pi r_{ij}^3}\left[\hat{\boldsymbol{\mu}}_i \cdot \hat{\boldsymbol{\mu}}_j - 3\left(\hat{\boldsymbol{\mu}}_i \cdot \hat{\mathbf{r}}_{ij}\right)\left(\hat{\mathbf{r}}_{ij} \cdot \hat{\boldsymbol{\mu}}_j\right)\right] \tag{6.1}$$

with $\hat{\mathbf{r}}_{ij} = \mathbf{r}_{ij}/r_{ij}$, and $\mathscr{U}_{ij}^{\mathrm{h}}$ as a sufficiently rigid isotropic hard sphere interaction. The total potential energy of N particles in an external magnetic field $\mathbf{B}$ is thus given by

$$\mathscr{U} = \frac{1}{2}\sum_{i \neq j}^{N}\mathscr{U}_{ij} - \sum_{i=1}^{N}\mathbf{B} \cdot \boldsymbol{\mu}_i. \tag{6.2}$$

Here, we neglect quantum mechanical effects which lead to, e.g., size-dependent magnetic moments in Fe particles with less than 1,000 atoms [18]. Due to the explicit use of magnetic interactions the calculation of the particle trajectory has to be extended by the orientations of the magnetic moments, $\boldsymbol{\mu}_i(t)$, of the particles under the influence of the force $\mathbf{F}_i = \mathbf{F}(\mathbf{r}_i, t)$ and the local magnetic field $\mathbf{B}_i = \mathbf{B}(\mathbf{r}_i, t)$ given by

$$\mathbf{F}_i = -\frac{\partial \mathscr{U}}{\partial \mathbf{r}_i}, \quad \mathbf{B}_i = -\frac{\partial \mathscr{U}}{\partial \boldsymbol{\mu}_i}. \tag{6.3}$$

As the magnetic moment $\boldsymbol{\mu}_i$ is assumed to be pinned within the particle by intrinsic anisotropies, the local magnetic field $\mathbf{B}_i$ will provide a torque $\mathbf{M}_i = \boldsymbol{\mu}_i \times \mathbf{B}_i$ to the particle. The resulting equations of motion have the form

Fig. 6.1 Kinetic energy e^{kin}, potential energy u, and total energy e^{tot} of a 3d system with $N = 9$ particles as function of temperature T for different system sizes $\tilde{L}$, as obtained from the molecular dynamics simulation. The transition temperature T_c is shown as *dotted line*. Figure taken from [17]

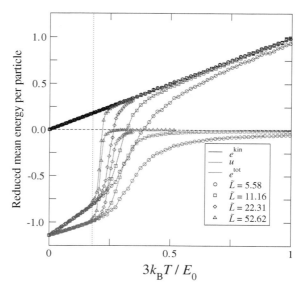

$$\dot{\mathbf{p}}_i = m\ddot{\mathbf{r}}_i = \mathbf{F}_i, \tag{6.4}$$

$$\dot{\mathbf{L}}_i = I\dot{\boldsymbol{\omega}}_i = \mathbf{M}_i = \boldsymbol{\mu}_i \times \mathbf{B}_i, \tag{6.5}$$

$$\dot{\boldsymbol{\mu}}_i = \boldsymbol{\omega}_i \times \boldsymbol{\mu}_i, \tag{6.6}$$

where $\mathbf{p}_i$ is the momentum, $I = m\sigma^2/10$ is the moment of inertia of spheres with diameter σ, and $\boldsymbol{\omega}_i$ is the angular velocity of the ith particle.

Due to the long-range dipole interaction the MD algorithm scales with N^2 and we restrict N to rather small values below 50 particles in order to get reliable equilibrium results. As we want to compare our results to experiments done in the gas phase [4, 6], we consider systems with very low particle densities and focus on the case system size $L \to \infty$ at fixed N. Figure 6.1 shows the reduced kinetic energy $e^{kin} = \langle E^{kin} \rangle / E_0 N$, the reduced potential energy $u = \langle \mathscr{U} \rangle / E_0 N$, and the reduced total energy $e^{tot} = \langle E^{tot} \rangle / E_0 N$ versus reduced temperature for a system consisting of $N = 9$ particles, with different linear size $\tilde{L} = L/\sigma$ in three dimensions, where the ground state energy of two particles in contact,

$$E_0 = \frac{\mu_0 \mu^2}{2\pi \sigma^3}, \tag{6.7}$$

denotes the energy scale. $N = 9$ is just one particular choice, $N > 9$ yields similar results but larger statistical errors. It can be seen that the potential energy u develops a jump with growing system size L, which results in a jump in the total energy e^{tot}, as the kinetic part is simply proportional to temperature in the canonical ensemble. The L-dependence of u stems from the fact that in a finite simulation volume particles

Fig. 6.2 Finite-size scaling plot of the potential energy u and specific heat c_V (inset) as function of the scaling variable x. Figure taken from [17]

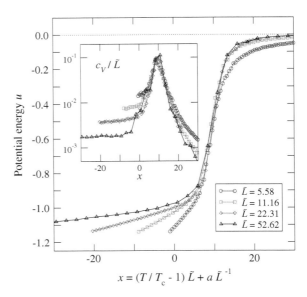

which leave the bound state, chain, or ring, are reflected back towards the structure by the boundaries. Thus, the bound state is more stable in small volumes than in large or even infinite volume. This effect broadens the width of the transition and shifts the effective transition temperature $T_c(L)$ to higher temperatures for smaller L. In order to determine the nature of the transition as well as the critical temperature T_c, a finite-size scaling plot of the potential energy $u(T)$ is shown in Fig. 6.2, together with the specific heat at constant volume and constant number of particles, $c_V = \partial u/\partial T|_{V,N}$. Using the scaling variable

$$x = \left(\frac{T}{T_c} - 1\right)\tilde{L} + \frac{a}{\tilde{L}}, \tag{6.8}$$

a data collapse both in $u(T)$ and in $c_V(T)$ can be achieved, leading to an estimation for the critical temperature T_c,

$$k_B T_c = 6.0(6) \times 10^{-2} E_0. \tag{6.9}$$

The constant a describes corrections to scaling and has the value $a = 25(3)$. The fact that no rescaling is necessary in u shows that the transition is of first order.

In our model E_0 depends on the size σ and magnetic moment μ of the particle via Eq. (6.7), leading to a simple dependence of the critical temperature T_c on the considered type of nanoparticles. For instance, Fe nanoparticles with a diameter of $\sigma \approx 6$ nm (10^4 atoms) have a saturation magnetization of $\mu \approx 2.2 \times 10^4 \mu_B$, where μ_B is the Bohr magneton. The critical temperature for the Fe nanoparticles follows from Eq. (6.9) to $T_c \approx 140$ K. A resulting phase diagram for magnetic nanoparticles

Fig. 6.3 Phase diagram of magnetic nanoparticles as function of particle size σ and temperature T. Shown are the melting temperature (*dotted black*), the Curie temperature (*dashed blue*), and the chain building temperature, both with constant moment μ (*long dashed pink*) as well as with temperature-dependent moment $\mu(T)$ (*thick red*), see text. Taken from [17]

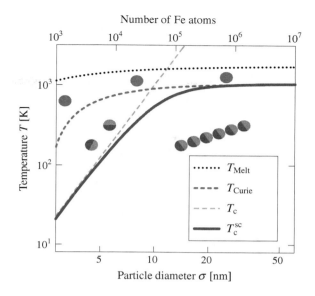

is depicted in Fig. 6.3, where $T_c(\sigma)$ is shown as a dashed red line. As to lowest order, $T_c(\sigma) \propto \sigma^3$, it has to be compared to other characteristic temperatures like the Curie temperature T_{Curie} and structural transition temperatures, e.g., the melting temperature T_{Melt}, in order to determine the validity of the theory. Hence these two temperatures, which are size dependent as well, are also depicted schematically for typical magnetic systems. The finite-size form of T_{Melt} can be approximated to lowest order using energetic arguments to give $T_{\text{Melt}}(\sigma)/T_{\text{Melt}}^\infty = 1 - \sigma_{\text{Melt}}/\sigma$ with material-specific constant σ_{Melt} [19]. In the case of the magnetic transition, standard finite-size scaling theory gives, again to lowest order, a temperature shift of the form $T_{\text{Curie}}(\sigma)/T_{\text{Curie}}^\infty = 1 - (\sigma_{\text{Curie}}/\sigma)^{1/\nu}$, with the exponent of magnetic correlations ν and constant σ_{Curie} [20]. As the magnetization μ vanishes at T_{Curie} as $\mu(T) = \mu(0)(1 - T/T_{\text{Curie}})^\beta$ with critical exponent β, and as E_0 is proportional to μ^2 (Eq. (6.7)), we can calculate a corrected $T_c^{\text{sc}}(\sigma)$ with temperature-dependent magnetic moments as solution of the self-consistency equation

$$\frac{T_c^{\text{sc}}(\sigma)}{T_c(\sigma)} = \left(1 - \frac{T_c^{\text{sc}}(\sigma)}{T_{\text{Curie}}(\sigma)}\right)^{2\beta}, \quad (6.10)$$

shown in Fig. 6.3 as thick red line. We used the parameters $T_{\text{Melt}}^\infty = 1{,}743$ K, $\sigma_{\text{Melt}} = 1$ nm, $T_{\text{Curie}}^\infty = 1{,}043$ K, $a_{\text{Curie}} = 2.5$ nm, $\nu = 0.7$, and $\beta = 0.33$. For small particles $T_c^{\text{sc}}(\sigma) \approx T_c(\sigma)$, as the rhs. of Eq. (6.10) is approximately unity. On the other hand, for big particles $T_c^{\text{sc}}(\sigma) \approx T_{\text{Curie}}(\sigma)$, as now the lhs. of Eq. (6.10) is vanishing.

In conclusion, Fe nanoparticles can undergo a phase transition from ferromagnetic chains/rings to a ferromagnetic gas for particle sizes below approximately

20 nm, while for larger particles the transition goes from ferromagnetic chains/rings directly to a paramagnetic gas. Although this transition disturbs the uniform growth of magnetic nanoparticles from the gas phase onto a substrate, it may serve as a new self-organized building mechanism of one-dimensional magnetic structures.

6.3 Molecular-Dynamics Simulations of Metal Cluster Agglomeration and Sintering

In this section we describe MD simulations of the agglomeration and sintering of metallic nanoparticle pairs. The aim of these simulations is to obtain a better understanding of the dynamics of structural changes and lattice defects during the coalescence of nanoparticles.

For the structural analysis of the results of our simulations we employed the common neighbor analysis (CNA) [21] which is a topological short-range analysis that can be used to assign a local crystalline order to individual atoms. It is most effective for the identification of close-packed structures like *fcc* and *hcp* but it can also be used to identify *bcc* order. This analysis is a highly effective method for the identification of structural defects in *fcc* metals since common defect structures have characteristic CNA signatures. Stacking faults, e.g., show up as a double layer of atoms with an *hcp* environment whereas a single layer of *hcp* atoms in a fcc matrix signals the presence of a twin boundary. Similarly this method can be used to identify certain low-energy surfaces of *fcc* and *bcc* lattices.

6.3.1 Sintering of Nickel Nanoparticles

In order to get insight into the evolution of structural defects during the sintering of nanoparticles, we have performed simulations of the sintering of two Ni nanoparticles with diameters of approximately 4 nm over a period of 100 ns. The inter-atomic forces in these simulations were calculated with the help of Cleri and Rosato's tight-binding second-moment potential for nickel [12] and the velocity-Verlet algorithm with a time step $\Delta t = 2$ fs for the integration of the equations of motion. Open boundary conditions were applied in all simulations.

For these simulations we used two similar initial configurations. Both configurations contained 8,562 Ni atoms and consisted of two identical nanoparticles that had been brought into close contact. All configurations shown in this section use a color coding to indicate the result of the CNA for the atoms. A blue (green) color represents atoms with a local *fcc* (*hcp*) environment. All other atoms, mainly surface atoms with a coordination number $Z \neq 12$ or atoms with an irregular environment, are shown in red. Figure 6.4 shows a cut through one of the initial configurations.

6 Simulation of Cluster Sintering, Dipolar Chain Formation

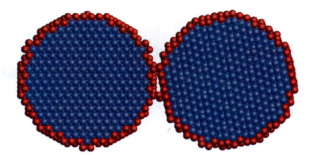

Fig. 6.4 Cut through an initial configuration for the simulations of the sintering of Ni nanoparticles; atoms are colored according to their local crystalline environment (see text for details)

For each of the two initial configurations we simulated the sintering of the nanoparticles at temperatures $T_S = 800, 900$ and $1,000$ K over a period of 100 ns (5×10^7 simulation steps). The high sintering temperatures have a negative impact on the structural analysis. The reason for this are the strong thermal vibrations which increase the rate of false identifications of the CNA. In order to avoid this problem we applied a minimization procedure of those configurations for which we wanted a structural analysis prior to the execution of the CNA. During the minimization step, we employed the conjugate-gradient method [22] to minimize the potential energy of the configurations. The minimization brings the configuration to a local minimum and thereby removes thermal noise.

In Fig. 6.5 we show the final configurations of our simulations after 100 ns of sintering. It can be seen from this figure that in all six cases, compact nanoparticles have formed. It is remarkable that most of the atoms in the final configurations show a perfect *fcc* ordering and that the only type of defects found inside the particles are stacking faults. In case of those configurations with a single stacking fault, the defects formed during the early stages of the simulations (within the first nanoseconds) and remained stable for the rest of the time. In contrast to this, the more complex defect structures shown in the panels (b) and (c) evolved much slower and continued to develop over much longer periods of time.

Figure 6.5 also illustrates the effect of temperature on the sintering process. Due to the slower diffusion rate (and maybe slower nucleation rates of new facets [23]) the particles obtained at a sintering temperature $T_S = 800$ K (top panels) are considerably more elongated and less compact than those obtained at the higher sintering temperatures. In contrast to this, our results show only small changes in the shape of the particles after 100 ns of sintering when T_S increases from 900 to 1,000 K. This can be explained by the fact that as the particles approach a spherical shape and develop more and more low energy facets, further improvements of the particles shape result in increasingly lower potential energy gains. This in turn reduces the driving force of the sintering process and allows particles sintered at lower temperatures to catch up with those compacted at higher temperatures.

Despite the differences in the structures of the final particles, we observed a similar sequence of stages during all of our simulations. In order to discuss these stages we use the simulation whose final configuration is shown in Fig. 6.5e as an example.

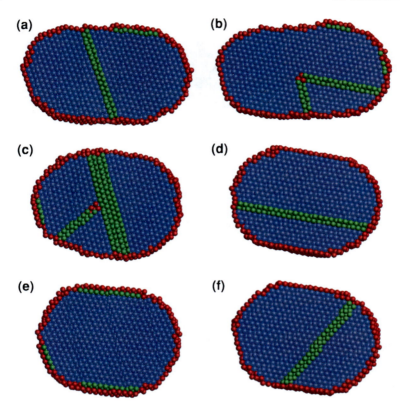

Fig. 6.5 Cuts through the final configurations of the coalesced nanoparticles after 100 ns of sintering. The *left*- and *right*-hand panels show the results from the two different starting configurations. Sintering temperatures T_S were 800 K (*top*), 900 K (*middle*), and 1,000 K (*bottom*)

The initial phase of our simulations is characterized by a very fast coalescence of the particles and the formation of a so-called sinter neck during the first picoseconds of the simulations. This phase is driven by the formation of new interatomic bonds when the particles are brought close together. These new bonds pull the particles together which leads to the creation of even more bonds. The fast progression of this initial coalescence is further facilitated by the potential energy released by the newly formed bonds which makes the atoms in the contact area highly mobile.

The initial fast coalescence of the particles comes to an end when a solid sinter neck has formed (cf. Fig. 6.6a). At this point, a disordered defect layer has formed at the interface between the two particles. The atoms in this interface layer are however still highly mobile and progressively minimize the energy of the interface. After a time of 0.1 ns the originally disordered interface has been transformed into a grain boundary, where the small misalignment between the crystal lattices gives rise to a grain boundary dislocation network (see Fig. 6.6b). Three such dislocations can be seen in the figure. The topmost of these dislocations has been split into two partials,

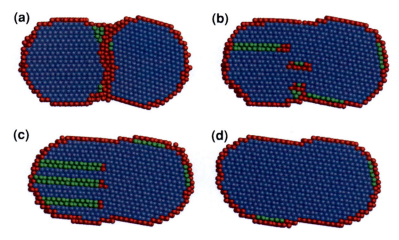

Fig. 6.6 Intermediate configurations after sintering times of 20 ps (**a**), 0.1 ns (**b**), 2.0 ns (**c**) and 2.5 ns (**d**)

one of which has left the system leaving behind the stacking fault in the upper left part of the particle.

During the further evolution of the system, all three dislocations visible in Fig. 6.6b split with one partial leaving the particle which lead to the defect pattern shown in Fig. 6.6c. Finally, the remaining three partial dislocations left the system to the left-hand side leaving behind a defect-free nanoparticle (Fig. 6.6c). In a similar case two partials of one dislocation left the particle on opposite side leaving behind the single stacking fault shown the middle right panel of Fig. 6.5. Later stages of the sintering process are then governed by surface diffusion processes. These processes do not affect the defect structure inside the nanoparticles but fill the sinter neck, replace high-energy facet by low-energy facets, and drive the particles toward a more spherical shape.

In conclusion, our simulations of the sintering of nickel nanoparticles show the formation of compact nanoparticles within a time of 100 ns. The only type of lattice defects in these particles are stacking faults. The necessary alignment of the lattices of the coalesced particles is not driven by diffusion processes but by grain-boundary dislocations that form at the interface between the particles. The interpretation of our results is however hampered by the small size of the particles which is due to the necessary long time scale. One might, for example, question whether partial dislocations will leave larger particles as easily as small particles. Simulations of the sintering of larger particles are therefore necessary in order to confirm to what extent the behavior observed in our simulations translates to larger particles.

6.3.2 Agglomeration of Icosahedral Iron Nanoparticles

Special numbers of atoms in a cluster give rise to the formation of high symmetric morphologies with closed atomic shells like the icosahedron. The number of atoms in a cluster is connected to the number of closed shells by $N = (10n^3 + 15n^2 + 11n + 3)/3$, where n is the number of shells and N the number of atoms. Recently, the shellwise Mackay [24] transformed cluster of Fe which was found in ab initio investigations could also be reproduced via MD-simulations with embedded-atom method [10, 11] potentials. This morphology is similar to the icosahedron but every atomic shell is partially transformed along the Mackay path which relates the icosahedron and the cuboctahedron. This partial transformation gives rise to a local *bcc* environment in some parts of the cluster.

From that starting point, we performed a more detailed analysis of Fe clusters in highly symmetric morphologies like the icosahedron or the cuboctahedron and their relative phase stability. During the study of the closed shell Fe clusters, structural phase transitions are found, which relate different morphologies. They are analogous to the bulk Fe transformations relating *bcc* and *fcc* structures and show a strong dependency of the transition temperature on the size of the cluster.

Additionally, we also took agglomeration processes between different Fe-clusters into account. In this case, induced structural transformations can be found when clusters agglomerate. For example, the formation of the sinter neck induces surface deformations and leads to the release of surface energy which may trigger a structural transformation. Additionally, the critical number of atoms, above which a particular morphology is no longer stable, can be exceeded in the resulting cluster as two smaller clusters start to agglomerate. Besides this, the initial collision of the clusters starting the agglomeration induces a shockwave which is known to be able to trigger structural transformations [25].

In summary, the agglomeration process affects the structural transformation significantly, especially the morphology and structure of the product cluster.

The other way round, the structural transformation can enhance the agglomeration due to the release of energy and also due to geometrical reasons. The latter means that the interface area is enlarged due to the change of the morphology which leads to a larger connectivity of the particles.

An example of this interplay between phase transformation and accompanied agglomeration can be seen in Fig. 6.7. Here, two shellwise Mackay transformed icosahedral Fe clusters of 3,871 atoms, each, agglomerate at a temperature of 60 K. The simulations have been performed in a canonical ensemble with open boundary conditions using time steps of 1 fs. This temperature is too low to induce the surface and subsurface diffusion of atoms which are the main mechanisms driving a sintering process. Normally, an agglomeration at such temperatures does not lead to a complete assembling of the clusters. For example, if two icosahedral Fe clusters are brought into contact, no phase transition appears and the agglomeration at comparable temperatures only leads to the formation of small sinter necks. For a complete agglomeration higher temperatures [26] or collisions with higher relative velocities

6 Simulation of Cluster Sintering, Dipolar Chain Formation   151

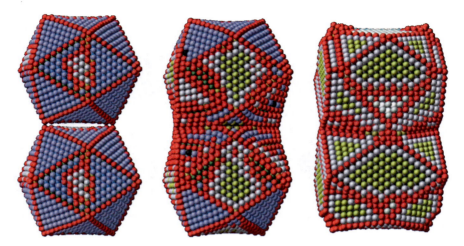

Fig. 6.7 Snapshots of the agglomeration of two shellwise Mackay transformed icosahedral Fe-clusters at 0, 40 and 85 ps. The atoms are colored according to their local crystalline environment identified by CNA, see Sect. 6.3. *Blue fcc*; *Gray bcc*; *Yellow and red* no local structure can be assigned

are needed, as complete or partial melting appears in the latter case leading to a strong aggregation.

In our simulation strong agglomeration arises on a time scale of less than a hundred picoseconds due to the structural transformation accompanied with the agglomeration. After the formation of a neck between the clusters a strong deformation of the surfaces starts. This surface deformation leads to a change in the contact area. While two edges of the clusters have been approaching first, the deformation leads to a contact area of two cluster facets and thus induces a much larger sinter neck. Although local structural transformations have already begun during the surface deformation, a fast and complete transformation of the whole aggregate starts afterwards. The transformation converts the *fcc* dominated lattice structure of the original clusters into *bcc*. The main reason for the transformation is the doubled number of atoms in the agglomerate. During the study of single Fe clusters we found that the shellwise Mackay transformed icosahedral morphology is the preferred structure for small clusters. But, with increasing cluster size and shrinking surface to volume ratio the shellwise Mackay transformed icosahedron becomes metastable with respect to morphologies that are complete *bcc*. This is due to the fact that the internal structure of icosahedral-like clusters is a multiply twinned *fcc* structure being higher in energy than the *bcc* structure which the ground state of bulk Fe, whereas the icosahedral clusters have a lower surface energy because they are more spherical.

In summary, our simulations show that strong agglomeration of clusters can be found at low temperatures if they are triggered by structural phase transformation.

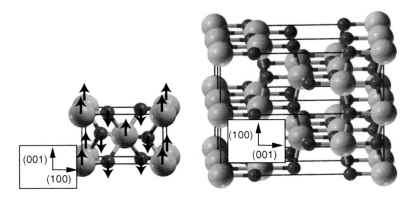

Fig. 6.8 *Left* atomic ground state structure of rutile. *Right* atomic ground state structure of anatase. Ti (*blue, light*), O (*red, dark*). Arrows mark the atomic displacement of the ferroelectric A_{2u} phonon mode. The relative atomic size is determined by the covalent radii

6.4 Density Functional Simulations of Dielectric Nanoparticles: Agglomeration and Ferroelectric Trends

In the following we want to discuss the properties of oxidic nanocomposite materials with a high polarizability as a first step toward magnetic-oxidic nanocomposites. Here, we concentrate our study on the properties of TiO_2-systems. Beside its *incipient* ferroelectric behavior [27], TiO_2 is a proper model system for the general behavior of metal oxides in the nano range as it is one of the few oxides which can be stabilized in different nanoscale sizes at ambient conditions. Although TiO_2 as such does not possess a ferroelectric phase transition, its local atomic structure is very similar to well-established ferroelectric materials as $BaTiO_3$. In both cases the Ti-ions are surrounded by the O-ions octahedrally with a similar bonding length. Figure 6.8 shows the atomic structure of TiO_2 in its most important morphologies: rutile and anatase. Although both structures show TiO_6 octahedra, their alignment varies. Due to its atomic structure and exceptionally large Born charges rutile shows a large polarizability and its dielectric constant along the c-axis shows an unexpected behavior with decreasing temperature [28]. Additionally, it has been shown theoretically that lattice expansion or strain on the system leads to a stable ferroelectric phase [29, 30]. Experimentally, an increase in the lattice constant of rutile nanoparticles in the range of 2–20 nm diameter has been found [31]. Such expansion would force the system to become ferroelectric and would therefore be of great technical importance.

We investigate the atomic structure of single TiO_2 nanoparticles, as well as the agglomeration of nanoparticle pairs using the plane-wave code VASP with PBE-potentials [32]. Hereby, Ti $3d^2 4s^2$ and O $3s^2 3p^4$ electrons are treated as valence electrons. Forces have been minimized to at least 0.05 eV/Å. Although we use periodic boundary conditions within our simulations we minimize interactions of the periodic pictures by separating vacuum layers of at least 9 Å.

6 Simulation of Cluster Sintering, Dipolar Chain Formation
153

The energetic ground state of small nanoparticles is a faceted nanoparticle of anatase morphology due to its lowest surface energy. Since, also spherical anatase or rutile nanoparticle can be stabilized, see [15] and references therein, we restrict our investigation to such spherical particles. The investigated $Ti_{41}O_{82}$ anatase particle has a mean diameter of 1.1 nm only, but, is already nearly spherical and systematic tests confirmed that this special choice of a stoichiometric configuration is quite stable, as the mean features of the anatase atomic structure in the center of the particle are already stable against surface relaxations, see Fig. 6.10a, c, and e. Taking into account additional O-atoms at the surface to saturate dangling Ti-bonds counteracts the stabilization of the structure because dimerization of the excess O-atoms appears.

In case of rutile particles, we use a $Ti_{101}O_{202}$ particle, see Fig. 6.12a, c, which approximately corresponds to a diameter of 1.5 nm. Due to the different arrangement of the TiO_6-octahedra a smaller rutile particle does not show a sufficient spherical shape in contrast to anatase. Systematic tests with even larger particles show that increasing the size does not alter the structure considerably, e.g., the O–Ti–O angles as well as the main surface relaxation stay constant. Like for anatase, a stoichiometric configuration is more stable than additional surface saturation with excess oxygen. The overall particle volume, given by the mean nearest Ti–Ti distances, is enlarged in comparison to bulk in agreement with experiment [31], as the distances in (100)- and (010)-direction expand by $+0.09$ Å on average which corresponds to 2 % of the lattice constant. However, in contrast to a uniform lattice expansion we find a large variety of modifications of the different bond lengths, e.g., the Ti–Ti-distance in (001)-direction shrinks on average by -0.04 Å. Thus, the tetragonal ratio is reduced which reduces the dielectric constant as well as ferroelectric trends [29, 30]. Additionally, the Ti–O bond lengths in the surface layers outside the shell of 5 Å are modified drastically due to the surface. As the size of the short Ti–O bond length is crucial for ferroelectric properties, since it is the main contribution to the short-range repulsion during a ferroelectric transition, a large modification of ferroelectric trends can be expected [30]. While the mean bond length in the subsurface shell increases by 0.04 Å, the Ti–O distance in the direct interface shrinks by -0.07 Å, which may prevent a ferroelectric transition. Additionally, the anisotropy of the Ti–O bond, which is the driving factor for the dynamical charge, is largely modified and thus the polarization during a ferroelectric transition may be reduced in comparison to bulk. Besides this, the system still shows a large polarizability, and therefore a magnetoelectric coupling at the surface to magnetic nanoparticles is still likely.

For rutile-TiO_2 surfaces, it has been shown by ab initio simulations [33] that the ionic character of the Ti–O bonds in the vicinity of the surface gains some covalent portion. This leads to the formation of a charge double-layer at the surface, as mainly charge from the O-atoms below the surface is transferred to the topmost Ti-atoms. For anatase and rutile nanoparticles similar charged double-layers have been found by variable-charge MD [15]. This charge double-layer can be reproduced by our ab initio simulations, as can be seen in Fig. 6.9. The charged particle surfaces may lead to an electrostatic repulsion, as two particles are brought into contact.

As a starting point for our agglomeration simulation the shortest distance between atoms in the different particles is assumed to be 4 Å unlike otherwise stated. One

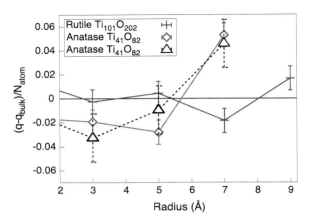

Fig. 6.9 Charge profiles of relaxed $Ti_{101}O_{202}$ rutile (*red*) and $Ti_{41}O_{82}$ anatase (*blue*) nanoclusters and two approaching anatase clusters (center of mass distance 13.55 Å) (*black, dashed*). $q - q_{bulk}$ is the difference in the charge transfer relative to bulk. The mean deviations over 2 Å shells of the particles are given. Error bars mark the accuracy of the charge partitioning

test simulation, which started with a distance of 10 Å and a gradual shrinking of the interparticle distance did not show any considerable particle–particle interaction before, see Fig. 6.11. After atomic relaxation the particles have been moved closer in compression steps of 1 Å (smaller steps at the end for some configurations).

For anatase particles attached in (001)-direction, the particles repel each other if the starting distance between the center of masses (COM) is 14 Å or smaller, which corresponds to interparticle O–O distance of 3.2 Å. As can be seen from Fig. 6.11, the distance between COMs of the particles increases after each further compression step. Figure 6.9 shows the corresponding charge profile for two relaxed anatase particles at a COM distance of 13.55 Å, which shows that even at this short distance the charge double-layer of the two nanoparticles is not substantially modified. Beside the resulting electrostatic repulsion, the interparticle O–O or Ti–Ti distances for this configuration are already in the same range of interparticle distances in bulk anatase. This obviously leads to additional repulsion due to short-range interactions, as Ti–Ti or O–O atoms are approaching in an unfavorable configuration. Nevertheless, further distance decrease leads to a formation of a sinter neck due to a sudden relative rotation of the particles, as the shortest interparticle O–O distance reaches 2 Å. Here, the main rearrangement consists of a relative particle rotation around (100) of about 11°, see Fig. 6.10b, which has been calculated for the Ti-sublattice as the positions of the oxygen atoms are modified by larger atomic relaxations. Due to the rotation, the attaching surface atoms can enlarge their coordination and distorted TiO_6- octahedra are formed. For particles attached along (010) or (100) the atomic rearrangement cannot be separated into such a simple rotation pattern. Instead, small relative rotations of the particles around different directions appear, which are superimposed by large atomic relaxations. Additionally, the COMs of the particles move against each other by 0.25/0.12 Å along (010)-/(001)-direction for the (100)-attachment, respectively 0.47/0.08 Å along (100)-/(001)-direction for (010)-attachment, see Fig. 6.10d, f. Due to this combination of relative particle shift and atomic rearrangement, an anatase- like structure can be built as a sinter neck

6 Simulation of Cluster Sintering, Dipolar Chain Formation 155

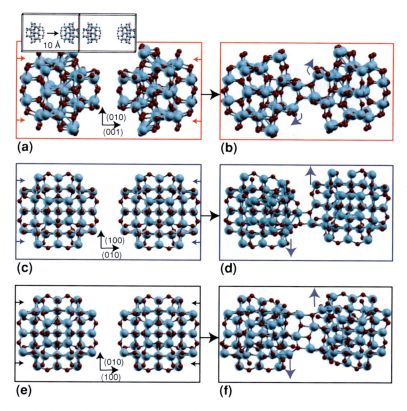

Fig. 6.10 Agglomeration of two anatase nanoclusters with 41 Ti and 82 O atoms. *Left* start configuration *Right* after repeated movement of the particles (**a**, **b**) (001)-attachment (**c**, **d**) (010)-attachment (**e**, **f**) (100)-attachment. Inset: Size of the real simulation cell, which contains at least 9 Å vacuum between periodic images in all simulations

in both cases. Here, the mean interparticle Ti–O bond length of 2.09/2.01 Å (for (100)/(010)) is only slightly longer than the corresponding bulk value of 2.00 Å. The starting configurations are nearly identical, as the (100)- and (010)-surfaces are very similar and thus, the same repulsion between the particles appears during the first simulation step. But, while the step size of the (010)-approach is reduced which leads to smaller repulsion and thus smaller rearrangement of the particles in each step, the particles in (100)-direction are strictly moved in 1 Å steps. Thus, the configuration is less favorable, as the particles approach to an interparticle O–O distance of 2.17 Å (COM distance of 14.9 Å), see energy curve in Fig. 6.11. This additional amount of energy pushes the system further away form the local energetic minimum of two undistorted clusters, which leads to a large rearrangement of the particles. As our simulation are at $T = 0$ K no further rearrangement or equilibration of the particles appear after a local energy minima is reached.

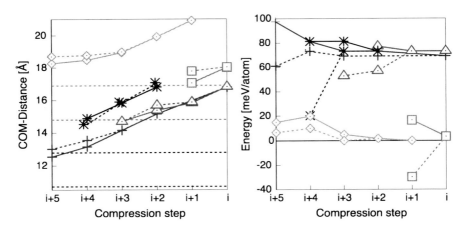

Fig. 6.11 *Left* center of mass (COM)'s distance between the approaching particles. Horizontal lines mark the diameter of the free particle. *Right* total energy of the particle system with arbitrary origin. For each compression step *i* initial (*solid lines*) and relaxed (*dashed lines*) configurations are compared. Anatase (001): *red, crosses*; Anatase (010): *blue, triangles*; Anatase (100): *black, stars*; Rutile (001): *magenta, squares*; Rutile (010): *cyan diamonds*

Basically, the attachment of rutile particles shows the same rearrangement mechanism. For the attachment of rutile particles in (010)-direction the nearest lying O-atoms form a dimer at a COM-distance of 18.94 Å, see Fig. 6.12a. Further decrease of the particle distance leads to an artificially short O–O bond. Because of this, the involved O-atoms are pushed into the nanoparticles and occupy an interstitial position (see Fig. 6.12). This configuration is a special case; however, the following simulation steps show the same behavior as in case of anatase, the atoms at the interface start to rotate relative to each other in the (100)/(010) plane, the interface atoms start to rearrange and the energy of the system decreases, see Fig. 6.12.

For particles attaching with (001)-orientation our simulation started with an interparticle distance of 3 Å. Within the second simulation step interparticle distances of about 2.9 Å are imposed, which are comparable with the bulk Ti–Ti distance in (001)-direction of 2.97 Å. This leads to a large interaction of the particles and the COMs start to shift against each other in (100)- and (010)-direction. Additionally, large atomic relaxations at the surface appear and interparticle Ti–O bonds with a similar bonding length like in bulk rutile are formed. This leads to a large energy gain.

The COM-distances for the different agglomeration steps in Fig. 6.11 show a similar distance of rutile or anatase surfaces for which the particles start to repel each other. Additionally, the increase in the COM-distances during atomic relaxation is of the same order for all investigated rutile and anatase configurations. In agreement with MD simulations by [15] our calculations show that anatase particles indeed form a sinter neck although they repel each other electrostatically. However, in contrast to a repulsion of rutile particles, which was also found in [15], we find similar

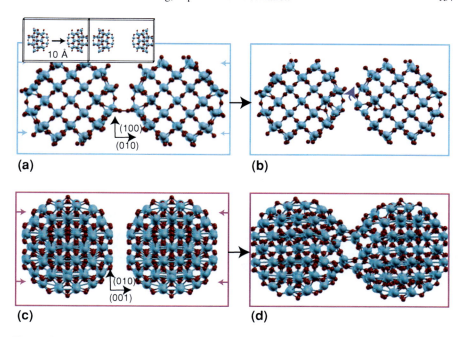

Fig. 6.12 Agglomeration of two rutile nanoclusters with 101 Ti and 202 O atoms. (**a, b**) (010)-attachment. **a** relaxed positions at a COM-distance of 18.94 Å **b** after repeated movement. (**c, d**) (001)-attachment. **c** start configuration. **b** *Right* intermediate relaxation step after repeated movement. Inset: size of the real simulation cell, which contains at least 9 Å vacuum between periodic images in all simulations

agglomeration trends for rutile particles. For both systems the energy barrier for particle rearrangement can be overcome if one repeatedly reduces the interparticle distance and a sinter neck appears. The main mechanism for the rearrangement in our simulation has been found to be a relative shift or rotation of the whole particles relative to each other. Additionally, the different agglomeration behavior between anatase particles attached in (100)- and (010)-direction show how critical the agglomeration process depends on the particular simulation history.

6.5 Summary

We have discussed the agglomeration of nanoclusters involving sintering processes of free Fe, Ni, and TiO_2 clusters as well as chain and ring configurations which appear due to the dipolar interactions between Fe nanoparticles. The sintering of Ni clusters is interesting because it finally leads to compact nanoparticles in the nanosecond range without any significant traces of defect structures which are observed during the sintering process. Here, molecular dynamics simulations of a larger array of Ni nanoclusters would be interesting. In the molecular dynamic simulations of Fe clusters it has become obvious that new structural motifs are involved in the agglomeration process. Chain and ring formation of Fe nanoparticles is particularly appealing as there is direct confirmation of the existence of such structures from experiment. Finally, the simulation of near-ferroelectric TiO_2 nanoclusters on an *ab initio* basis has revealed the importance of the high polarizability of the clusters in the sintering process.

Acknowledgments We thank M. Fendrich who prepared the start configuration for the sinter simulations of Ni nanoparticles, and S. Buschmann who worked on the dipolar nanoparticles. Computation time was granted by the John-von-Neumann Institute of Computing.

References

1. B. Bhushan, *Springer Handbook of Nanotechnology* (Springer, Berlin, 2006)
2. A.S. Edelstein, R.C. Cammarata, *Nanomaterials: Synthesis Properties and Applications* (Institute of Physics, Bristol, 1996)
3. C. Guet, P. Hobza, F. Spiegelmann, F. David, Atomic clusters and nanoparticles. Agregats atomiques et nanoparticules: Les Houches Session LXXIII 2–28 July 2000 (Les Houches - Ecole d'Ete de Physique Theorique), Vol. 73, (Springer, Berlin, 2001)
4. J. Knipping, H. Wiggers, B.F. Kock, T. Huelser, B. Rellinghaus, P. Roth, J. Nanotechnol. **15**, 1665 (2004)
5. V. Salgueiriño-Maceira, M.A. Correa-Duarte, A. Hucht, M. Farle, J. Magn. Magn. Mater. **303**, 163 (2006)
6. T. Huelser, H. Wiggers, P. Ifeacho, O. Dmitrieva, G. Dumpich, A. Lorke, J. Nanotechnol. **17**, 3111 (2006)
7. R. Ramesh, N.A. Spaldin, Nat. Mater. **6**, 21–29 (2007)
8. M.P. Allen, D.J. Tildesley, *Computer Simulations of Liquids* (Clarendon, Oxford, 1991)
9. R. Car, M. Parrinello, Phys. Rev. Lett. **55**, 2471–2474 (1985)
10. M.S. Daw, M.I. Baskes, Phys. Rev. B **29**, 6443 (1984)
11. M.S. Daw, M.I. Baskes, Phys. Rev. Lett. **50**, 1285 (1983)
12. F. Cleri, V. Rosatto, Phys. Rev. B **48**, 22–33 (1993)
13. W.G. Hoover, Phys. Rev. A **31**, 1695–1697 (1985)
14. H.C. Andersen, J. Chem. Phys. **72**, 2384 (1980)
15. S. Ogata, H. Iyetomi, K. Tsurutu, F. Shimojo, A. Nakano, R.K. Kalia, P. Vashishta, J. Appl. Phys. **88**, 6011–6015 (2000)
16. S. Ogata, H. Iyetomi, K. Tsurutu, F. Shimojo, R.K. Kalia, A. Nakano, P. Vashishta, J. Appl. Phys. **86**, 3036–6041 (1999)
17. A. Hucht, S. Buschmann, P. Entel, Europhys. Lett. **77**, 57003 (2007)
18. I.M.L. Billas, J.A. Becker, A. Châtelain, W.A. de Heer, Phys. Rev. Lett. **71**, 4067 (1993)
19. P. Pawlow, Z. Phys. Chem. **65**, 1 (1909)
20. P.V. Hendriksen, S. Linderoth, P.-A. Lindgard, Phys. Rev. B **48**, 7259 (1993)
21. J.D. Honeycutt, H.C. Andersen, J. Phys. Chem. **91**, 4950 (1987)
22. R. Barrett, M. Berry, T.F. Chan, J. Demmel, J. Donato, J. Dongarra, V. Eijkhout, R. Pozo, C. Romine, H. Van der Vorst, *Templates for the Solution of Linear Systems: Building Blocks for Iterative Methods*, 2nd edn. (SIAM, Philadelphia, 1994)
23. N. Combe, P. Jensen, A. Pimpinelli, Phys. Rev. Lett. **85**, 110–113 (1985)
24. G. Rollmann, M.E. Gruner, A. Hucht, R. Meyer, P. Entel, M. Tiago, J.R. Chelikowsky, Phys. Rev. Lett. **99**, 083402 (2007)

6 Simulation of Cluster Sintering, Dipolar Chain Formation

25. K. Kadau, T.C. Germann, P.S. Lomdahl, B.L. Holian, Science **296**, 1681–1684 (2002)
26. S. Hendy, S.A. Brown, M. Hyslop, Phys. Rev. B **68**, 241403 (2003)
27. C. Lee, P. Ghosez, X. Gonze, Phys. Rev. B **50**, 13379–13387 (1994)
28. J.G. Traylor, H.G. Smith, R.M. Nicklow, M.K. Wilkinson, Phys. Rev. B **3**, 3457 (1971)
29. B. Montanari, N.M. Harrison, J. Phys. Condens. Matter. **16**, 273 (2004)
30. A. Grünebohm, C. Ederer, P. Entel, First-principles investigation of incipient ferroelectric trends of rutile TiO_2 in bulk and at the (110) surface, submitted to Phys. Rev. B (2012)
31. G. Li, J. Boerio-Goates, B.F. Woodfield, L. Li, Appl. Phys. Lett. **85**, 2059 (2004)
32. G. Kresse, D. Joubert, Phys. Rev. B **59**, 1758 (1999)
33. P. Reinhardt, B.A. Heß, Phys. Rev. B **50**, 12015 (1994)

Chapter 7
Nanopowder Sintering

D.E. Wolf, L. Brendel, M. Fendrich and R. Zinetullin

Abstract We define nanopowder sintering as the conversion of a loose agglomerate of nanoparticles into a *nanostructured* solid. This means that grain boundaries between the particles must survive the sintering process to a large extent. The key issue here is structural self organization; external control is limited to macroscopic parameters like temperature and pressure, while the desired structure on the particle scale should then form by itself. This chapter reviews, how the early, intermediate, and late stages of sintering are influenced by the presence of grain boundaries, with a special focus on particle sizes in the nanometer range. A new, efficient computer simulation model is presented and its applicability to Ni- and to ITO-particles is briefly discussed.

7.1 Introduction

In essence, sintering converts a powder into a solid at temperatures below the melting point. This happens in pottery, which belongs to the cultural achievements of early mankind. However, although clay, the raw material of pottery, contains nanoparticles, it would be misleading to date the beginnings of Nanotechnology back to the stone age: nanoscale features are usually lost, when pottery is fired. Sintering cannot be regarded as part of Nanotechnology just because it starts out from nanoparticles, but only, if it results in a solid that is still nanocrystalline. Reaching this goal is a nontrivial task which requires a thorough understanding of the processes down to the atomic level. New experimental techniques like high resolution transmission

---

[1] Not always sintering is used to create a nanocrystalline solid of macroscopic dimension. Sintering of mass selected agglomerates of a few nanoparticles has also been used as a fabrication tool of monodisperse, spherical particles [1].

D. E. Wolf (✉) · L. Brendel · M. Fendrich · R. Zinetullin
Department of Physics and CENIDE (Center for Nanointegration Duisburg-Essen),
University of Duisburg-Essen, Lotharstraße 1, 47057 Duisburg, Germany
e-mail: dietrich.wolf@uni-due.de

A. Lorke et al. (eds.), *Nanoparticles from the Gas Phase*, NanoScience and Technology,
DOI: 10.1007/978-3-642-28546-2_7, © Springer-Verlag Berlin Heidelberg 2012

electron microscopy (TEM) provide information at this level. At the same time, larger and larger systems can be simulated with atomic resolution due to increasing computer power. The combination of both experimental and computational studies has improved the microscopic understanding of nanopowder sintering enormously during the last decade.

The current understanding of sintering has been summarized in textbooks by German [2], Rahaman [3] and Ring [4]. The progress in modeling and simulation was documented in a recent review by Barnard [5]. This chapter tries to complement those reviews by focussing on aspects that are particularly relevant for securing nanocrystallinity for sintered samples.

Direct manipulation at the nanoscale is expensive. External control of the sintering process should therefore be limited to macroscopic parameters like temperature, mechanical stress, electric or magnetic field. Fortunately, under certain conditions nature does the desired nanocrystalline consolidation all by itself. To find out such conditions was the goal of the research reviewed in this chapter.

Sintering begins with the preparation of the green body. In pottery this is usually processed from a water-based paste of clay and other particles with a complicated chemical composition into a dry body. By contrast, in the following only the sintering of nano*powders* will be considered. The green body is a dry agglomerate of essentially round nanoparticles, all of the same material. The van-der-Waals cohesion between two such particles with radii R_1 and R_2 is

$$F = \frac{A_\mathrm{H}}{6a^2} R_\mathrm{red} \quad \text{with} \quad R_\mathrm{red} = \frac{R_1 R_2}{R_1 + R_2}. \tag{7.1}$$

A_H denotes the Hamaker constant, and a is a typical separation between atoms [6]. All other forces increase with a higher power of the particle radii and hence become negligible, when at least one of the particles is small. Therefore, nanopowders, which are deposits of nanoparticles or aerosol flakes from the gas phase, are extremely porous, and usually have a fractal substructure [7–9]. Their own weight leads to hardly any compaction [10].

In the following, it will be assumed that initially the contacts between the particles are fragile, i.e., no solid necks have formed yet. One calls this a soft agglomerate. Sintering converts it first into a hard agglomerate, where particles cannot be separated easily any more, because they are connected by solid necks. The pore volume changes very little in this first stage. During the later stages of sintering the pores shrink so that ultimately a polycrystalline solid is formed.

Thermally activated atomic diffusion processes are responsible for neck growth and pore shrinkage. The diffusion bias is such that the agglomerate lowers its free energy (including surface, interfacial, and strain contributions), thereby relaxing toward thermal equilibrium. If thermal equilibrium could be reached, all particles would have merged into a single crystal with Wulff shape (see e.g. [11]), which minimizes the surface free energy for a given volume.

Surface diffusion redistributes mass from one part of the surface to a nearby one. This leads to fast neck growth, because the diffusion length is small of the order of a

7 Nanopowder Sintering

particle diameter. However, pores shrink only slowly by surface diffusion, because redistribution of mass around a pore does not reduce the pore volume. This requires surface diffusion all the way from the outer boundary of the sample to the pore surfaces inside the powder, which takes considerably more time than neck growth.

In general, atoms in grain boundaries or interfaces are less mobile than at surfaces.[2] Nevertheless, grain boundary diffusion plays an important role for neck growth and shrinkage, because the distance between sources and sinks of the mass transport is small of the order of the neck diameter. Grain boundary diffusion erodes the adjacent crystal bulk, so that the two lattices move toward each other. The mass is transported out into the pore, where it is released along the contact line, at which the grain boundary and the free surfaces meet. Hence, the contact line acts as a source for adatoms diffusing on the pore surface.

A special form of biased diffusion along grain boundaries is Coble creep. It describes the slow, quasi-viscous yield of a polycrystalline solid to an applied stress [13]. Atoms get redistributed along the grain boundaries such that the grains grow on the faces with tensile stress and erode at the other ones.

All diffusional processes, being thermally activated, can be suppressed by avoiding high temperatures. Then one has to resort to a second type of irreversible mass transport in order to reduce the porosity of a nanopowder: by applying a sufficiently high pressure, the particle agglomerate can be compactified (for a recent review see [14]). This requires breaking of existing necks so that particles on opposite sides of a pore can move toward each other. Considering compaction (pressure control) and sintering (temperature control) together opens up new possibilities to optimize the structure of a nanopowder. Current assisted sintering works this way: Joule heat release and pressure are combined to obtain nanocrystalline bulk samples [15] (see also Chap. 10 in this book).

In the present chapter, however, we focus on thermal sintering. After a short recollection of the current understanding of particle coalescence and late stage sintering in Sect. 7.2, mainly the effect of grain boundaries on the early stages of sintering will be considered (Sect. 7.3).

7.2 Particle Coalescence

The most elementary sintering process is the one, where two spherical particles— generally of different size—come into contact, the crystalline structure of which may be ignored. This is the case, if they are amorphous or fluid-like (at least at the surface). Also, if crystalline particles deform so easily that they adopt a coherent or twinned orientation immediately, and if the temperature is high enough that their surfaces are rough, the crystal structure has negligible effect on sintering.

A somewhat more general case is an agglomerate of several spherical particles, all of approximately the same size and material. In particular the agglomerate should

---

[2] Mobility in crystalline bulk is in general still lower [12].

not be supported by a substrate. It should be thermostated by a surrounding inert gas so that the potential energy released, while the particles fuse to become a single sphere, does not lead to a substantial temperature increase of the agglomerate. According to [16] these conditions are approximately met for instance in the experiment reported in [1].

7.2.1 Phenomenological Theory

Diffusive mass transport, the basis of sintering, is driven by gradients of the local chemical potential, which is a function of the local mean curvature of the surface. This raises the question, whether a porous medium, as it may arise from sintering of a powder, can become metastable by adopting a constant mean curvature throughout. The answer is "no", as has recently been proven [17]. Any configuration of constant mean curvature is unstable with respect to small perturbations, apart from the final equilibrium configuration.

Phenomenologically, the time evolution of the total surface area A of an agglomerate with fixed volume V is usually described by the Koch–Friedlander equation [18]

$$\frac{\mathrm{d}A}{\mathrm{d}t} = -\frac{A - A_{eq}}{\tau}. \tag{7.2}$$

The relaxation time τ depends on the microscopic mechanism of mass transport (see [11], Chap. 8 for an excellent review). Equation (7.2) means that the driving force changing the surface area A is its deviation from the equilibrium value, $A_{eq} = (6\sqrt{\pi}V)^{2/3}$. From a microscopic point of view this is somewhat disturbing; how can the (local) change of surface area of a large agglomerate possibly be influenced by the global size? The answer is simple; locally meaningful is only the *relative* change,

$$A^{-1}\frac{\mathrm{d}A}{\mathrm{d}t} = \frac{1}{\tau}\left(1 - \frac{A_{eq}}{A}\right) \approx \frac{1}{\tau}, \tag{7.3}$$

where the influence of the global size may be neglected as long as $A \gg A_{eq}$. In fact, for an agglomerate of $\mathcal{N}$ particles, $A \propto \mathcal{N} \propto V$. If the particles are densely packed, $A_{eq}/A \propto \mathcal{N}^{-1/3}$ is a small correction proportional to the surface to volume ratio of the sample. For large agglomerates with a fractal dimension d_f the correction should be proportional to $\mathcal{N}^{-1/d_f}$. Then Eq. (7.2) is not applicable.

Equation (7.2) has been interpreted and applied in several different ways in the literature. Often τ is regarded as independent of A [18, 19]. Then Eq. (7.2) is a linear equation describing an exponential decay:

$$A(t) = A_{eq} + \left(A_0 - A_{eq}\right)e^{-t/\tau}. \tag{7.4}$$

7 Nanopowder Sintering

Table 7.1 Exponent α, Eq. (7.5), for different transport mechanisms

Mechanism	α	References
Surface diffusion	4	[11], Chap. 8
Grain boundary diffusion	4	[3], Chap. 8.5.1
Bulk diffusion	3	[11], Chap. 8
Evaporation-recondensation	2	[11], Chap. 8
Viscous flow	1	[24]

It has been shown in [1, 20] (see Sect. 7.2.3) that this assumption works only for the final stage of coalescence, when the typical distance between sources and sinks of the diffusing adatoms is of the same order of magnitude as the radius of the final sphere, $R_{eq} = 3V/A_{eq}$.

A more appropriate model takes into account that the speed of relaxation depends on the typical distance R between sources and sinks of the mass transport, which in general grows with time. Using a thermodynamic continuum description of mass transport during relaxation toward equilibrium, one finds that the relaxation time on length scale R is an atomic time constant t_0, which is independent of R, times a power law of R/a,

$$\tau = t_0 \left(\frac{R}{a}\right)^\alpha, \tag{7.5}$$

where the exponent α depends on the transport mechanism [21–24], see Table 7.1.

For diffusional processes the atomic time constant t_0, up to a dimensionless prefactor, is given by

$$1/t_0 \propto B\gamma. \tag{7.6}$$

The surface tension (surface free energy per unit area) γ provides the driving force for the diffusive drift. The mobility B, which is related to the diffusion coefficient D and temperature T by the Einstein relation $B = D/k_B T$, describes, how easily an atom responds to the driving force.

In practice, several of the mechanisms listed in Table 7.1 are active simultaneously. Which diffusion process dominates for a certain length scale, depends on the ratio R^α/B, if one compares particles with the same surface tension γ and lattice constant a. This argument shows that, for instance, the coalescence of two particles with radius R is dominated by surface diffusion rather than bulk diffusion, as long as

$$R/a < B_{surface}/B_{bulk}, \tag{7.7}$$

whereas for larger particles volume diffusion determines the coalescence time. Since atoms are much less mobile in the bulk than at the surface, $B_{bulk} \ll B_{surface}$ [12], one may neglect bulk diffusion for nanoparticle sintering.

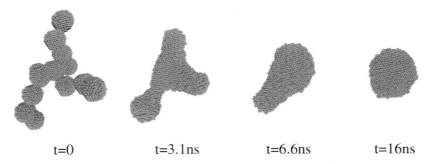

Fig. 7.1 Snapshots of the coalescence of an agglomerate of initially 12 particles without grain boundaries between them. From [20]

Simulation snapshots of the coalescence of a larger agglomerate are shown in Fig. 7.1. Obviously, the distance R between sources and sinks of surface diffusion (convex, respectively concave surface parts) increases between the first and the second snapshot. Hence, the reduction of surface area slows down as coalescence proceeds. In order to estimate the typical value of $R(t)$, one may replace the agglomerate at time t by one that consists of spherical particles with radius $R(t)$. Then $R(t)$ can be calculated from the total surface area $A(t)$:

$$R(t) = \frac{3V}{A(t)}. \tag{7.8}$$

Together with Eq. (7.5), Eq. (7.2) becomes nonlinear:

$$\frac{dA}{dt} = -\frac{1}{C(3V)^4} A^4 (A - A_{eq}). \tag{7.9}$$

The solution of this equation is

$$t = CR_{eq}^4 \left(f\left(\frac{A_{eq}}{A_0}\right) - f\left(\frac{A_{eq}}{A}\right) \right), \tag{7.10}$$

where A_0 is the total surface area of the agglomerate at $t = 0$. This formula expresses the time t that elapses while the surface area decreases from A_0 to A in terms of a function $f(A_{eq}/A)$ given by

$$f(x) = \ln(1-x) + x + \frac{1}{2}x^2 + \frac{1}{3}x^3. \tag{7.11}$$

For $A \gg A_{eq}$ it is enough to take the leading order of $f(x) \approx -\frac{1}{4}x^4$. Then, with Eq. (7.8),

$$t \approx \frac{C}{4}(R(t)^4 - R(0)^4), \tag{7.12}$$

which implies that the surface area decays initially as a power law, in contrast to the exponential decay, Eq. (7.4). Equation (7.12) agrees with the empirical relation derived in [25].

In [1] size selected Ni nanoparticle agglomerates were sintered at different temperatures for a fixed time, while being carried through a tube furnace by an Ar-gas stream. The higher the temperature, the more the agglomerate had coarsened, and at the highest temperatures complete coalescence was reached within the fixed sinter time. This can be explained by the temperature dependence of the diffusion constant $D = k_{\mathrm{B}} T B$ in Eq. (7.6), $D = D_0 \exp(-E_{\mathrm{a}}/k_{\mathrm{B}} T)$. The surface area of the Ni-particle agglomerates could be determined experimentally. Applying Eq. (7.9) then allowed to extract the activation energy of surface diffusion, $E_{\mathrm{a}} = 0.6 \pm 0.1 \mathrm{eV}$, from these data.

7.2.2 Atomistic Modeling

The phenomenological theory has been compared to computer simulations of the atomic mass transport during the equilibration of agglomerates. In Molecular Dynamics simulations (MD) the equations of motion are solved for all atoms. The Debye frequency of phonons is of the order of $10^{13} s^{-1}$ so that at least femtosecond time resolution is required to calculate the atomic trajectories. Sintering proceeds much more slowly, in particular at temperatures, where the surface is not molten. Then mass transport happens by thermally activated diffusion processes. This is why MD-simulations consume a lot of computation time and are limited to the early stages of sintering or small systems [19, 26].

Kinetic Monte Carlo simulations (KMC) [27, 28] replace atomic dynamics by stochastic transitions (kinetics) between configurations. Hence, the time resolution is of the order of the inverse rate of these transitions. Even a radically simplified model of the transition rates allows a *quantitative* description of the slow processes, provided the kinetic bottlenecks are captured correctly [29]. The results presented in this chapter were obtained with this method.

In the model used here, only the solid particles are considered, not the surrounding gas. All atoms are located on a perfect face centered cubic (fcc) lattice. This excludes topological crystal defects like dislocations or stacking faults. Elastic deformations could be taken into account [20, 30] but are ignored in the following. As mentioned in the beginning of Sect. 7.2, such a simplified model is appropriate in situations, where the sinter process is largely independent of the detailed atomic structure.

In order to reduce the number of configurations that have to be considered and to suppress sublimation, all atoms must be at least twofold coordinated. If an atom has less than the maximal number of 12 neighbor atoms, it can move to one of the empty, at least twofold coordinated neighbor sites with a certain configuration dependent rate. The rates ν of the individual processes are given by the Arrhenius law determined by the activation energies E_{act},

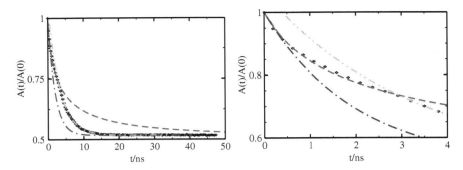

Fig. 7.2 Fits (see text) of the time dependent surface area (*symbols*). *Left* long-time behavior, *right* short-time behavior. From [20]

$$\nu = \nu_0 \exp\left(-\frac{E_{act}}{k_B T}\right) \quad (7.13)$$

with attempt frequency ν_0, temperature T, and Boltzmann constant k_B. The activation energies were calculated from a simple model described in the appendix.

All allowed diffusional processes and their rates have to be known at any time. In each KMC-step one particular of these processes is chosen at random with a probability $\nu/\Sigma^{(\nu)}$, where $\Sigma^{(\nu)}$ is the sum of all rates. The chosen process is carried out, and time is incremented by

$$\Delta t_{KMC} = -\frac{1}{\Sigma^{(\nu)}} \ln\left(\text{Rnd}(0,1)\right), \quad (7.14)$$

where Rnd(0,1) denotes a (pseudo-)random number between 0 and 1. The sum t_{KMC} corresponds to real time [31].

7.2.3 Coalescence of Agglomerates

Coming back to Fig. 7.1 the time evolution of the surface area of the agglomerate (number of atoms with less than 12 neighbors) was evaluated. The simulation data are shown in Fig. 7.2 together with three different fits. The dash-dotted line is the solution of the linear Koch–Friedlander equation (7.4), assuming that τ is a constant. As expected it describes a relaxation that is too fast compared to the data. The dashed line is the solution, Eq. (7.10) describing successive equilibration on larger and larger length scales, which leads to a (shifted) power law decay of the surface area. The enlarged early time regime (right part of Fig. 7.2) shows that it represents the data perfectly up to a time of about 3 ns.

Later on, the power law decay approaches A_{eq} too slowly compared to the data. The late stage of coalescence is better described by an exponential decay of $A - A_{eq}$, which starts at 3.5 ns (dash-dot-dotted line). Figure 7.1 gives a clue why this is the case: after 3.1 ns there remain three bumps on the surface of a central volume part. All of them have approximately the same size. The distance between the convex and the concave parts of the surface (sources respectively sinks of surface diffusion current density) has reached a maximum. Therefore, the effective time constant τ does no longer increase, but becomes a bit smaller, which explains the increase of the slope of $A(t)$ in Fig. 7.2b at $t \approx 3$ ns, and then stays roughly constant, which explains the exponential decay for $t > 3.5$ ns.

7.2.4 Coalescence of Two Particles of Different Size

The phenomenological theory reviewed in Sect. 7.2.1 assumes that the sintering is dominated by a single, possibly time dependent length scale R, which describes the typical distance between sources and sinks of the mass transport. However, generically two spherical, gas-borne particles that meet will have different radii, R_1 and R_2. Then it is not clear, what the characteristic length scale R should be. This question was addressed by atomistic simulations in [20] and [19]. As these two papers come to different conclusions, it is important to discuss precisely what was analyzed.

In [20] the coalescence of two spherical particles sharing the same fcc lattice was simulated by the KMC method. The deviation from a spherical shape was quantified by the squared radius of gyration,

$$R_{\mathrm{g}}^2 = \frac{1}{N} \sum_{i=1}^{N} \vec{r}_i^{\,2} - \left(\frac{1}{N} \sum_{i=1}^{N} \vec{r}_i \right)^2 . \tag{7.15}$$

$\vec{r}_i$ is the position of the ith atom, and $N \approx 5.924 \left[(R_1/a)^3 + (R_2/a)^3 \right]$ is the total number of atoms, which ranged from 6,300 to about 53,000. R_{g}^2 was averaged over many statistically independent simulation runs all starting from the same initial configuration.

The average squared radius of gyration R_{g}^2 decreases monotoneously and, for long times, approaches a minimal value, $R_{\mathrm{g,eq}}^2$. The fraction of the distance from this value,

$$Q(t) = \frac{R_{\mathrm{g}}^2(t) - R_{\mathrm{g,eq}}^2}{R_{\mathrm{g}}^2(0) - R_{\mathrm{g,eq}}^2}, \tag{7.16}$$

decays from the initial value 1 and fluctuates around zero for $t \rightarrow \infty$. The coalescence time t_c was defined as the time at which $Q(t_c) = 0.02$ for the first time.

Assuming the decay of $Q(t)$ is dominated by a single characteristic time τ, as in Eq. (7.4), one might expect that $Q(t)$ behaves similar to

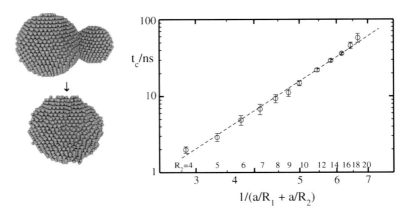

Fig. 7.3 Coalescence of two particles with radii $R_1 = 10a$ and $4 \leq R_2/a \leq 20$ (a is the nearest neighbor distance of the fcc lattice). Double-logarithmic plot of coalescence time t_c versus reduced radius in units of a (*dashed line* has slope 4). From [20]

$$\frac{A(t) - A_{eq}}{A_0 - A_{eq}} = e^{-t/\tau}. \quad (7.17)$$

This would imply that

$$t_c \propto \tau. \quad (7.18)$$

Figure 7.3 shows that

$$t_c \propto R_{red}^4, \quad \text{with} \quad R_{red} = \frac{R_1 R_2}{R_1 + R_2}. \quad (7.19)$$

Therefore, it had been concluded in [20] that the characteristic relaxation time τ for the coalescence of two spherical particles is proportional to their reduced radius R_{red} to the fourth power. Using the total volume $V = V_1 + V_2$ and the initial volume ratio $x = V_1/V_2$ as variables (instead of R_1 and R_2), one finds

$$\tau(V, x) \propto R_{red}^4 \propto V^{4/3} \left[(1+x)^{1/3} + (1+x^{-1})^{1/3} \right]^{-4}. \quad (7.20)$$

In [19] a different conclusion was drawn. The coalescence times obtained from a MD simulation seemed to be consistent with the exponential decay, Eq. (7.4), if one assumed that τ depends only on V, but not on x. In the following, a possible reason for this discrepancy will be given.

In [19] six particle pairs with different volume ratios x were considered, all having the same total volume V ($N = 3{,}200$ atoms). The simulation begins with a head-on collision of the particles. The ratio

$$\tilde{Q}(t) = \frac{I_\perp(t)}{I_\parallel(t)} \tag{7.21}$$

of the moments of inertia belonging to the axes perpendicular respectively parallel to the direction of the collision was taken as a measure of asphericity. This quantity has a maximum for the initial configuration and equals 1, when the two spheres have fused into a single one. A coalescence time $\tilde{t}_c$ was determined from the condition $\tilde{Q}(\tilde{t}_c) = 1.1$.

In [19] $\tilde{t}_c(x)/\tilde{t}_c(1)$ was fitted with the theoretical function

$$\frac{\tilde{t}_\epsilon(x)}{\tilde{t}_\epsilon(1)} = \frac{\ln\left(\frac{A_0(x)}{A_{eq}} - 1\right) - \ln \epsilon}{\ln\left(\frac{A_0(1)}{A_{eq}} - 1\right) - \ln \epsilon} \tag{7.22}$$

calculated from Eq. (7.4) with the condition $A(\tilde{t}_\epsilon)/A_{eq} = 1 + \epsilon$ and assuming that τ does not depend on x. A reasonable fit was obtained for $\epsilon = 0.05$. Hence the conclusion was reached that the x-dependence of the normalized coalescence time $\tilde{t}_c(x)/\tilde{t}_c(1)$ is consistent with the exponential decay, Eq. (7.4), with a characteristic time τ that depends only on V, but not on the initial volume ratio x.

This conclusion may be questioned, as the starting values,

$$\tilde{Q}(0) = 1 + \frac{5}{2}\left[\frac{1+x}{(1+x^{-1/3})^2} + \frac{1+x^{-1}}{(1+x^{1/3})^2}\right]^{-1} \tag{7.23}$$

and

$$\frac{A(0)}{A_{eq}} = \frac{1+x^{2/3}}{(1+x)^{2/3}} \tag{7.24}$$

have completely different x-dependencies. The values of $\tilde{t}_c$, respectively $\tilde{t}_\epsilon$ depend on these initial values. Hence, a fit of $\tilde{t}_c(x)/\tilde{t}_c(1)$ by (7.22) is not justified.

Both papers, [19] and [20], assume that the coalescence of the two particles can approximately be described by a single, constant characteristic time τ. This assumption, which seems to be justified in view of the result for late stage sintering in the previous Sect. 7.2.3, may actually be too simple, as both quantities, Q and $\tilde{Q}$, which were used to characterize the asphericity, do not have a purely exponential decay (see the curve labeled $\Sigma = 1$ in Fig. 7.7).

7.3 The Effect of Grain Boundaries

If crystalline particles without amorphous or fluid-like surfaces agglomerate, their crystallographic orientations usually do not match. Then the contact area is a grain boundary. These grain boundaries must survive the sintering process, if nanocrystalline properties should be maintained. The surface tension γ now depends on the local orientation of the surface with respect to the lattice axes. For high temperatures this dependence becomes weak. Therefore, in the literature on sintering, it is usually neglected. Here, too, Eqs. (7.25) and (7.26) are only given for a constant γ. The faceted shape of the simulated particles shows, however, that the applicability of these formulas is limited.

Grain boundaries strongly affect the sinter dynamics of nanoparticle agglomerates [32]. This is partly due to the interfacial free energy per unit grain boundary area, γ_{gb}. It limits neck growth, which increases the grain boundary area. The neck grows only as long as the gain in surface free energy overcompensates the cost of interfacial energy. Balance between gain and loss is reached, when the free surface formes the so-called dihedral angle ϑ at its intersection with the grain boundary. ϑ is given by Young's equation

$$2\gamma \cos(\vartheta/2) = \gamma_{gb}. \tag{7.25}$$

It is clear that only grain boundaries exist, for which $\gamma_{gb} < 2\gamma$. Otherwise, two particles with the corresponding orientational mismatch would not stick together.

If atoms on one side of the grain boundary accommodate to the crystal orientation on the other side, the grain boundary moves. This recrystallization is suppressed by the constriction, which exists due to the dihedral angle, provided the particle radii are not too different. A simple geometrical argument [33] for a dumbbell configuration with particle radii $R_1 > R_2$ shows that the grain boundary is pinned to a constriction as long as

$$\frac{R_1 - R_2}{R_1} < 1 + \cos\vartheta = \frac{\gamma_{gb}^2}{2\gamma^2}. \tag{7.26}$$

This means that narrow size distributions of the particles are favorable for maintaining grain boundaries as desired in order to get a nanocrystalline material as a result of the sintering.

Any size difference between two particles leads to an extra contribution to the chemical potential gradient. It is due to inhomogeneous elastic deformations induced by the surface tension and leads to Ostwald ripening (growth of the larger particle at the expense of the smaller one) [34]. It is conceivable that grain boundaries provide a higher activation barrier for surface diffusion from one particle onto the neighboring one, and hence slow down the Ostwald ripening.

Atomic mobility is enhanced in grain boundaries compared to the crystalline bulk, because (with few exceptions) the atoms are less densely packed. This is another reason why grain boundaries have a strong effect on sinter dynamics. One might expect that a strong misorientation leads to a large volume of the defected zone between the particles and to a high atomic mobility, thus to fast neck growth. However, such a grain boundary would also have a high interfacial energy per unit area, hence a small dihedral angle. Consequently, the neck must remain small, because the driving force for neck growth is lacking.

7 Nanopowder Sintering

This example shows that the orientational mismatch between the two grains is an important degree of freedom for sintering, because the interfacial free energy per unit area, as well as the grain boundary diffusion coefficient depend on it. A useful way to quantify the misorientation is by the coincidence site lattice (CSL) index Σ [35]. One imagines that both lattices extend throughout space, penetrating each other. Together they form a periodic lattice with a larger unit cell than for each individual lattice. Σ is the ratio of the large unit cell volume to the original one. It is always an odd integer. $\Sigma = 1$ means that both lattices have the same orientation, a twin boundary separates grains with relative orientation $\Sigma = 3$.

7.3.1 Rigid Body Dynamics Combined with KMC

As long as no solid neck has formed between them, two particles have six degrees of freedom to move as rigid bodies relative to each other: three components of the vector connecting the centers of mass, and three Euler angles for the relative crystallographic orientation.

In order to take atomic diffusion and particle reorientation into account simultaneously, a hybrid simulation technique has been developed [32]. It combines KMC steps for the atomic diffusion (as described in Sect. 7.2.2) with the numerical integration of the equations of motion for the six rigid body degrees of freedom [36, 37] (RBD).

In this simulation model, both particles have an fcc lattice of their own, with individual orientation and position of the origin. Each lattice extends beyond the occupied region into free space. A diffusion step takes an atom either to a free neighboring site of the same lattice or to a nearby free site of the other lattice. How the activation energies of these diffusion steps are modeled in the present calculations, is described in the appendix. (In [32] the activation energies were modeled differently, however, the conclusions drawn from those simulations were the same.)

The new ingredient of the hybrid simulation scheme, the rigid body dynamics (RBD), will be described now. Both particles are treated as rigid bodies. At first, their centers of mass as well as the tensors of their moments of inertia are calculated. Next, the total force exerted by the atoms of one particle on the atoms of the other one is evaluated as well as the total torque. Finally, the rigid translation and rotation of the particles is calculated by integrating the corresponding equations of motion [36]. This defines a second time scale, t_{RBD}, besides t_{KMC}, Eq. (7.14). Both times, t_{KMC} and t_{RBD}, are kept approximately synchronous [32] by alternating KMC and RBD steps.

This simulation model allows to analyze the reorientation of the two crystal lattices prior to the merging process, see Fig. 7.4. Generalization to larger agglomerates is straightforward.

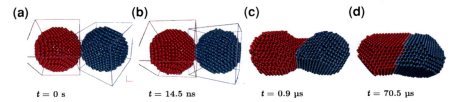

$t = 0$ s $\quad\quad\quad t = 14.5$ ns $\quad\quad\quad t = 0.9$ μs $\quad\quad\quad t = 70.5$ μs

Fig. 7.4 Snapshots of the coalescence of two particles of the same crystalline material, but different lattice orientations. Atoms are colored to indicate the fcc lattice, on which they are located. Color changes, when an atom moves over to the other lattice. Lattice axes are indicated by the perspective frames in **a** and **b**. In **c** the particles locked into a $\Sigma = 3$ configuration with a twin boundary, which is not yet planar. The neck prevents further reorientation. The RBD dynamics was switched off to speed up the simulation. In **d** coalescence has proceeded, the twin boundary has become flatter. Adapted from [32]

7.3.2 Competition Between Reorientation and Neck Growth for Two Particles

Figure 7.4 shows four snapshots of a typical simulation run. It starts with two crystalline particles of random relative orientation (a). In the course of time, they change their orientation (b), until—in the case shown—they adopt a $\Sigma = 3$ configuration (c), which does not change any more and allows that the neck grows until the dihedral angle is formed (d) (for a twin boundary $\vartheta \approx \pi$, because γ_{gb} is very small).

The initial configuration was generated as follows: all sites of an fcc lattice within a certain distance from the origin were filled by atoms, giving a compact (nearly) spherical particle. This particle was "cloned" to obtain a second, independent one. The shapes of both particles were equilibrated separately for a sufficiently large number of KMC diffusion steps (depending on the particle size) at a fixed temperature. Then one of the particles was rotated by three random Euler angles uniformly distributed between 0 and π. Finally, the origins of both lattices were placed on the [110]-axis of one particle in such a way that the particles just touched at a point. Then the KMC–RBD hybrid simulation started and ran for typically 200,000–500,000 KMC steps (corresponding to a real time of 40 ns).

More than 1,000 such simulation runs were performed for each particle size. In all cases the particles changed their relative orientation, generally pausing at some intermediate orientations for a while before rotating further. An example is given in Fig. 7.5a.

The simulation time was chosen long enough that in about 99 % of the runs the two particles finally locked into a fixed orientation, where the growing neck prevented further rotation of the lattices. The duration of the orientation phase is proportional to the initial particle radius, see Fig. 7.5b.

The final orientation was evaluated statistically in terms of the CSL index Σ, using the criterion derived in [39]. The histograms, Fig. 7.6, have a remarkable dependence on the particle size: the smaller the primary particles, the more likely are orientations

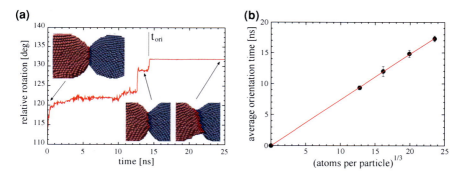

Fig. 7.5 a The "red" lattice is rotated with respect to the "blue" one. The rotation axis (not shown) and the angle (shown) depend on time in a complex way. In this example a neck has formed at an exceptionally early time (*left* configuration). This leads almost to a stationary orientation. However, thermal fluctuations are strong enough to trigger two major reorientation events at later times (*middle* and *right* configuration). At the orientation time t_{ori} the orientation locks in. **b** The orientation time depends linearly on the radius of the primary particles

with low Σ-values as a result of the sintering process. This is in accordance with [40], where MD simulations showed that small Pt particles on a Pt substrate rotate to adjust to the underlying crystal orientation, while larger ones do not. Figure 7.6 shows on the left the histogram for particles with a radius R, which is seven times the nearest neighbor distance a of the fcc lattice, and on the right the one for $R = 13a$. Whereas for $R = 7a$ about 11% of the configurations had lost the grain boundary, and about 17% had developed a twin boundary, these two probabilities were less than 5% for $R = 13a$. The fraction of configurations with $\Sigma < 47$ is 56% ($R = 7a$), respectively 33% ($R = 13a$). Also shown is the histogram for the Σ-values obtained by analyzing the initial configuration: Only 5.8% had $\Sigma < 47$. This confirms that single crystalline nanoparticles, when sintering pairwise, preferentially adopt configurations with a large number of coherent sites in both lattices.

These results can be explained by the competition between neck growth and reorientation. An orientation with a large Σ-value in general implies a large grain boundary energy, hence a small driving force for neck growth. The contact between the particles remains fragile. At the same time, the grain boundary position fluctuates thermally, due to atoms that leave the lattice sites on one to adopt the crystal structure on the other side. The positional fluctuations are the stronger the smaller the grain boundary area. Together with thermal shape fluctuations of the particles themselves, they trigger the reorientation moves of the particles, because they lead to a fluctuating torque. Large particles move more slowly than small ones, because their thermal fluctuations are weaker and their inertia is higher. Therefore, they have less chance to find an orientation favorable for neck growth, leading to a larger orientation time (time until orientation locks in), Fig. 7.5, and to a less frequent realization of low-Σ boundaries, Fig. 7.6.

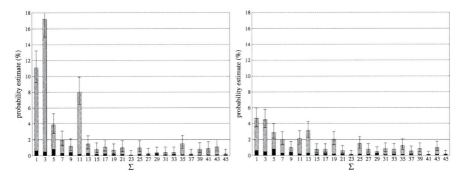

Fig. 7.6 Normalized histograms of final orientations (characterized by the coincidence site lattice index Σ) of two coalescing particles. *Left* initial particles had radius $R = 7a$ (2,071 atoms per particle). *Right* $R = 13a$ (13,045 atoms per particle). The *error bars* indicate the 95% confidence interval (Agresti–Coull interval [38]). The *black bars* indicate the probabilities in the case of random orientations

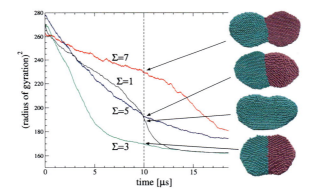

Fig. 7.7 Coalescence after the orientation has locked in. Time axis begins for each orientation individually at t_{ori} defined in Fig. 7.5. Squared radius of gyration evolves qualitatively differently for different values of the coincidence site lattice index Σ. Only single runs are shown, but the same qualitative behavior has been observed for a large number of runs. All particles have the same size

For a dumbbell configuration of nanoparticles the orientation phase is very short compared to the subsequent stages of the sinter process. When the neck has grown big enough that a further reorientation becomes unlikely, the orientational fluctuations decrease significantly. Then the rigid body dynamics can be switched off. This allows to extend the simulations to much longer times (20 ms), in order to study the later stages of coalescence in the presence of a grain boundary. In Fig. 7.7 the time evolution of the radius of gyration is shown starting at the time, at which the orientation of the particles locked in. Snapshots of the configurations were taken after 10 μs. Clearly, the development of a near-spherical shape is prevented by the formation of a dihedral angle for $\Sigma = 5$ and $\Sigma = 7$.

7 Nanopowder Sintering

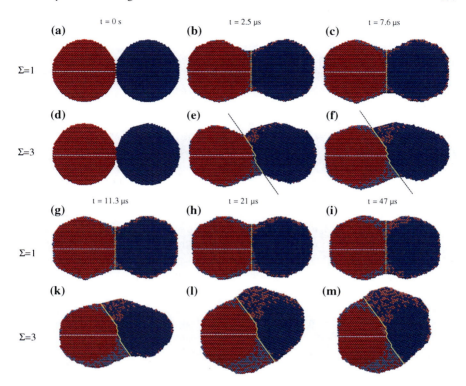

Fig. 7.8 Central cross-sections through a coherent ($\Sigma = 1$) and a twinned system ($\Sigma = 3$) compared at six stages of coalescence. Primary particle radii $R = 17a$ (29,035 atoms per particle). *Dashed white lines* are fixed to the bulk region of the left particle, where atoms do not change their positions. They serve as an orientation to identify corresponding atoms in the snapshots. For $\Sigma = 1$ an additional line marks the initial symmetry plane. For $\Sigma = 3$ the additional lines mark the grain boundary

Unexpectedly, the radius of gyration decreases faster for $\Sigma = 3$ than for $\Sigma = 1$. The snapshots in Fig. 7.7 show that the coherent configuration ($\Sigma = 1$) is still more elongated after 10 μs than the one with the twin boundary ($\Sigma = 3$). Obviously, the neck grows faster for $\Sigma = 3$ than for $\Sigma = 1$. This surprising result can be explained by a look into the interior structure of the particles, Fig. 7.8.

The CSL index Σ only describes the relative orientation of the particle bulk lattice structures. It does not determine the crystallographic orientation of the grain boundary in between, whose normal vector can point into any direction. The atomic structure of the grain boundary, and hence its specific interfacial energy γ_{gb} depends both on Σ and the normal direction.[3] Generically, when the crystal lattices lock into

---

[3] In addition, the structure of the grain boundary depends also on a translation vector that fixes the positions of the origins of the lattices within the unit cell of their common CSL-superlattice. In the case considered here, the origins can be chosen to coincide.

a $\Sigma = 3$ orientation, the normal vector of the grain boundary must rotate in order to establish a densely packed twin boundary, which is no longer perpendicular to the line connecting the centers of the primary particles.

Let us assume for simplicity that $\gamma_{gb} = 0$ for the twin boundary. Then local equilibrium is reached, when the twin boundary intersects the free surface at an angle of 90 degree. The mass transport along the surface has to be such that the centers of mass of the two lattices move parallel to the densely packed twin boundary. This is different for $\Sigma = 1$, where the centers of mass move toward each other along the symmetry axis of the dumbbell. As a consequence, the cross-sectional area perpendicular to this axis is smaller than the area of the twin boundary, when the concave surface parts vanish. This means that chemical potential gradients along the surface remain strong for $\Sigma = 3$, until a rather round shape is reached. For $\Sigma = 1$ the chemical potential gradients become weak while the shape is still cigar-like. Therefore, the mass transport is faster for $\Sigma = 3$ than for $\Sigma = 1$.

Grain boundary diffusion, which in principle could also speed up the neck growth compared to the case $\Sigma = 1$, played no role in the present simulation. However, it is interesting to note, that the grain boundary is dynamical due to atom exchange between the two lattices. In Fig. 7.8e one can see seven (111)-facets. At later times their size changed.

An in situ TEM study of the sintering of Indium Tin Oxide (ITO) nanoparticle agglomerates [32] shows that grain boundaries between the particles exist so that the model described in this section is applicable. There is clear experimental evidence that the ITO particles must adopt a suitable orientation, before coalescence takes place. Moreover, after the coalescence of small particles one observes a preference for twin boundaries compared to larger misorientation [1, 32]. That particle reorientations are important for the dynamics of coalescence has also been confirmed recently in MD simulations of a two dimensional Lennard–Jones model [41].

7.3.3 Reorientation Effects in Porous Agglomerates

According to the above picture, one expects that coalescence of an agglomerate of randomly oriented crystalline nanoparticles can be suppressed, if the particles cannot freely change their relative orientations. It has been predicted [42] that pores surrounded by sufficiently many particles (sufficiently large "pore coordination number") become stable. However, in that study it was assumed that all grain boundaries have the same dihedral angle, which determines the critical pore coordination number. For randomly oriented particles this is, of course, not the case.

This raises the question, how a large agglomerate of randomly oriented nanoparticles can be formed adding particle by particle, in view of the previous result, that the reorientation phase lasts only for about 10 ns (see Fig. 7.5), if a newly added particle can freely rotate. One possibility is that the primary particles are covered by a layer of surfactants, as is usually the case, if they are produced by wet chemistry. When agglomerating, the crystalline core of the particles is hidden behind this layer

7 Nanopowder Sintering

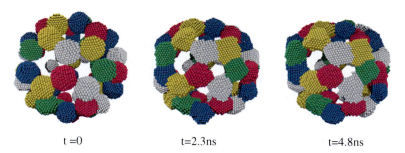

t =0   t=2.3ns   t=4.8ns

Fig. 7.9 59 randomly oriented round particles with fcc structure initially arranged in a buckyball configuration with one defect (the sixtieth particle is missing). Different colors are used to distinguish different lattice orientations

so that it cannot promote reorientation. Presumably this is the way, how a random orientation of the ITO particles was obtained in the initial agglomerates used for the experiments mentioned above [32]. Raising the temperature then removes the surfactant layer, so that the reorientation dynamics can set in.

In the following some preliminary results will be shown, which demonstrate that it is feasible to study the influence of grain boundaries on sinter dynamics of porous agglomerates with the model developed in Sect. 7.3.1. Figure 7.9 shows 3 snapshots of such a simulation [29] for 59 randomly oriented particles placed on the vertices of a bucky ball. One vertex was left empty in order to perturb the symmetry, as it might have an additional stabilizing effect intervening with the one of interest here. Whereas the previous simulations were based on a Lennard–Jones pair potential between the atoms, here a many body interaction (tight binding second moment) was used. Also the temperature was different. Therefore, the timescale is not comparable to the ones of the previous figures.

After 2.3 ns those neighboring particles that could rotate their lattices into a coherent or a twinned orientation have formed larger necks than the others. Obviously, the result that reorientation dynamics and neck growth mutually influence each other does not depend on the type of interatomic potential used. The new aspect here is purely geometric: particles cannot rotate freely as in Sect. 7.3.2, because they have more than one neighbor. Adapting to one of them in general leads to no improvement of the relative orientation with respect to the other neighbors.

Shrinkage and configurational changes are much weaker in the subsequent 2.5 ns (from b to c in Fig. 7.9). Sintering seems to stagnate in agreement with the ideas put forward in [42]. The reason is that the necks, which have formed, prevent further reorientation and have adopted local equilibrium configurations: the agglomerate is frozen into a metastable, nanocrystalline state. Further shrinkage is slow for three reasons: the particles have similar sizes so that Ostwald ripening due to atom migration across the grain boundaries is slow; at all grain boundaries the local equilibrium dihedral angle has been adopted; the positional fluctuations of grain boundaries, which are pinned at constrictions, are weak, so that recrystallization by sweeping a grain boundary through a whole particle is unlikely.

7.4 Conclusion and Outlook

In this review on nanopowder sintering we have focused on atomic scale effects that may help to preserve nanocristallinity. Smallness of the primary particles in many ways promotes the opposite, i.e., coarsening processes that tend to destroy nanocrystallinity. Reasons are:

- the enormous potential for lowering the free energy by reducing the surface and interfacial area, combined with
- short relaxation pathways, hence fast relaxation,
- the frequent and fast adoption of a coherent configuration, when two nanoparticles collide and stick together,
- enhanced thermal fluctuations making structural changes easier,
- size dependent lowering of the melting temperature bearing the risk that the smaller particles melt and merge with, respectively recrystallize as a larger single crystal.

The remedies preserving nanocristallinity in the sintering process largely rely on kinetic constraints leading to a frozen-in metastable nonequilibrium state:

- A narrow size distribution helps by reducing Ostwald ripening.
- When collecting the powder, the crystalline primary particles should be hindered to adjust their crystallographic orientations . This could be achieved e.g., by covering them by a surfactant layer.
- The transition from a soft to a hard agglomerate should be performed in a way that preserves grain boundaries between the particles. For this it is an advantage, if each particle has several differently oriented neighbors such that a collective alignment becomes unlikely.
- Grain boundaries with larger interfacial free energy per unit area are more helpful, because local equilibrium will be reached with a small dihedral angle and a narrow neck. This helps to pin the grain boundary. Moreover, it makes it easier to compactify the porous powder structure by applying a pressure.
- The metastable state only remains frozen in, if temperatures stay low enough.

Recently, particle-based phenomenological simulations of sintering have been presented for instance in [43] and in [44]. The atomic scale effects described in this review should next be incorporated into such phenomenological simulations on the particle scale in order to study the evolution of large assemblies of particles. For applications it is, for example, important to know, how the porosity and possible fractal substructures of a powder evolve.

The recipes listed above to preserve nanocrystallinity in general also favor a porous sintered structure, as discussed in Sect. 7.3.3. A very promising method to reduce the porosity, while maintaining the nanocrystallinity, is current assisted sintering

[15]. This method is discussed in the Chap. 10 in this book. After pores have been eliminated, one obtains a nanocrystalline solid with a network of grain boundaries. Further ripening kinetics is similar to the coarsening of foam (the grain boundaries corresponding to the fluid lamellas separating the gas bubbles of a foam), with the difference that grain rotations may assist the elimination of grain boundaries [45].

Acknowledgments We thank F. Westerhoff for his contribution to the research work presented here. Many fruitful and inspiring discussions with our colleagues within the Collaborative Research Center (SFB) 445 "Nanoparticles from the Gas Phase" funded by the German Research Foundation (DFG) are gratefully acknowledged, in particular with P. Entel, J. Kästner, E. Kruis, R. Meyer, B. Rellinghaus, G. Schierning, R. Schmechel, R. Theissmann, H. Wiggers, and M. Winterer.

Appendix: Activation Energies Used in the KMC-RBD Hybrid Model

A possible diffusion process leads from an occupied initial site to a nearby unoccupied site on either lattice, which must have at least two occupied nearest neighbors. If there are only two neighbor atoms, it is not a binding site but a saddle point. Then the atom is immediately moved on to a randomly chosen neighbor site exhibiting three or more direct neighbors.

In case that the initial and the destination site belong to the same lattice, they must be nearest neighbors. Then the activation energy in our simulation model is defined by

$$E_{\text{act}} = \begin{cases} E_s - E_{b,i}, & \text{for more than 2 final neighbors} \\ E_{b,f} - E_{b,i}, & \text{for exactly 2 final neighbors} \end{cases}, \qquad (7.27)$$

where $E_{b,i}$ and $E_{b,f}$ are the binding energies at the initial, respectively the final site calculated from a Lennard–Jones potential, if not stated otherwise. The saddle point energy E_s is assumed to be constant for simplicity. Its value is chosen of the same order of magnitude as the saddle point energies on high symmetry surfaces of the Lennard–Jones crystal.

For diffusion processes of atoms to a site on the other lattice, we have to address the fact that the hopping distance is no longer constant. We allow only distances between the starting and the destination site which do not exceed $1.5a$, where a is the nearest neighbor distance on the fcc lattices. Assuming that energy barriers drop with decreasing hopping distance we approximate the local energy landscape by the radial parabolic potential $E(r) = E_b + \kappa r^2$. Given the initial site on one lattice and an allowed destination site on the other lattice, the saddle point energy E_S between them is defined by the intersection of both sites' parabolas and hence depends on the distance between the minima. The activation energy is $E_{\text{act}} = E_s - E_{b,i}$. The parameter κ is determined by the requirement that simulation results should match if either a single-lattice model or two exactly coinciding lattices are used.

References

1. S. Tsyganov, J. Kästner, B. Rellinghaus, T. Kauffeldt, F. Westerhoff, D.E. Wolf, Analysis of Ni nanoparticle gas phase sintering. Phys. Rev. B **75**, 045421 (2007)
2. R.M. German, *Sintering Theory and Practice* (Wiley-Interscience, New York, 1996)
3. M.N. Rahaman, *Ceramic Processing and Sintering*, 2nd edn. (Marcel Dekker, New York, 2003)
4. T. Ring, *Fundamentals of Ceramic Powder Processing and Synthesis*. (Academic Press, San Diego, 1996)
5. A.S. Barnard, Modelling of nanoparticles: approaches to morphology and evolution. Rep. Prog. Phys. **73**, 086502 (2010)
6. J.N. Israelachvili, *Intermolecular and Surface Forces* (Academic Press, London, 1991)
7. Lutz Mädler, Anshuman A. Lall, Sheldon K. Friedlander, One-step aerosol synthesis of nanoparticle agglomerate films: simulation of film porosity and thickness. Nanotechnology **17**, 4783 (2006)
8. T. Schwager, D.E. Wolf, T. Pöschel, Fractal substructure of a nanopowder. Phys. Rev. Lett. **100**, 218002 (2008)
9. D.E. Wolf, T. Pöschel, T. Schwager, A. Weuster, L. Brendel, Fractal substructures due to fragmentation and reagglomeration, in *Powders and Grains 2009 (AIP CP 1145)*, ed. by M. Nakagawa, S. Luding (AIP, Melville, 2009), p. 859
10. D. Kadau, H.J. Herrmann, Density profiles of loose and collapsed cohesive granular structures generated by ballistic deposition. Phys. Rev. E **83**, 031301 (2011)
11. A. Pimpinelli, J. Villain, *Physics of Crystal Growth*. (Cambridge University Press, Cambridge, 1998)
12. A. Suzuki, Y. Mishin, Atomic mechanisms of grain boundary diffusion: low versus high temperatures. J. Mater. Sci. **40**, 3155 (2005)
13. R.L. Coble, A model for boundary diffusion controlled creep in polycrystalline materials. J. Appl. Phys. **34**, 1679 (1963)
14. M. Morgeneyer, J. Schwedes, K. Johnson, D.E. Wolf, L. Heim, Microscopic and macroscopic compaction of cohesive powders, in *Powders and Grains*, ed. by R. Garcia-Rojo, H.J. Herrmann, S. McNamara (A. A. Balkema, Rotterdam, 2005), pp. 569–573
15. D. Schwesig, G. Schierning, R. Theissmann, N. Stein, N. Petermann, H. Wiggers, R. Schmechel, D.E. Wolf, From nanoparticles to nanocrystalline bulk: percolation effects in field assisted sintering of silicon nanoparticles. Nanotechnology **22**, 135601 (2011)
16. P. Song, D. Wen, Surface melting and sintering of metallic nanoparticles. J. Nanosci. Nanotechnol. **10**, 8010 (2010)
17. Benny Davidovitch, Deniz Ertaş, Thomas C. Halsey, Ripening of porous media. Phys. Rev. E **70**, 031609 (2004)
18. W. Koch, S.K. Friedlander, The effect of particle coalescence on the surface area of a coagulating aerosol. J. Colloid Interface Sci. **140**, 419 (1990)
19. T. Hawa, M.R. Zachariah, Coalescence kinetics of unequal sized nanoparticles. J. Aerosol Sci. **37**, 1 (2006)
20. F. Westerhoff, R. Zinetullin, D.E. Wolf, Kinetic Monte-Carlo simulations of sintering, in *Powders and Grains*, ed. by R. Garcia-Rojo, H.J. Herrmann, S. Mc-Namara (A. A. Balkema, Rotterdam, 2005), pp. 641–645
21. C. Herring, Effect of change of scale on sintering phenomena. J. Appl. Phys. **21**, 301 (1950)
22. F.A. Nichols, W.W. Mullins, Morphological changes of a surface of revolution due to capillarity-induced surface diffusion. J. Appl. Phys. **36**, 1826 (1965)
23. F.A. Nichols, Coalescence of two spheres by surface diffusion. J. Appl. Phys. **37**, 2805 (1966)
24. J. Frenkel, Viscous flow of crystalline bodies under the action of surface tension. J. Phys. (USSR), **9**, 385 (1945)
25. A.S. Edelstein, R.C. Cammarata, *Nanomaterials: Synthesis, Properties and Applications* (Institute of Physics, London, 1996)

7 Nanopowder Sintering

26. R. Meyer, J.J. Gafner, S.L. Gafner, S. Stappert, B. Rellinghaus, P. Entel, Computer simulations of the condensation of nanoparticles from the gas phase. Phase Transitions Multinatl. J. **78**, 35 (2005)
27. M. Kotrla. Numerical simulations in the theory of crystal growth. Comput. Phys. Commun. **97**, 82 (1996)
28. A.F. Voter, F. Montalenti, T.C. Germann, Extending the time scale in atomistic simulation of materials. Annu. Rev. Mater. Res. **32**, 321 (2002)
29. R. Zinetullin, Zur Kalibrierung Kinetischer Monte-Carlo Simulationen durch Molekulardynamik. Ph.D. thesis, Universität Duisburg-Essen, 2010
30. L. Brendel, A. Schindler, M. von den Driesch, D.E. Wolf, Different types of scaling in epitaxial growth. Comput. Phys. Commun. **147**, 111 (2002)
31. A.B. Bortz, M.H. Kalos, J.L. Lebowitz, A new algorithm for Monte Carlo simulation of Ising spin systems. J. Comput. Phys. **17**, 10 (1975)
32. R. Theissmann, M. Fendrich, R. Zinetullin, G. Guenther, G. Schierning, D.E. Wolf, Crystallographic reorientation and nanoparticle coalescence. Phys. Rev. B **78**, 205413 (2008)
33. F.F. Lange, B.J. Kellett, Thermodynamics of densification.2. grain-growth in porous compacts and relation to densification. J. Am. Ceram. Soc. **72**, 735 (1989)
34. J. Pan, H. Le, S. Kucherenko, J.A. Yeomans, A model for the sintering of spherical particles of different sizes by solid state diffusion. Acta Materialia **46**, 4671 (1998)
35. G.A. Chadwick, D.A. Smith, *Grain Boundary Structure and Properties* (Academic Press, New York, 1976)
36. N.S. Martys, R.D. Mountain, Velocity Verlet algorithm for dissipative-particle-dynamics-based models of suspensions. Phys. Rev. E **59**, 3733 (1999)
37. D.C. Rapaport, *The Art of Molecular Dynamics Simulation* (Cambridge University Press, Cambridge, 1995)
38. L.D. Brown, T. Tony Cai, A. DasGupta, Interval estimation for a binomial proportion. Stat. Sci. **16**, 101 (2001)
39. G. Palumbo, K.T. Aust, E.M. Lehockey, U. Erb, P. Lin, On a more restrictive geometric criterion for 'special' CSL grain boundaries. Scripta Materialia **38**, 1685 (1998)
40. Y. Ashkenazy, R.S. Averback, K. Albe, Nanocluster rotation on Pt surfaces: twist boundaries. Phys. Rev. B **64**, 205409 (2001)
41. Lifeng Ding, Ruslan L. Davidchack, Jingzhe Pan, A molecular dynamics study of sintering between nanoparticles. Comput. Mater. Sci. **45**, 247 (2009)
42. B.J. Kellett, F.F. Lange, Thermodynamics of densification.1. Sintering of simple particle arrays, equilibrium-configurations, pore stability, and shrinkage. J. Am. Ceram. Soc. **72**, 725 (1989)
43. B. Henrich, A. Wonisch, T. Kraft, M. Moseler, H. Riedel, Simulations of the influence of rearrangement during sintering. Acta Materialia **55**, 753 (2007)
44. Max L. Eggersdorfer, Dirk Kadau, Hans J. Herrmann, Sotiris E. Pratsinis, Multiparticle sintering dynamics: from fractal-like aggregates to compact structures. Langmuir **27**, 6358 (2011)
45. D. Moldovan, V. Yamakov, D. Wolf, S.R. Phillpot, Scaling behavior of grain-rotation-induced grain growth. Phys. Rev. Lett. **89**, 206101 (2002)

Chapter 8
Material and Doping Contrast in III/V Nanowires Probed by Kelvin Probe Force Microscopy

Sasa Vinaji, Gerd Bacher and Wolfgang Mertin

Abstract We have studied the local surface potential and the voltage drop along individual VLS grown GaAs nanowires using Kelvin probe force microscopy. With the obtained information, we identify a core–shell structure in GaAs/GaP heterostructure nanowires, which we attribute to the difference in radial and vertical growth between the two semiconductor materials. In p-doped GaAs nanowires, qualitative and quantitative doping levels are estimated. Furthermore, we find a better incorporation of the zinc compared to the carbon to realize doping in partially p-doped GaAs nanowires by localizing the doping transitions and estimating the width of their depletion layers. Additionally, the p–n junction can be localized with a resolution better than 50 nm and the bias dependence of the depletion layer width can be studied.

8.1 Introduction

III–V semiconductor nanowires are believed to have enormous benefits in the field of advanced electronic [1], optoelectronic [2], and thermoelectric devices [3] exhibiting improved or even new properties. First, the improved electrostatic charge control in a nanowire by omega-shaped gate electrodes and the possibility to design heterostructures lead to a transconductance of 2 S/mm in InAs nanowire field effect transistors [4] and to on currents of 1 mA/μm in InAs/InP radial nanowire heterostructure devices [5]. Second, the enhanced light collection efficiency compared to conventional thin film devices [6, 7] together with the freedom of integrating multiple p–n junctions with optimized spectral response to the solar spectrum [8] make III–V nanowires quite attractive for the development of high performance photovoltaic devices. A few successful examples of GaAs nanowire photovoltaic devices have been demonstrated.

S. Vinaji · G. Bacher · W. Mertin (✉)
Faculty of Engineering and CENIDE, Electronic Materials and Nanodevices,
University of Duisburg-Essen, Bismarckstraße 81, 47057 Duisburg, Germany
e-mail: wolfgang.mertin@uni-due.de

A. Lorke et al. (eds.), *Nanoparticles from the Gas Phase*, NanoScience and Technology, DOI: 10.1007/978-3-642-28546-2_8, © Springer-Verlag Berlin Heidelberg 2012

Efficiencies ranging from 0.83 % for coaxial nanowire diode arrays [9] up to 4.5 % for a single coaxial nanowire diode [10] have been reported. Finally, due to e.g. increased boundary scattering [11, 12], the thermal conductivity in nanowires can be decreased, while the electrical conductivity remains relatively high. This leads to ZT values largely enhanced as compared to bulk material.

The widely used vapor–liquid–solid (VLS) technique [13] allows the growth of nanowires with controlled properties like morphology and crystallinity [14], material composition, and doping concentration [15, 16]. These parameters define the device functionality [17]. Therefore, it is crucial to determine these parameters, especially the local variation of composition and doping, because of their high impact on the electronic, optoelectronic, and thermoelectric behavior of the device. Although first proof of principle devices have been demonstrated it is obvious that a further improvement of the growth of nanowires with controlled composition and doping is needed to optimize the properties of future nanowire devices.

In contrast to bulk semiconductors where the technology for obtaining high quality doping and material transitions is well established, this is quite challenging for semiconductor nanowires. That is especially true for III–V nanowires where a detailed model of the VLS growth mechanism is still under discussion [18]. Thus, nondestructive experimental techniques are required, which simultaneously provide information on the electrical (e.g. doping, conductivity, potential distribution) and structural (e.g. size, tapering, core–shell structure) properties of nanowires with an inherently high spatial resolution. Standard tools for extracting information on structure and composition are, for example, scanning electron microscopy, transmission electron microscopy or energy dispersive X-ray spectroscopy [15, 19]. These techniques, however, do not allow one to locally extract electrical properties, like doping levels and transitions or interface resistances. More sophisticated techniques like high spatially resolved atom probe tomography measure directly the dopant profile, even in the cross-section of a nanowire [20]. As this technique is highly destructive, it cannot be applied for investigating operating nanowire devices. Very recently, scanning spreading resistance microscopy was used to trace the profile of the carrier concentration in an individual Si nanowire [21], whereas an access to the voltage drop along the nanowire or at the contacts in a nanowire device cannot be obtained.

A scanning force microscopy (SFM) technique that has shown its application potential to semiconductor device research is Kelvin probe force microscopy (KPFM). Hereby the electrostatic force between an atomically sharp tip and the sample beneath is measured [22–24]. In KPFM, there is no direct mechanical contact between the tip and the sample, thus making this technique ideally suited for the study of sensitive samples like nanowires. Furthermore, it is material sensitive and can work under ambient conditions even for operating devices. Most importantly, the surface potential can be measured with a high voltage resolution of only a few mV, even on samples with poor conductivity [25]. KPFM was used e.g. for measuring the work function of metals [22] as well as the surface potential of conventional semiconductor diodes [26–28]. Its application to biased semiconductor devices allows a spatially resolved determination of the voltage drop even quantitatively, thus helping to understand and optimize the device functionality [29–31].

8 Material and Doping Contrast in III/V Nanowires 187

During the last years, KPFM has gained importance in the characterization of nanowires. KPFM was used to investigate the contact properties between nanowires and metal contacts by measuring the voltage drop in the contact area for carbon nanotubes [32]. KPFM was also used to visualize the influence of covalent attachments to the conductance of single-walled carbon nanotubes [33]. More recently, Koren et al. probed the dopant distribution in a phosphorous-doped silicon nanowire [34]. By using a successive etching technique, they could even measure the radial doping distribution and found that the doping concentration decreases from the surface to the core of the nanowire by a factor of two [35, 36].

In contrast to KPFM experiments done on carbon nanotubes and elementary semiconductors, similar experiments on compound semiconductor nanowires with their potential for electronic, optoelectronic, and thermoelectric applications are quite rare. Potential drops along ZnO [37], CuO [38] or CdSe [39] nanowires have been measured. A more device-related question was investigated by Minot et al., who measured the voltage drop along an InP/InAsP nanowire and subsequently calculated the depletion zone in this diode-like structure [40]. However, no such data is reported up to now for nanowires based on the technologically relevant GaAs system.

8.2 Instrumental Setup

A Veeco Innova^{TM} atomic force microscope was used for all the experiments. A schematic diagram of the measurement setup is depicted in Fig. 8.1. For the KPFM experiments, the system was equipped with an external homebuilt Kelvin controller. The Si tips used for the measurements have a conductive Cr/Pt coating, a nominal radius of about 25 nm, and a nominal resonance frequency ω_{res} of about 75 kHz (BudgetSensors ElectriMulti75). To enhance the sensitivity, the externally applied voltage V_{DC} to the tip was modulated by an AC voltage with amplitude V_{AC} and frequency ω_{Ref}. All measurements were performed in the non-contact mode with a distance of about 20 nm between nanowire and tip. This allows fully nondestructive measurements, which makes KPFM ideally suited for sensitive samples like nanowires. V_{DC} is controlled via the Kelvin loop system in order to compensate the contact potential difference V_{cpd} between tip and sample, and thus corresponds to the Kelvin voltage V_{Kelvin} [22].

V_{cpd} can thus be written as

$$V_{\text{CPD}}(x, y) = -e^{-1} \cdot (\Phi_{\text{tip}} - \Phi_{\text{S}}(x, y)) = -V_{\text{Kelvin}}(x, y) + V_{\text{drop}}(x, y) \quad (8.1)$$

with $V_{\text{drop}}(x, y)$ being the local voltage drop in the device with respect to ground. Here, Φ_{tip} is the work function of the tip material and $\Phi_{\text{S}}(x, y)$ is the local work function of the material under investigation, which is given by [41]

$$\Phi_{\text{S}}(x, y) = \chi(x, y) + \Delta E_{\text{F}}(x, y) + \Delta \Phi_{\text{S}}(x, y) \quad (8.2)$$

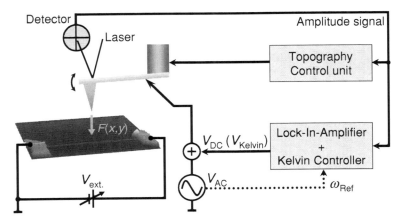

Fig. 8.1 Schematic diagram of the KPFM system used. The applied voltages V_{DC} and V_{AC} are needed for the measurement of the Kelvin voltage $V_{Kelvin}(x, y)$

$\chi(x, y)$ is the local electron affinity of the semiconductor, $\Delta E_F(x, y)$ the local energy difference between the Fermi energy, and the conduction band edge, and $\Delta \Phi_S(x, y)$ is the local surface band bending. The value of V_{DC} is controlled via a Kelvin loop system in such a way that the electric force $F(x, y)$ between the tip and the sample surface becomes zero. In the unbiased case ($V_{ext} = 0$ V and therefore $V_{drop}(x, y) = 0$ V), V_{DC} is equal to $-V_{cpd}(x, y)$ and in the case of a bias voltage ($V_{ext} \neq 0$ V and therefore $V_{drop}(x, y) \neq 0$ V), V_{DC} corresponds to $-(V_{cpd}(x, y) + V_{drop}(x, y))$. In both cases V_{DC} is called Kelvin voltage $V_{Kelvin}(x, y)$.

Using the amplitude modulated noncontact mode it is possible to measure simultaneously the topography and the local work function $\Phi_S(x, y)$ of a sample surface via the $V_{cpd}(x, y)$ [23, 31]. During the KPFM measurements we used an amplitude of 1 V for the modulation voltage V_{AC}, a frequency $\omega_{Ref} = 44$ kHz, and a scan speed of 1/3 Hz. With this technique, we are able to measure the local Kelvin voltage distribution of the sample surface with a spatial resolution down to few tens of nanometer and a sensitivity of a few mV. For the application of a bias voltage V_{ext}, the nanowires are contacted with electrodes, created by ebeam lithography. The substrate was grounded for all measurements.

For the evaluation of the achievable spatial resolution for the determination of both, material and doping transitions, an InP/InGaAs butterfly structure with wellknown growth conditions was investigated (Fig. 8.2a). The structure was grown lattice matched on InP substrate. The In-content of the InGaAs layers was 53 %. The InP layers were n-doped with a concentration of 1×10^{19} cm^{-3}. The InGaAs layers were undoped except from two layers (marked by red arrows in Fig. 8.2a), which were p-doped with a concentration of 3×10^{19} cm^{-3}. The smallest layer with a width of 5 nm was located in the center of the structure while the width of the subsequent layers increases up to 70 and 90 nm, respectively, toward both sides. In the scanning

8 Material and Doping Contrast in III/V Nanowires

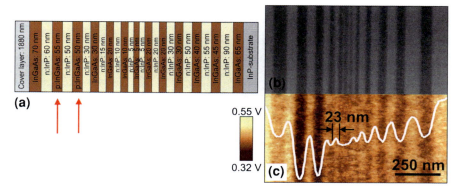

Fig. 8.2 InP/InGaAs butterfly structure for the determination of the electrical spatial resolution of the KPFM setup. **a** Schematic layer structure with indicated layer thicknesses. The two p-doped layers are indicated by *red arrows*. **b** Scanning transmission electron microscopy Z-contrast image of the structure. **c** 2D-KPFM-image. The color code represents the Kelvin voltage. The *white line* is a line scan obtained by averaging 20 lines [42]

transmission electron microscopy Z-contrast image of the structure, the InP layers appear bright while the InGaAs layers appear dark (Fig. 8.2b).

A 2D image of a cross-section KPFM measurement on the InGaAs/InP butterfly structure is shown in Fig. 8.2c. The bright areas represent the InP layers and the dark ones the InGaAs layers. The difference in the work function between InGaAs and n-InP is visualized by the color contrast. Due to the larger work function, the two p-doped InGaAs layers can be clearly identified as dark layers on the left side demonstrating the high sensitivity regarding different doping levels. For the evaluation of the spatial resolution, 20 lines were averaged (white line in Fig. 8.2c). It is obvious that the difference in the Kelvin signal between InP and InGaAs layers decreases toward smaller layers. This can be explained by the averaging effect of the electrostatic fields for small areas [43, 44]. The smallest resolvable layer is a n-doped InP layer with a nominal width of 15 nm, which shows a Kelvin signal with a FWHM of 23 nm [42]. This demonstrates the very good electrical spatial resolution of our KPFM setup.

The voltage sensitivity of our system was determined using a gold evaporated GaAs substrate. For that experiment, an AC voltage with an amplitude of 0.5 V and a frequency of 27 kHz was used for modulation. The voltage V_{ext} applied to the sample was stepwise changed from 0 V to 150 mV with decreasing step sizes. The corresponding Kelvin voltages (in each case 20 line scans were performed) are shown in Fig. 8.3a as a 3D plot. From this plot it can be seen that with decreasing applied voltage the Kelvin signal decreases too. The 20 line scans of the Kelvin voltage for each applied voltage were averaged and plotted versus the applied voltage as shown in Fig. 8.3b. The slope Δ of the fitted linear curve is nearly one. Differences in the applied voltage as small as 5 mV can be resolved [45] representing the voltage sensitivity of the system.

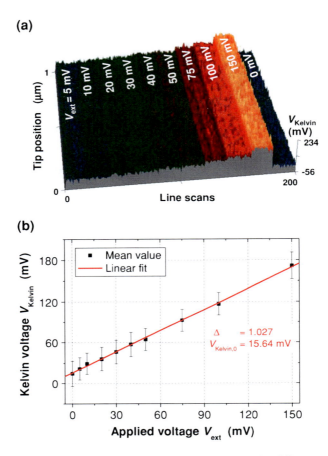

Fig. 8.3 Estimation of the voltage sensitivity. **a** 3D image of V_{Kelvin} for different applied voltages. **b** Measured V_{Kelvin} versus applied voltages V_{ext} [45]

8.3 Material and Doping Contrast in Single GaAs Based Nanowires

8.3.1 Material Transitions in Single GaAs Based Nanowires

Nanowires consisting of a nominally undoped GaAs/GaP heterostructure were grown with Metal–organic vapor phase epitaxy [46]. The growth was performed under vapor–liquid–solid (VLS) conditions on a (111) GaAs substrate. As seeds for the VLS growth, monodisperse Au particles from the gas phase (Ted Pella Inc.) were deposited on the substrate. While trimethylgallium (TMGa) was used as group-III precursor, tertiarybutylarsine (TBAs), and tertiarybutylphosphine (TBP) were utilized as group V precursors for the growth of the GaAs and the GaP nanowire

Fig. 8.4 Schematic illustration and scanning electron microscope image of a GaAs/GaP heterostructure nanowire. Reprinted with permission from [24]. Copyright 2009, IOP Publishing

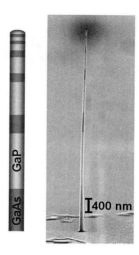

sections, respectively. The substrate was annealed at 600 °C for 10 min before the nanowires were grown at a constant temperature of about 480 °C [24].

A schematic diagram along with a scanning electron microscope (SEM) image of the heterostructure nanowire is depicted in Fig. 8.4. From the bottom to the top of the nanowire, the length of the GaAs and the GaP sections was reduced stepwise from 2000 to 30 nm by reduction of the growth time. As shown in the schematic diagram in Fig. 8.4, ten heterojunctions have been realized along the axial direction of the nanowire. To obtain sharper material transitions, at each interface all precursors have been switched off for a 1 min growth interruption. Due to the material contrast, the different sections of the heterostructure nanowire can be distinguished clearly in the SEM image. The bright areas correspond to GaP, while the darker ones represent the GaAs layers.

The atomic force microscope topography measurement of the heterostructure nanowire is displayed in Fig. 8.5a. The upper image shows the 2D data plot of the topography, while in the lower graph an averaged line measurement extracted from the 2D image is depicted.

Over the total length of the nanowire (∼6.5 µm), the diameter of the nanowire changes from about 40 nm close to the Au seed nanoparticle to approximately 100 nm at the bottom of the nanowire. As visualized in Fig. 8.5a, the increase in nanowire diameter shows a steplike behavior along the axial direction of the nanowire. A comparison with the SEM measurement shown in Fig. 8.4 reveals that the nanowire diameter increases only during GaAs growth (blue solid line), while it remains constant for the GaP regions (red dashed lines). In our measurements, all layers can be resolved except the thinnest GaAs part, because of the influence of the Au nanoparticle close to it [24].

From the topography measurements, we are able to extract the vertical and the lateral growth rates r_V and r_L, respectively (see Fig. 8.5b). This is done by determining

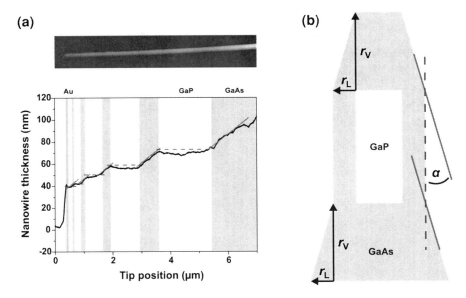

Fig. 8.5 Lateral versus radial growth. **a** 2D measurement (*top*) and averaged line scan (*bottom*) of the topography measurement of the heterostructure nanowire with the Au particle on the left side. The color scale represents 116 nm for the 2D plot. **b** Schematic drawing of the core–shell structure. Reprinted with permission from [24]. Copyright 2009, IOP Publishing

the angle α, which represents the growth ratio r_V/r_L. The value we obtain for the GaAs parts is about 1°, resulting in a growth ratio of $r_V/r_L \approx 55$, which is in good agreement with literature data [19]. This ratio is constant over the whole nanowire length as indicated by the slope of the blue solid lines in Fig. 8.5a. In contrast, the lateral growth can be neglected for the GaP growth. From our data it can be seen, that the radial GaAs overgrowth is not limited to the GaAs regions. GaP regions of the nanowire are also covered by GaAs, resulting in a core–shell structure as schematically shown in Fig. 8.5b.

The corresponding Kelvin voltage, measured simultaneously with the topography (see Fig. 8.5a), is shown in Fig. 8.6. The Kelvin voltage was measured with a modulation voltage of 4 V to enhance the sensitivity, and therefore the contrast of the measurement [24]. The dark spot in the 2D plot (top of Fig. 8.6) corresponds to a low Kelvin voltage and indicates the Au nanoparticle at the top of the nanowire. Additionally, a distinct change of the Kelvin voltage along the nanowire axis can be seen in the vicinity of the nanoparticle, while the Kelvin voltage remains nearly constant at larger distances from the gold nanoparticle.

In the bottom of Fig. 8.6, a line scan extracted from the Kelvin voltage data is plotted. The positions of the GaAs and GaP layers have been taken from the topography measurement that has been done simultaneously on the same nanowire. According to Eq. (8.1), the Kelvin voltage depends on the work function difference between the tip and the sample. The positive value for the Kelvin voltage obtained

Fig. 8.6 2D KPFM measurement (*top*) and extracted line scan averaged (*bottom*) across the heterostructure nanowire as shown in Fig. 8.5. The color scale represents a Kelvin voltage scale of 130 mV. The inset shows an enlarged data plot in the vicinity of the Au nanoparticle. Reprinted with permission from [24]. Copyright 2009, IOP Publishing

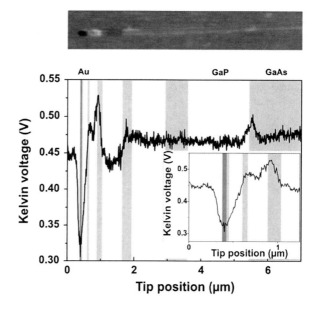

here agrees with the fact, that the work function of the used tip (Pt-coated) is expected to be higher compared to all materials studied (Au, GaAs and GaP) [47, 48]. The higher value of the Kelvin voltage in the semiconductor area of the nanowire indicates a lower work function as compared to the seed Au nanoparticle. This is again in good agreement with literature, where a work function of about 5.1–5.26 eV is expected for Au [47], which is higher than 4.78 and 4.93 eV for GaAs and GaP, respectively [48].

Like in the topography measurements (see Fig. 8.5a), the thinnest layers cannot be resolved, most probably due to long range electrostatic forces from the nearby Au nanoparticle as discussed in Sect. 8.2. However, at some distance from the particle a material contrast can be seen and a maximum work function difference of about 75 meV is measured between GaAs and GaP. This is half of the value expected from the literature data as discussed above. The measured work function is influenced by band bending, defect states or contaminations at the surface, as well as from the averaging effect of nearby layers due to the finite spatial resolution. Moreover, a possible background doping of the semiconductor sections cannot be excluded. The radial overgrowth of GaAs, as discussed above, reduces the contrast in the KPFM measurement more and more as the GaAs shell is getting thicker toward the bottom of the nanowire. This results in a nearly constant Kelvin voltage when the shell thickness reaches approximately 5 nm.

8.3.2 KPFM on Single p-Doped GaAs Nanowires

As discussed before, one of the major advantages of KPFM is that it is sensitive toward doping variations in semiconductor materials. In order to study the doping efficiency during the growth process, two differently doped GaAs nanowires have been analyzed with KPFM. While the first sample was p-doped with zinc in situ during growth, the doping of the second one was realized by zinc ion implantation subsequently after growth.

8.3.2.1 Axial Doping Gradient in a Single p-Doped GaAs Nanowire

GaAs nanowires have been grown using polydisperse Au nanoparticles as seeds. The growth was similar to what is described in Sect. 8.3.1. In contrast, however, the nanowires have been grown at a lower temperature of about 400 °C to prevent additional radial overgrowth. For p-type doping, diethylzinc (DEZn) was introduced in addition to TMGa and TBAs during growth. A detailed description of the growth parameters and corresponding SEM images of the grown nanowires can be found elsewhere [46, 49]. Macroscopic current/voltage measurements at different positions along the nanowire indicated a variation of the wire resistance along the growth direction, which was attributed to a local change of the doping concentration [49].

In order to study the axial change of the doping concentration in more detail, the nanowires have been removed from the substrate and transferred to an insulating substrate carrier. KPFM measurements have been performed on unbiased nanowires in order to get access to local variations of the work function and thus the doping level.

Figure 8.7 shows the result of the KPFM measurement on a typical nanowire. The measurement reveals a graded Kelvin voltage along the whole nanowire in axial direction (blue curve), while no contrast change is visible either in the topography of the nanowire (inset in the top of Fig. 8.7) or in the Kelvin voltage on the substrate close to the nanowire (black curve in Fig. 8.7). From the macroscopic current/voltage measurements performed on similar structures, the hole concentration is expected to increase toward the nanowire tip (left side in Fig. 8.7). This should result in an increase of the local work function and thus in a decrease of the Kelvin voltage in this direction. This is exactly what is found in the KPFM experiments, confirming the increasing p-doping level toward the nanowire tip.

The change of the Kelvin voltage is about 80 mV along the $\sim 9\,\mu$m long part of the nanowire. Note that a quantitative estimation of the doping level and the change in doping concentration along the nanowire cannot be extracted from the data due to band bending effects related to surface states and surface charges [41, 51].

Fig. 8.7 Kelvin voltage versus tip position measured on the nanowire (*blue*) and on the substrate (*black*) as a reference. The inset in the top depicts the corresponding 2D topography measurement [50]

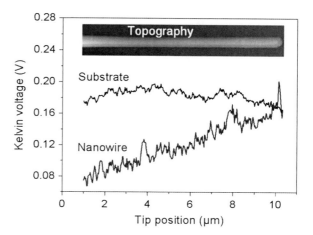

8.3.2.2 Effective Carrier Concentration in a Single GaAs Nanowire Doped by Zinc Ion Implantation

Nominally undoped GaAs nanowires have been grown on a (100) GaAs substrate using dispersed Au particles with nominal diameters of about 150 nm as seeds. The preferred growth axis of the GaAs nanowires is in the (111) direction [46], resulting in tilted nanowires with an angle of about 35° with respect to the substrate surface. As shown schematically in Fig. 8.8a, the as-grown nanowires were subsequently doped with Zn by ion implantation to reach a nominal doping density of $3 \times 10^{19}\,\text{cm}^{-3}$ [52].

The nanowires have been transferred to an insulating substrate for further electrical characterization. Electron beam lithography has been used to realize standard p-type contacts (Pt/Ti/Pt/Au) resulting in contacted nanowires as shown in the inset of Fig. 8.8b. The linear macroscopic current/voltage characteristic of the measured nanowire (Fig. 8.8b) reveals the Ohmic character of the contacts and an overall macroscopic resistance of about 130 kΩ. The latter is determined by both, the contact resistances and the intrinsic resistance of the nanowire in between.

In order to extract the intrinsic resistance of the nanowire and subsequently estimate the carrier concentration in the nanowire, KPFM has been performed to gain knowledge about the local voltage drop along the nanowire. The Kelvin voltage along the nanowire has been measured with and without an externally applied bias. The local voltage drop was deduced quantitatively by subtracting the KPFM signal obtained in the unbiased case from the KPFM signal measured when the nanowire is biased [29, 31].

The Kelvin voltage measurements are depicted in Fig. 8.9a for the unbiased (open circles) and the biased (full circles) case. Without an external bias, the KPFM signal shows a relatively homogeneous Kelvin voltage along the whole nanowire. In contrast, an applied bias of 1 V at the left contact results in a pronounced change of the local Kelvin voltage. The measured Kelvin voltage is determined by both, the

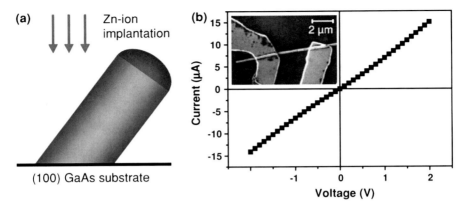

Fig. 8.8 Schematic illustration and I–V characteristic of the ion-implanted nanowire **a** Schematic drawing of a tilted nanowire. **b** Macroscopic current/voltage characteristic of the measured nanowire. The inset shows a SEM image of a typical contacted nanowire. Reprinted with permission from [53]. Copyright 2009, American Institute of Physics

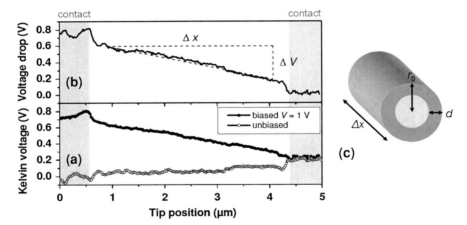

Fig. 8.9 Determination of the local voltage drop along the nanowire. **a** Kelvin voltage measured for the unbiased (*open circles*) and for the biased (*full circles*) nanowire. **b** Local voltage drop along the nanowire and the contacts calculated from the Kelvin voltage difference of the biased and the unbiased device. The contacts are drawn for clarity. **c** Schematic drawing of the assumed model for estimating the carrier concentration. Reprinted with permission from [53]. Copyright 2009, American Institute of Physics

contact potential difference $V_{cpd}(x)$ and the local voltage drop $V_{drop}(x)$. Subtracting the zero-bias KPFM signal reveals the local voltage drop $V_{drop}(x)$ along the nanowire (Fig. 8.9b), as outlined above. While the measured value of 0 V on the right contact is exactly the expected one, the voltage on the left contact is only 0.8 V instead of the applied 1 V. This difference can be explained most probably by the averaging effect of the long ranging electrostatic field [43]. However, this measurement accuracy is sufficient enough for an estimation of the carrier concentration.

In between the contacts, the voltage drops linearly with a change of $\Delta V \sim 0.4$ V over a length of $\Delta x \sim 3.2 \mu$m. The voltage drop at the left and the right contact is 0.26 and 0.19 V, respectively. With the measured current of about 7μA this results in an internal resistance of about 57 kΩ for the nanowire and contact resistances of 38 and 27 kΩ, respectively, can be extracted. The corresponding total resistance of 122 kΩ is very close to the macroscopic measured value of 130 kΩ.

A cylindrical symmetry shown schematically in Fig. 8.9c is assumed for the calculation of the carrier concentration. The resistance of the nanowire in between the contacts is given by $R = \Delta x/[\sigma \pi (r_0-d)^2]$ with the conductivity $\sigma = qp\mu_h$, a depletion width $d = (2\varepsilon_0\varepsilon_r V_S/qN_A)^{1/2}$ at the nanowire surface, and an average wire radius r_0 of about 120 nm. V_S is the surface potential and the hole mobility can be calculated by $\mu_h = \mu_0/[1 + (N_A/10^{18}$ cm$^{-3})^{1/2}]$. With typical values for p-GaAs of $\mu_0 = 450$ cm^2/Vs, $V_S = 0.45$ V and $\varepsilon_r = 13.1$ taken from literature [52], an effective carrier concentration of $p \sim 6 \times 10^{17}$ cm^{-3} can be estimated from our measurements [53].

Like mentioned before, the particles used as seeds for the nanowire growth had a diameter of about 150 nm. Therefore, for the implantation process ion energies were adjusted to provide a homogeneous doping profile across the nominal diameter of the nanowires. Our topography measurements on the investigated structure, however, showed a tapering of the nanowire, resulting in an average diameter of about 240 nm. This causes a lower and a more inhomogeneous doping concentration than intended by the implantation parameters used here. Together with the fact that not all acceptors are ionized at room temperature [52] the value extracted from our measurements is in reasonably good agreement with what could be expected for the effective carrier concentration.

8.3.3 Localization of Doping Transitions in Single p-Doped GaAs Nanowires

Because KPFM is very sensitive toward doping concentrations and the type of doping, this technique is able to localize doping transitions, e.g., transitions from a p-doped area to an intrinsic one, of individual nanowires. GaAs nanowires with a nominal change in doping type were grown similar as described before. A 2.5 nm thin Au film was evaporated on a (111) GaAs substrate. To form seeds for the VLS growth, the substrate was annealed at 600 °C for 5–10 min, resulting in polydisperse Au droplets with diameters ranging from 50 to 350 nm. Two different sources have been added to the gas phase to realize p-type doping. For sample A (Fig. 8.10a) tetrabromide (CBr$_4$) has been used, while DEZn has been included for the p-doping of sample B (Fig. 8.10b).

Fig. 8.10 SEM images and schematic drawings of (**a**) a GaAs nanowire with a doping transition from a C-doped to an undoped part (sample A) and (**b**) a GaAs nanowire containing a transition from an undoped stump to a Zn-doped nanowire part at the top (sample B). Reprinted with permission from [24]. Copyright 2009, IOP Publishing

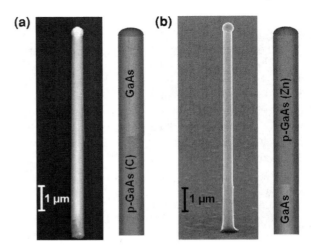

Figure 8.10a shows a SEM micrograph and a schematic drawing of a nanowire from sample A grown at a constant temperature of 480°C. First, the p-doped part was grown by adding CBr_4 to the gas phase. As C is found not to dissolve in Au, the doping is attributed to a side wall incorporation rather than trough doping via the VLS mechanism [24]. The doping source was switched off after about half of the growth time of about 6 min. Hence, the top parts of the A-type nanowires are nominally undoped. In sample B (Fig. 8.10b), DEZn was used as the precursor for p-type doping. A growth temperature of about 400 °C was applied for the p-doped part of the nanowires preventing radial overgrowth at the nanowire sidewalls. Before the growth of the p-doped layer, a nominally undoped stump was grown at a temperature of about 450 °C for better crystal quality [46, 49]. Like described before, the nanowires have been transferred to another substrate for performing KPFM measurements.

The results of the simultaneously measured topography and KPFM signal of sample A are depicted in Fig. 8.11a–c. The 2D image of the topography (Fig. 8.11a) shows no specific features and, in particular, no structural variations in the center of the nanowire, i.e., in the vicinity of the doping transition. In contrast, a pronounced change in contrast close to the center of the nanowire axis can be recognized in the KPFM signal (Fig. 8.11b). The lower Kelvin voltage at the right side of the image (bottom of the nanowire) is consistent with an increasing work function as expected for the p-doped part.

This local change in the Kelvin voltage is analyzed in more detail by extracting a line scan from the 2D KPFM image and plotting the Kelvin voltage versus tip position in Fig. 8.11c. Starting from the bottom of the nanowire (right side) the Kelvin voltage shows a strong increase of about 80 mV in the center of the structure where the p-doping has been switched off. To localize the exact position of the doping transition more accurately [27], the derivative of the Kelvin voltage has been extracted from the measured data and plotted in the inset in Fig. 8.11c. The doping transition can be localized at a tip position of approximately 5.2 μm and a full width half maximum

8 Material and Doping Contrast in III/V Nanowires

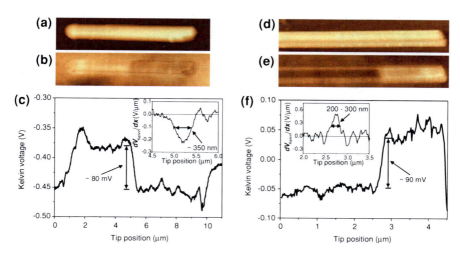

Fig. 8.11 Investigation of doping transitions. 2D measurements of the topography **a** and the Kelvin voltage **b** for sample A with the color scale representing 370 nm and 260 mV, respectively. **c** Line scan of the Kelvin voltage extracted from the 2D image. The inset shows the derivative of the Kelvin voltage in the vicinity of the doping transition. **d** 2D images of the topography (color scale: 80 nm) and **e** the Kelvin voltage (color scale: 210 mV) of sample B. **f** Kelvin voltage versus tip position with the corresponding derivative in the inset. Reprinted with permission from [24]. Copyright 2009, IOP Publishing

(FWHM) of about 350 nm is obtained. Ideally, this width could be used to estimate the depletion width and thus the achieved doping level. Note however, that surface states will result in a deviation of the depletion width in the bulk and at the surface [54]. Therefore, a quantitative discussion of the depletion width and a deduction of the average doping level in the nanowire are not conducted.

Figure 8.11d–f show analog data for sample B. Again, no contrast change can be seen in the 2D image of the topography (Fig. 8.11d), whereas a clear contrast between the two differently doped areas is visible in the Kelvin voltage (Fig. 8.11e). In contrast to sample A, the nanowire B consists of a nominally undoped area at the right side and a p-doped area at the left side, which is the top of the nanowire. Hence, the expected contrast change in the Kelvin voltage image should be exactly opposite to the one measured for sample A, in agreement with the experimental data. The extracted line scan in Fig. 8.11f shows a change in the Kelvin voltage of about 90 mV at the doping transition. From the derivative depicted in the inset, the position of the doping transition can be localized at a tip position of approximately 2.75 μm and a transition width between 200 and 300 nm is estimated.

From the schematic energy band model shown in Fig. 8.12 it can be seen, that the work function difference $\Phi_p - \Phi_i$ and therefore the change in the Kelvin voltage between a nominally p-doped and an ideally intrinsic semiconductor is expected to be in the order of half the bandgap, i.e., about 0.7 V for GaAs. This value is much larger than the values of 80 and 90 mV measured in our KPFM experiments discussed

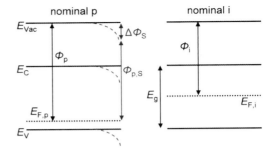

Fig. 8.12 Schematic models of the energy bands for a p-doped (*left*) and an intrinsic (*right*) semiconductor. The surface band bending $\Delta\Phi_S$ results in a depletion zone (*red dotted lines*) and a decreased surface work function $\Phi_{p,S}$

above. In similar studies on bulk GaAs p–n junctions with a doping concentration of $5 \times 10^{18}\,\text{cm}^{-3}$ for both, p and n doping, voltage differences between 25 and 40 mV have been measured at the heterojunction with KPFM [28]. In order to explain this discrepancy, surface band bending effects $\Delta\Phi_S$ caused by surface states and surface charges [28, 51] have to be considered.

As can be seen on the left side of Fig. 8.12 for p-doped GaAs, the surface band bending (red dashed lines) results in a decreased work function at the surface, and therefore reduces the Kelvin voltage difference between the p-doped and the intrinsic area measured at the surface. In addition, the depletion width suffers from the finite lateral resolution of KPFM. Hence, it is not possible to derive quantitative data for the concentration change at the doping transition. Nevertheless, the position of the doping transition can be localized with a precision of ∼50 nm. In addition, the reduced value of the depletion width in sample B indicates a better incorporation of the Zn as compared to the C into the nanowire for p-type doping.

8.4 GaAs p–n Junction Nanowire Devices

To realize GaAs nanowires with a p–n junction [46, 55], polydisperse Au seeds were prepared as described before (see Sect. 8.3.3). The precursors for the wire growth were TMGa and TBAs, while diethylzinc (DEZn) and tetraethyltin (TESn) were used for p- and n-type doping, respectively. The nanowires were transferred onto an insulating substrate and contacts were defined via electron beam lithography. Contacts for the p-doped nanowire part were realized with a Ti/Pt/Ti/Au metallization, while Pd/Ge/Au was used to form an ohmic contact to the n-doped area of the nanowire.

The schematic structure of the nanowire p–n junction is shown on the left side of Fig. 8.13. First, a nominally undoped GaAs nanowire of a few micrometer length was grown at a higher temperature (450 °C) to provide better crystal quality at the beginning of the growth [46, 55]. Afterwards, the temperature was reduced to 400 °C and the n- and p-doped parts of the nanowires were grown. The nominal doping concentrations were extracted from macroscopic current/voltage characteristic measurements performed on purely n- and purely p-doped reference nanowires, respectively.

8 Material and Doping Contrast in III/V Nanowires

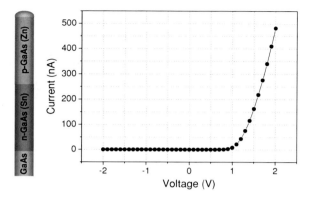

Fig. 8.13 Schematic illustration of the nominal structure of the p–n nanowire (*left*). Macroscopic current/voltage characteristic showing a diode-like behavior for the measured nanowire (*right*)

For these nanowires, a maximum doping concentration of about $10^{18}\,\text{cm}^{-3}$ has been achieved for the purely n-doped, while a nominal value of about $1.6 \times 10^{19}\,\text{cm}^{-3}$ is obtained for purely p-doped nanowires [46, 49, 56]. These values are assumed as the nominal values for the n- and p-doped part of the investigated p–n nanowire discussed here, because the growth parameters were similar in these experiments.

Figure 8.13 shows the macroscopically measured current/voltage characteristics of the investigated p–n nanowire. A clear diode-like behavior can be seen for the applied voltages between -2 and $2\,\text{V}$. The built-in voltage extracted from the curve ($\sim 1.4\,\text{V}$) is in good agreement with the expected value for a GaAs p–n heterojunction device.

To further analyze the diode-like behavior, KPFM measurements have been performed on the p–n nanowire. In Fig. 8.14, the 2D images of the topography (a) and the simultaneously measured Kelvin voltages for various applied biases (b–d) are depicted. In all images, the n- and p-contacts are on the left and on the right side, respectively.

The topography (Fig. 8.14a) shows the nanowire with a constant diameter of about 100 nm in between the contacts. While no structural features are measured in the topography, a relatively sharp transition in the contrast can be seen in the Kelvin voltage image along axial direction of the nanowire in Fig. 8.14b. This already indicates that the diode-like behavior originates from a p–n junction inside the nanowire and not from possible Schottky characteristics of the contacts.

Starting from the n-doped part of the nanowire on the left side, the Kelvin voltage decreases abruptly at the junction toward the right side, which is consistent with an increasing work function for the p-doped wire part. The contrast is enhanced even more when applying a bias in reverse direction, e.g., $-1\,\text{V}$ to the p-contact (Fig. 8.14c). While the Kelvin voltage remains constant for the grounded n-side on the left, the Kelvin voltage decreases on the p-doped nanowire part due to the additional negative potential. If the nanowire is biased in forward direction with $+1\,\text{V}$ at the p-contact (Fig. 8.14d), the Kelvin voltage increases with respect to the grounded side of the nanowire. In both cases the voltage apparently drops mainly at

Fig. 8.14 2D images of the **a** topography (color scale: ~400 nm), **b** the Kelvin voltage for the unbiased nanowire and the nanowire biased in **c** reverse and in **d** forward direction. The color scale is the same for all KPFM images and represents approximately 2 V. As depicted in **a**, the n-contact is on the left side, while the p-contact is on the right side. The scan size is 17.1 × 17.1 μm² for all images

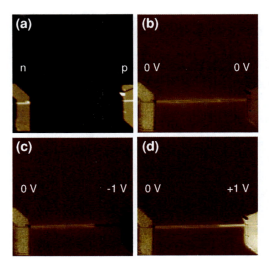

the p–n junction inside the nanowire proving the good ohmic quality of the metallic contacts.

To analyze the p–n junction in more detail, a KPFM measurement in the vicinity of the doping transition (inset in Fig. 8.15a) has been performed with −1 V applied to the p-contact. An averaged line scan of the Kelvin voltage has been extracted and plotted versus the tip position in Fig. 8.15a. The Kelvin voltage decreases by approximately 0.8 V from the left to the right side. Note that this value results from the difference in the local work function superimposed by the potential due to the externally applied voltage. To estimate the depletion length of the p–n junction, the derivative of the Kelvin voltage has been determined and depicted versus tip position in Fig. 8.15b. As the derivative of the Kelvin voltage represents the electric field in the p–n junction [54], this procedure allows an estimate of the depletion zone of about 410 nm under the reverse bias of −1 V.

The Kelvin voltage has been measured for varying reverse biases ranging from −2 to 0 V and the corresponding depletion lengths have been estimated like described above. The result is depicted in Fig. 8.16 where the depletion length is plotted versus the applied bias. The voltage was again applied at the p-contact for all measurements. The estimated depletion length increases with increasing reverse bias from ∼300 nm for the unbiased wire to approximately 530 nm for a bias of −2 V. The observed behavior is consistent with the theoretical model of a p–n junction, where the depletion length x_d is expected to increase when the p–n junction is biased in reverse direction [57].

The depletion length as a function of the applied voltage and of the acceptor concentration has been calculated and depicted in Fig. 8.16 (solid lines). For the calculation an abrupt p–n junction has been assumed. Hence, the depletion length is given by $x_d = [2\varepsilon_0\varepsilon_r(N_A+N_D)(V_{bi}-V_{ext})/(qN_AN_D)]^{1/2}$ [57], where ε_0 and ε_r are

8 Material and Doping Contrast in III/V Nanowires

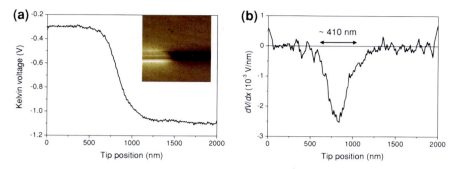

Fig. 8.15 Estimation of the depletion zone. **a** Line scan of the Kelvin voltage plotted versus tip position in the vicinity of the p–n junction for a bias of -1 V at the p- contact (*right side*). The inset shows the corresponding 2D image of the Kelvin voltage with a size of $2 \times 2\,\mu m^2$ **b** The derivative of the Kelvin voltage for the same position at -1 V

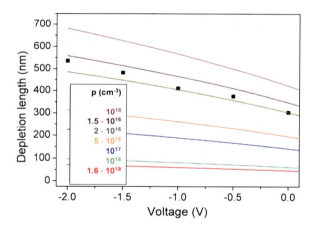

Fig. 8.16 Measured depletion length (*black squares*) of the p–n junction versus applied bias in reverse direction. For the calculation of the depletion length (*solid lines*) a nominal concentration for n-doping of $10^{18}\,cm^{-3}$ was used and an abrupt p–n junction was assumed. The p-doping concentration was varied as shown in the legend

the permittivity of the vacuum and the material, N_A and N_D are the acceptor and donor concentrations, and V_{bi} and V_{ext} are the built-in and the externally applied voltage, respectively. As described before, the n-doped part of the nanowire has been grown previous to the p-doped part. Therefore, a fixed nominal n-doping concentration of $10^{18}\,cm^{-3}$ has been assumed for the calculations, while the concentration for the p-doping was varied from the nominal value of $1.6 \times 10^{19}\,cm^{-3}$ down to $10^{16}\,cm^{-3}$. It can be seen in Fig. 8.16, that a relatively good agreement is achieved for a p-doping level between 1.5 and $2 \times 10^{16}\,cm^{-3}$, which is three orders of magnitude smaller than the nominal one.

The main reason for the low value extracted for the p-doping concentration is attributed to a deviation from an abrupt p–n junction. As discussed in Sect. 8.3.2.1 and elsewhere [46, 49, 50], the p-doping level along the nanowire is not constant. In fact, the incorporation of the Zn into the nanowire is delayed resulting in an increasing doping concentration with growth time. Moreover, when the doping is switched from n- to p-type, some amount of Sn is still present due to the memory effect of the Au-seed during VLS growth. This is believed to cause a compensated region at the junction [46, 58]. Thus, a continuous increase of the effective p-doping along the nanowire is expected rather than an abrupt junction. Note that in addition surface states and surface charges cause a surface band bending resulting in an increased depletion region at the GaAs semiconductor surface as compared to the nanowire center [28, 51]. Moreover, the finite spatial resolution of KPFM results in a broadening of the measured depletion length [44].

8.5 Conclusion

The potential of Kelvin probe force microscopy for probing material and doping contrast in III/V nanowires with a $\sim$50 nm spatial resolution is demonstrated. The simultaneous measurement of topography and contact potential difference reveals new insights into both, core–shell nanowires, and nanowires with axial doping transitions.

In GaP/GaAs core–shell nanowires, the experiments yield a ratio between vertical and lateral growth rate of $\sim$55 for GaAs while the lateral growth is found to be negligible for GaP. Applying our technique to p-doped GaAs nanowires, we are able to monitor local variations of the doping level and to extract the local potential drop, if the nanowire is externally biased. This finally allows an estimation of the effective doping concentration in the nanowire.

Possibly, the largest benefit of the Kelvin probe force microscopy is obtained if p–i or p–n nanowire junctions are studied. The position of the junction is localized with an accuracy of 50 nm or less and the width of the depletion zone is estimated. The experiments have been performed under external bias to directly visualize the bias dependence of the depletion width.

Acknowledgments We gratefully acknowledge the financial support by the German Research Foundation (DFG) through the collaborative research centre SFB 445. The authors also thank F.-J. Tegude, W. Prost, I. Regolin, C. Gutsche, and A. Lysov from the Institute of Solid-State Electronics of the University Duisburg-Essen, for the preparation of the nanowires and for helpful discussions.

References

1. L.-E. Wernersson, C. Thelander, E. Lind, L. Samuelson, Proc. IEEE **98**, 2047 (2010)
2. W. Lu, C.M. Lieber, J. Phys. D Appl. Phys. **39**, R387 (2006)
3. X. Zianni, Appl. Phys. Lett. **97**, 233106 (2010)
4. Q.-T. Do, K. Blekker, I. Regolin, W. Prost, F.J. Tegude, IEEE Electr. Dev. Lett. **28**, 682 (2007)
5. X. Jiang, Q. Xiong, S. Nam, F. Qian, Y. Li, C.M. Lieber, Nano Lett. **7**, 3214 (2007)
6. S.L. Diedenhofen, G. Vecchi, R.E. Algra, A. Hartsuiker, O.L. Muskens, G. Immink, E.P.A.M. Bakkers, W.L. Vos, J.G. Rivas, Adv. Mat. **21**, 973 (2009)

8 Material and Doping Contrast in III/V Nanowires

7. E. Garnett, P. Yang, Nano Lett. **10**, 1082 (2010)
8. M.T. Borgström, J. Wallentin, M. Heurlin, S. Fält, P. Wickert, J. Leene, M.H. Magnusson, K. Deppert, L. Samuelson, IEEE J. Sel. Top. Quant. DOI: 10.1109/JSTQE.2010.2073681
9. J.A. Czaban, D.A. Thompson, R.R. LaPierre, Nano Lett. **9**, 148 (2009)
10. C. Colombo, M. Heiß, M. Grätzel, A. Fontcuberta i Morral, Appl. Phys. Lett. **94**, 173108 (2009)
11. A.I. Hochbaum, R. Chen, R.D. Delgado, W. Liang, E.C. Garnett, M. Najarian, A. Majumdar, P. Yang, Nature **451**, 163 (2008)
12. A.I. Boukai, Y. Bunimovich, J. Tahir-Kheli, J.-K. Yu, W.A. Goddard III, J.R. Heath, Nature **451**, 168 (2008)
13. R.S. Wagner, W.C. Ellis, Appl. Phys. Lett. **4**, 89 (1964)
14. K.A. Dick, P. Caroff, J. Bolinsson, M.E. Messing, J. Johansson, K. Deppert, L.R. Wallenberg, L. Samuelson, Semicond. Sci. Technol. **25**, 024009 (2010)
15. M.S. Gudiksen, L.J. Lauhon, J. Wang, D.C. Smith, C.M. Lieber, Nature **415**, 617 (2002)
16. I. Regolin, D. Sudfeld, S. Lüttjohann, V. Khorenko, W. Prost, J. Kästner, G. Dumpich, C. Meier, A. Lorke, F.J. Tegude, J. Cryst. Growth **298**, 607 (2007)
17. Y. Li, F. Qian, J. Xiang, C.M. Lieber, Mater. Today **9**, 18 (2006)
18. K.A. Dick, K. Deppert, L.S. Karlson, L.R. Wallenberg, L. Samuelson, W. Seifert, Adv. Funct. Mater. **15**, 1603 (2005)
19. M.A. Verheijen, G. Immink, T. de Smet, M.T. Borgström, E.P.A.M. Bakkers, J. Am. Chem. Soc. **128**, 1353 (2006)
20. D.E. Perea, E.R. Hemesath, E.J. Schwalbach, J.L. Lensch-Falk, P.W. Voorhees, L.J. Lauhon, Nat. Nanotechnol. **4**, 315 (2009)
21. X. Ou, P.D. Kanungo, R. Kögler, P. Werner, U. Gösele, W. Skorupa, X. Wang, Nano Lett. **10**, 171 (2010)
22. M. Nonnenmacher, M.P. O'Boyle, H.K. Wickramasinghe, Appl. Phys. Lett. **58**, 2921 (1991)
23. A. Lochtofen, W. Mertin, G. Bacher, L. Hoeppel, S. Bader, J. Off, B. Hahn, Appl. Phys. Lett. **93**, 022107 (2008)
24. S. Vinaji, A. Lochtofen, W. Mertin, I. Regolin, C. Gutsche, W. Prost, F.J. Tegude, G. Bacher, Nanotechnology **20**, 385702 (2009)
25. V. Palermo, M. Palma, P. Samori, Adv. Mater. **18**, 145 (2006)
26. T. Meoded, R. Shikler, N. Fried, Y. Rosenwaks, Appl. Phys. Lett. **75**, 2435 (1999)
27. A. Doukkali, S. Ledain, C. Guasch, J. Bonnet, Appl. Surf. Sci. **235**, 507 (2004)
28. T. Mizutani, T. Usunami, S. Kishimoto, K. Meazawa, Jpn. J. Appl. Phys. **38**, 4893 (1999)
29. G. Lévêque, P. Girard, E. Skouri, D. Yarekha, Appl. Surf. Sci. **157**, 251 (2000)
30. A.V. Ankudinov, V.P. Evtikhiev, E.Y. Kotelnikov, A.N. Titkov, R. Laiho, J. Appl. Phys. **93**, 432 (2003)
31. Kl.-D. Katzer, W. Mertin, G. Bacher, A. Jaeger, K. Streubel, Appl. Phys. Lett. **89**, 103522 (2006)
32. X. Cui, M. Freitag, R. Martel, L. Brus, P. Avouris, Nano Lett. **3**, 783 (2003)
33. B.R. Goldsmith, J.G. Coroneus, V.R. Khalap, A.A. Kane, G.A. Weiss, P.G. Collins, Science **315**, 77 (2007)
34. E. Koren, Y. Rosenwaks, J.E. Allen, E.R. Hemesath, L.J. Lauhon, Appl. Phys. Lett. **95**, 092105 (2009)
35. E. Koren, N. Berkovitch, Y. Rosenwaks, Nano Lett. **10**, 1163 (2010)
36. E. Koren, J.K. Hyun, U. Givan, E.R. Hemesath, L.J. Lauhon, Y. Rosenwaks, Nano Lett. **11**, 183 (2011)
37. Z. Fan, J.G. Lu, Appl. Phys. Lett. **86**, 032111 (2005)
38. G. Cheng, S. Wang, K. Cheng, X. Jiang, L. Wang, L. Li, Z. Du, G. Zou, Appl. Phys. Lett. **92**, 223116 (2008)
39. Y.-J. Doh, K.N. Maher, L. Ouyang, C.L. Yu, H. Park, J. Park, Nano Lett. **8**, 4552 (2008)
40. E.D. Minot, F. Kelkensberg, M. van Kouwen, J.A. van Dam, L.P. Kouwenhoven, V. Zwiller, M.T. Borgström, O. Wunnicke, M.A. Verheijen, E.P.A.M. Bakkers, Nano Lett. **7**, 367 (2007)
41. P.M. Bridger, Z.Z. Bandić, E.C. Piquette, T.C. McGill, Appl. Phys. Lett. **74**, 3522 (1999)

42. A. Lochthofen, Mikroskopische Strom- und Spannungsverteilung in GaN-Lichtemittern. Dissertation Universität Duisburg-Essen, Fakultät für Ingenieurwissenschaften, 2009
43. H.O. Jacobs, P. Leuchtmann, O.J. Homan, A. Stemmer, J. Appl. Phys. **84**, 1168 (1998)
44. U. Zerweck, C. Loppacher, T. Otto, S. Grafström, L.M. Eng, Phys. Rev. B **71**, 125424 (2005)
45. Kl.-D. Katzer, Rasterkraftmikroskopie zur elektrischen Charakterisierung von innovativen Bauelementen und Nano-Strukturen. Dissertation Universität Duisburg-Essen, Fakultät für Ingenieurwissenschaften, 2008
46. All nanowires investigated in this paper were prepared at the Institute of Solid-State Electronics of the University Duisburg-Essen. For details see also: C. Gutsche, I. Regolin, A. Lysov, K. Blekker, Q.-T. Do, W. Prost, F.-J. Tegude, III/V Nanowires for electronic and optoelectronic applications in this volume
47. C. Calandra, G. Chiarotti, U. Gradmann, K. Jacobi, F. Manghi, A.A. Maradudin, S.Y. Tong, R.F. Wallis, in *Landolt-Börnstein New Series Physics of Solid Surfaces: Electronic and Vibrational Properties III/24b* ed. by G. Chiarotti (Springer, Berlin, 1994), pp. 64–66
48. M. Levinshtein, S. Rumyantsev, M. Shur, *Handbook Series on Semiconductor Parameters 1* (World Scientific Publisher, Singapore, 1996), p. 77 and pp. 104–105
49. C. Gutsche, I. Regolin, K. Blekker, A. Lysov, W. Prost, F.J. Tegude, J. Appl. Phys. **105**, 024305 (2009)
50. I. Regolin, C. Gutsche, A. Lysov, W. Prost, M. Malek, S. Vinaji, W. Mertin, G. Bacher, M. Offer, A. Lorke, F.-J. Tegude, in Proceedings of the EW MOVPE XIII Ulm, pp. 111, 7–10 June 2009
51. S. Saraf, Y. Rosenwaks, Surf. Sci. **574**, L35 (2005)
52. D. Stichtenoth, K. Wegener, C. Gutsche, I. Regolin, F.J. Tegude, W. Prost, M. Seibt, C. Ronning, Appl. Phys. Lett. **92**, 163107 (2008)
53. S. Vinaji, A. Lochthofen, W. Mertin, G. Bacher, I. Regolin, K. Blekker, W. Prost, F.J. Tegude, AIP Conf. Proc. **1199**, 329 (2010)
54. R. Shikler, T. Meoded, N. Fried, B. Mishori, Y. Rosenwaks, J. Appl. Phys. **86**, 107 (1999)
55. I. Regolin, C. Gutsche, A. Lysov, K. Blekker, Z.-A. Li, M. Spasova, W. Prost, F.-J. Tegude, J. Cryst. Growth **315**, 143 (2011)
56. C. Gutsche, A. Lysov, I. Regolin, K. Blekker, W. Prost, F.-J. Tegude, Nanoscale Res. Lett. **6**, 65 (2011)
57. S.M. Sze, *Physics of Semiconductor Devices*, 2nd edn. (Wiley, New York, 1981), pp. 77–79
58. A. Lysov, M. Offer, C. Gutsche, I. Regolin, S. Topaloglu, M. Geller, W. Prost, F.-J. Tegude, Nanotechnology **22**, 085702 (2011)

Part III
Properties and Applications

Chapter 9
Optical Properties of Silicon Nanoparticles

Cedrik Meier and Axel Lorke

Abstract This chapter reviews recent results on optical spectroscopy on silicon nanoparticles. The quantum confinement effect causing a spectral shift of the photoluminescence together with an intensity enhancement is discussed. The small spatial dimensions lead not only to a change of the electronic states, but affect also the vibronic spectrum as is seen in results on first- and second-order Raman scattering. Using time-resolved spectroscopy, the excitonic fine structure of silicon nanoparticles is investigated and a crossover of bright and dark exciton states is found. The analysis of the recombination dynamics allows to determine the size-dependence of the oscillator strength, which is in the order of 10^{-5} and increases with decreasing particle size. Finally, we demonstrate an electroluminescence device based on silicon particles using impact ionization.

9.1 Introduction

Due to the fact that most physical, chemical, and mechanical properties of nanostructured materials are determined by their structure, especially by geometry and size, many applications have been developed based on fundamental research in this field. The reasons for the novel effects in such nanomaterials are manifold, ranging from surface effects, e.g., in nanoparticle based gas sensors, to quantum confinement effects in sufficiently small structures. Among the inorganic materials used

---

C. Meier (✉)
Physics Department and CeOPP, University of Paderborn, Warburger Str. 100,
33098 Paderborn, Germany
e-mail: cedrik.meier@uni-paderborn.de

A. Lorke
Faculty of Physics and CENIDE, University of Duisburg-Essen, Lotharstraße 1,
47057 Duisburg, Germany
e-mail: axel.lorke@uni-due.de

A. Lorke et al. (eds.), *Nanoparticles from the Gas Phase*, NanoScience and Technology,   209
DOI: 10.1007/978-3-642-28546-2_9, © Springer-Verlag Berlin Heidelberg 2012

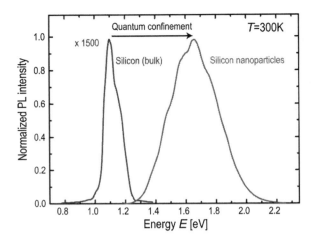

Fig. 9.1 Photoluminescence of bulk silicon (*red*) and silicon nanoparticles (*blue*) at room temperature. Note that the bulk photoluminescence signal is 1,500 times weaker than that of the nanoparticles

in electronic and/or optoelectronic applications, silicon plays a unique role. As the second most abundant element in earth's crust it is widely available and large single crystals can be pulled with extremely low defect densities. Together with the highly developed technology, these facts have caused the wide success of silicon in modern electronics and partly in optoelectronics. For the latter, however, the applications of bulk silicon are mostly limited to detector devices, as the electronic band structure of silicon exhibits an indirect bandgap, requiring the involvement of phonons for interband transitions due to the momentum conservation rule.

The report of optical emission from porous silicon in 1990, however, made silicon attractive also for optoelectronic applications as an emitter material [1]. Two years earlier, Furukawa and Miyasato had already studied the optical absorption of silicon nanocrystals and shown that with decreasing particle diameter an increase of the optical bandgap was observed, which they attributed to effects of quantum confinement [2]. Since then, many groups have contributed to the study of optoelectronic properties of silicon nanocrystalline structures. Figure 9.1 shows the photoluminescence of bulk silicon in comparison to the photoluminescence of silicon nanocrystals from the gas phase. In the beginning phase of the rapidly developing research activities, especially porous silicon fabricated by electrochemical etching was studied; later, other fabrication methods for such nanomaterials have been developed, such as ion implantation in SiO_2 [3] films and subsequent thermal annealing, nanocrystal growth from SiO_2/SiO superlattices [4] as well as the fabrication of silicon nanoparticles from the gas phase [5, 6]. While these processes differ in the degree of control, the obtained nanostructured are similar with respect to their optical properties.

In this article, we will focus mostly on results obtained using gas phase synthesis [5, 6]. The benefit of this method over the other processes mentioned above is that the resulting particles can—after proper chemical surface treatment—be dispersed in solution and easily processed using, e.g., printing or roll-to-roll methods. Moreover,

9 Optical Properties of Silicon Nanoparticles

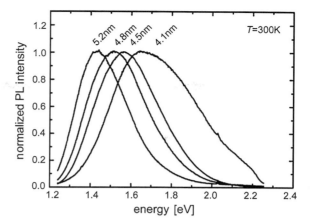

Fig. 9.2 Size-dependence of the room temperature photoluminescence of silicon nanoparticles

the temperatures reached during the nucleation and growth process are very high, leading to a high degree of crystallinity.

The photoluminescence obtained from gas-phase synthesized silicon nanoparticles exhibits significant size dependence, as shown in Fig. 9.2. With a change of mean particle diameter from $d = 5.2$ nm to $d = 4.1$ nm, the photoluminescence peak shifts by about $\Delta E = 300$ meV, about one-third of the bandgap energy of bulk silicon. However, the origin of the observed size dependence of the silicon nanoparticle photoluminescence has been under debate for quite a while. As mentioned above, already in 1988 the origin of the size dependent optical bandgap was attributed to quantum confinement. Furukawa and Miyasato [2] suggested a 3D confinement model to explain their experimental results. However, many authors have argued quantum confinement to not be the only effect responsible for the observed behavior. Wolkin et al. have studied the influence of the surface passivation of silicon nanoparticles on the optical bandgap and the emission properties. They found that for particles above $d = 3$ nm, quantum confinement is the dominant effect for the increasing bandgap with decreasing particle size. For smaller particles, the recombination mechanisms are different for oxygen or hydrogen terminated particle surfaces. The observation of a large Stokes shift is indicative of electron or exciton trapping near the surface, leading to size-independent recombination [7]. In 2008, Godefroo et al. [8] reported on photoluminescence experiments in high magnetic fields up to $B = 50$ T, in which they could show that properly terminated silicon nanoparticles show a clear diamagnetic shift as expected for quantum confinement dominated recombination, while untreated samples were governed by localization effects.

It is worth noting that the nanoscale dimensions do not only affect the optical bandgap. Other effects observed for silicon nanoparticles include phonon confinement effects, leading to alteration of the vibrational states as observed in Raman spectroscopy [9, 10]. Moreover, the nonlinear optical properties can change when going from bulk silicon to nanoparticles. Bulk silicon, which crystallizes in the centrosymmetric diamond structure, has a vanishing $\chi^{(2)}$ optical susceptibility. For silicon

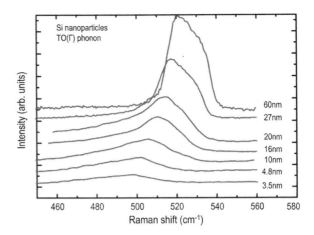

Fig. 9.3 Raman spectra from silicon nanoparticles with diameters down to 3.5 nm

nanoparticles, however, second harmonic generation could be observed, indicative of a break of the inversion symmetry of the lattice for very small particles ($d \sim 1$ nm) [11].

For device applications, the recombination dynamics are of greatest importance. In this context, questions of interest are: What is the internal quantum efficiency for interband transition? How are the radiative and nonradiative decay times? What is the oscillator strength for the involved transitions? In the first part of this chapter, we will discuss the recombination dynamics of nanocrystalline silicon. In the second part, we will demonstrate carrier injection across the oxide interface leading to electroluminescence in a silicon nanoparticle-based light emitting device.

9.2 Vibrational Properties

Light scattering methods such as Raman or Brillouin scattering are valuable to gain insights in the interplay of the impact of nanoscale dimensions on the lattice properties, which, in turn, affect the electronic structure. Especially for the case of silicon, Raman scattering allows to determine particle sizes as well as to distinguish between crystalline and amorphous material. Therefore, Raman spectra have been recorded for first and second-order light scattering processes on silicon nanoparticles with diameters ranging from $d = 60$ nm down to 3.5 nm [10]. The material was fabricated from a low-pressure microwave plasma using silane (SiH_4) as a precursor [5]. Higher order scattering processes are of great interest, as they allow to gather information not only from the Γ-point in the reciprocal space, but also from other symmetry points with $\vec{k} \neq 0$ as long as the vibrational density of states (VDOS) has a maximum.

Figure 9.3 shows the results for the TO(Γ) phonon obtained in first-order scattering. One can see that for the largest particle diameters investigated in these

9 Optical Properties of Silicon Nanoparticles

experiments, the frequency of the TO-phonon line is nearly identical to that of bulk silicon ($\tilde{v} = 522\,\text{cm}^{-1}$). With decreasing particle diameter, however, the Raman shift reduces down to around $497\,\text{cm}^{-1}$ for the smallest particles investigated in this study. At the same time, a broadening of the peak is observed.

The phonon confinement model introduced by Richter and Campbell [12, 13] explains the decrease of the Raman shift by the spatial confinement of the phonon wavefunction, leading in turn to an increased contribution from parts of the phonon band structure with $\vec{k} \neq 0$. As the phonon dispersion of the TO phonon has a local maximum at the Γ-point, the confinement effect can only lead to lower frequencies contributing to the scattered intensity, and thus, to a softening of the TO phonon. However, using the assumption of a Gaussian envelope for the phonon wavefunction, which is spatially confined to the region of the nanoparticle, one obtains significant frequency shifts only for nanoparticles with $d < 10\,\text{nm}$. The fact that in the experiments already for particle sizes around $d = 27\,\text{nm}$ a sizeable shift was observed might be related to the fact that the method used for size measurement is probing a different quantity than what is relevant for the confinement of the phonon wavefunction. Indeed, the size is measured via the specific surface area using the Brunauer-Emmett-Teller (BET) method [14]. As this measurement is purely sensitive to the surface, it cannot distinguish between crystalline silicon and a (possibly amorphous) surface layer. Thus, the phonon confinement method is only sensitive to the crystalline core and thus phonon confinement induced shifts are already observed for particles that are comparably large. It should be noted that the presence of a substoichiometric silicon oxide layer can also be determined using Auger spectroscopy, as shown in separate experiments [9].

Second-order Raman processes occur by scattering of a photon $\hbar\omega$ with two phonons. Thus, the frequency of the scattered light is increased or decreased by 2Ω, for Stokes and anti-Stokes observation. However, as the wavevector of the photon is negligible in comparison to that of the phonons, the total wavevector of the phonons must add up to zero. Thus, the phonons must have equal but opposite wavevectors to fulfill this momentum conservation rule. The highest scattering signal is therefore obtained for the frequencies with the highest VDOS, namely in the region of the van-Hove-singularities. Moreover, for single crystalline samples, the scattered intensity strongly depends on the experimental geometry due to symmetry considerations. For the case of a macroscopic sample containing randomly oriented nanoparticles, however, these symmetry conditions do not play a role. The scattering experiment averages over all possible orientations. Taking these considerations into account, the second- order Raman scattering on nanoparticle samples should rather direct to be a measure for the VDOS.

The experimental results for second-order Raman scattering are shown in Fig. 9.4. One can see that for the largest particles a peak due to the TO(X) phonon at $490\,\text{cm}^{-1}$ is found. With decreasing particle size, this peak shifts to lower frequencies down to $460\,\text{cm}^{-1}$. In general, the observed decrease in frequency is very similar to what is observed at the Γ-point using first-order scattering. However, the same explanation via the relaxation of the momentum conservation does not apply in this case. At the symmetry point involved here, the TO phonon branch has a local minimum, so

Fig. 9.4 Second-order Raman scattering on silicon nanoparticles of different sizes. Note that the frequency axis has been divided by two, so that the frequency of a single phonon is directly obtained

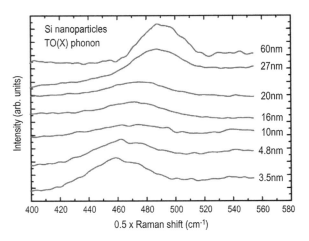

that an increase in frequency would be expected in this simple model. However, one should note that large wavevectors correspond to short length scales in real space. Thus, it is very plausible that atom–atom interaction effects are responsible for the softening of the TO phonon at the X point. Using molecular dynamics calculations it was recently shown by Meyer and Comtesse, that the observed redshift in the VDOS regardless from the symmetry point can be explained by the increasing number of surface atoms [15].

9.3 Recombination Dynamics

When we compare the photoluminescence emission traces for bulk silicon and the nanoparticle sample with a mean particle diameter of $d = 4.5$ nm in Fig. 9.1, we immediately find significant differences: Firstly, the luminescence from the silicon nanoparticles is significantly enhanced over the signal from the bulk crystal. Secondly, the photoluminescence peak for the nanostructures is much broader than that of the bulk sample. The reason for the latter is quite obvious and not directly related to the dynamics of the recombination process: Due to the large number of different nanocrystal sizes that are sampled in a single photoluminescence measurement, different optical bandgaps are present in the ensemble. Indeed, as shown in Fig. 9.2, the size of the optical bandgap in the nanoparticles is very sensitive even to slight changes in crystal size. These facts cause the comparably broad photoluminescence emission spectrum of the nanoparticles over the bulk crystal. On the other hand, the count rate one can obtain from a nanoparticle sample are—taking the instrument function into account—approximately 1,500 times higher than for bulk. The reason for this intensity enhancement is mostly the strong spatial localization of electron and hole in the nanoparticle, increasing the dipole transition matrix

9 Optical Properties of Silicon Nanoparticles

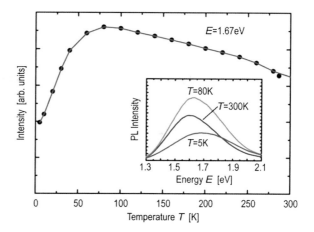

Fig. 9.5 Photoluminescence of silicon nanoparticles at $T = 300$, $T = 80$ and $T = 5\,\text{K}$

element. One might speculate, however, if the reduction in size relaxes the $\Delta k = 0$ selection rule due to Heisenberg's uncertainty principle. Following this idea, Trani et al. [16] have used the tight-binding method to compute absorption spectra for silicon nanoparticles of different sizes in order to assess the nature of the bandgap. They found that for particle diameters as discussed here with $d > 2.0\,\text{nm}$, one can safely assume that the band structure remains indirect. We will later analyze this question by evaluating absorption spectra.

Apart from the particle dimensions, another parameter that influences the recombination dynamics is the temperature. Temperature effects are well-known to affect dynamics in quantum dots or nanocrystals, mostly due to the competition between thermal energy $k_B T$ and the exciton binding energy E_{exc}. This effect usually leads to a steady decrease of the intensity of interband transitions when increasing the temperature of the sample [17–19]. For silicon nanostructures, however, the temperature dependence of the luminescence intensity is different. When the sample is cooled starting at room temperature, the photoluminescence signal first starts to increase until a maximum is reached at about $T = 80\,\text{K}$, as shown in Fig. 9.5. When the temperature is lowered even further, the photoluminescence intensity decreases again without being fully quenched when approaching $T \to 0\,\text{K}$ [20]. A similar behavior was reported for silicon nanocrystals in a SiO_2 layer by Brongersma et al. [21]. The origin for this remarkable temperature dependence is the fact that the exciton in silicon nanoparticles exhibits a fine structure. Therefore, the excited state is split in more than one state whose occupation is influenced by their respective relaxation rates and the thermal relaxation.

To analyze the situation in more detail, we will firstly assume that the conduction band states in silicon nanoparticles are mostly s-like, while the valence band is constituted of p-like states. With these assumptions, we obtain a simple model for the electronic band structure as shown in Fig. 9.6. It should be noted that these assumptions do not hold strictly for the discussed case. One important point is that

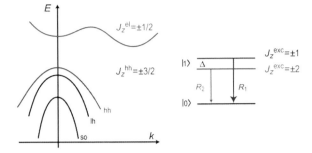

Fig. 9.6 *Left* schematic band structure of silicon nanoparticles. Due to quantum confinement, the degeneracy of heavy holes (hh) and light holes (lh) is lifted. Spin-orbit interaction is responsible for the split-off (so) band. *Right* excitonic states

while the holes involved for interband transitions originate from the Γ-point, the conduction band has its minimum at about 0.7 in the Γ-X direction. While for semiconductors with a direct gap, where only states with $k = 0$ are involved, the character of the orbitals is fairly well-characterized as p-like for the valence band and s-like for the conduction band, for indirect bandgaps one needs to consider mixing of p-like character into the s-like orbitals describing the conduction band. As suggested by Dovrat et al., this leads to some further adjustment of the fine structure model discussed in Ref. [22]. The principal results shown here, however, remain correct. As seen in Fig. 9.6, we describe the s-like conduction band and the p-like valence band by their angular momenta $J_{el} = 1/2$ and $J_{hh} = 3/2$. For an exciton formed of an electron and a hole, one obtains four possible configurations due to the fact that for the z-components of the angular momenta one needs to account for two parallel and two antiparallel configurations. For the exciton, this leads to two configurations with twofold degeneracy each, one with $J_{exc} = 1$ (for the antiparallel configurations) and the other one with $J_{exc} = 2$ (for the parallel configurations), as shown in the right part of Fig. 9.6. Considering the selection rules for optical transitions, one would expect only the $J_{exc} = 1$ states to be able to recombine radiatively, as only such an emission process would fulfill the conservation of angular momentum rule for circularly polarized photons with $J = 1$. For this reason, this state is labeled as "bright exciton", while the $J_{exc} = 2$ state is referred to as "dark exciton". However, as we will see later, these selection rules do not hold strictly for the nanoparticles as in solid-state based systems—in contrast to, e.g., ions or atoms—more degrees of freedom are present that allow to alleviate such selection rules.

As discussed above, the optical bandgap is strongly dependent on the particle size. Ledoux et al. have given a simple power law to relate the bandgap energy to the mean particle diameter [23]. The particle diameters discussed here are very small, leading to large confinement energies. On the other hand, the electronic barrier posed by either the SiO_2 shell surrounding the silicon core or the vacuum is extremely high with respect to the ground state energy shift. Thus, we can safely assume that only the first state in a quantum confinement model is relevant here and higher states are not involved in the optical transitions. These facts allow selecting distinct particle sizes from the optical spectra by restricting the observation to a narrow energy interval.

9 Optical Properties of Silicon Nanoparticles

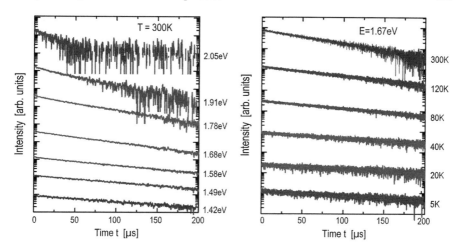

Fig. 9.7 *Left* Time-resolved photoluminescence for silicon nanoparticles for $E = 1.67$ eV at room temperature. *Right* Temperature-dependent decay curves for the recombination at $E = 1.67$ eV

The left part of Fig. 9.7 shows the time-resolved photoluminescence at room temperature for a sample with a mean particle diameter of $d = 4.8$ nm. However, due to the broad particle size distribution we can observe dynamics from different particle size within one sample by looking at different energies, as shown in Fig. 9.7. The energy range of from 1.42 to 2.05 eV means spectrally selected particle diameters between 2.7 and 6.1 nm. The observed decay curves can be described using a single exponential decay model. In contrast to other samples with a higher degree of disorder and defects, this indicates a very high crystalline quality of the individual nanoparticles. For disordered systems, typically a stretched exponential decay is observed [24]. It should also be noted that the observed time constants are very large, being in the range of tens to hundreds of microseconds. For quantum dots or nanoparticles made from direct bandgap semiconductors, often recombination times of $\tau_{PL} \approx 0.3$–1 ns are found [25, 26].

From Fig. 9.7, it is clearly visible that the significant difference in particle diameter has a strong effect on the recombination dynamics. With reduced particle diameter, the observed decay times are reduced from $\tau_{PL} = 125\,\mu$s for the largest particles to $\tau_{PL} = 25\,\mu$s for the smallest particles. This enhancement of recombination rate is caused by two contributions: Firstly, as the particle size decreases, the electron–phonon interaction is enhanced, as shown by Kovalev et al. [27]. Therefore, the contribution from phonon-assisted transitions increases. Secondly, with decreasing particle size also phonon-less transitions start to play a role (zero-phonon-line) [28]. These transitions are enhanced due to the increased overlap of electron and hole wavefunctions. Both effects lead to increased radiative recombination rates.

As discussed above, the photoluminescence intensity of silicon nanoparticles exhibits a non-monotic temperature behavior. To further assess this observation,

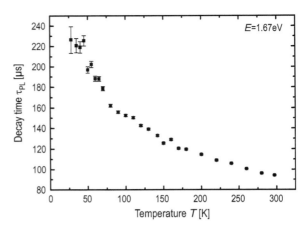

Fig. 9.8 Temperature dependence of the photoluminescence decay time τ

one needs to study the temperature dependence of the photoluminescence decay. In the following, we will analyze the dynamics of the recombination from nanoparticles at $E = 1.67$ eV. Using Ledoux' relation [23], this means we restrict observation to particles with a size of $d = 3.96$ nm.

The experimental results are shown in the right part of Fig. 9.7. One can see that the fastest photoluminescence decay, and thus, the largest recombination rates are observed at room temperature. When the sample is cooled down, the decay time τ_{PL} increases from 100 μs at room temperature to about 220 μs at low temperature. The temperature behavior is summarized in Fig. 9.8. In contrast to the temperature behavior of the intensity, the photoluminescence decay time exhibits a monotonic dependence on the temperature.

The photoluminescence decay rate $R_{PL} = 1/\tau_{PL}$ measured in the time-resolved experiments has contributions from radiative recombinations ($R_R = 1/\tau_R$) and nonradiative recombinations ($R_{NR} = 1/\tau_{NR}$). Thus, we obtain for the decay rate $R_{PL} = R_R + R_{NR}$. The intensity, on the other hand, is directly proportional to the quantum efficiency η, which is the ratio of radiative recombinations and all recombinations, radiative and nonradiative:

$$I \propto \eta = \frac{R_R}{R_R + R_{NR}}$$

From these equations we obtain that the radiative recombination rate R_R is proportional to the product of the intensity $I(T)$ and the measured photoluminescence decay rate $R_{PL} = 1/\tau_{PL}$:

$$R_R(T) \propto I(T) \cdot R_{PL}$$

Following this route, we obtain the temperature dependence of the radiative recombination rate, as plotted in Fig. 9.9.

9 Optical Properties of Silicon Nanoparticles

Fig. 9.9 Temperature dependence of the radiative recombination rate obtained via intensity and decay rate measurements

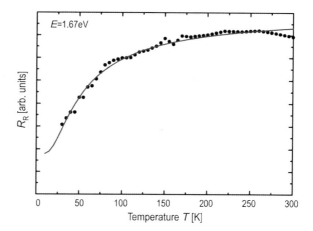

As discussed above, for quantum dots in direct bandgap semiconductors, one expects an increase of the radiative recombination rate at low temperatures due to the contribution from excitonic recombinations. For silicon nanoparticles, however, the contrary is observed. With decreasing temperature, the radiative recombination rate is steadily reduced. The reason for this—at first sight—unexpected result is the contribution from the above mentioned "dark excitons". In fact, as shown in the right part of Fig. 9.6, the "dark" state is lower in energy than the bright state and therefore, while being dipole forbidden, energetically favored. Such dark states, of course, are by no means unique to silicon, but are present in all semiconductors with s-like conduction band and p-like valence band. The difference is, however, that in silicon the lifetimes are significantly longer than in semiconductors with direct bandgaps. Therefore, in silicon nanoparticles thermalization of the dark and the bright states occurs, such that we can assume a thermal occupation for the states given by Maxwell–Boltzmann statistics. This means, that there must be spin-flip processes involved to reach the thermal equilibrium. Julsgaard et al. have directly observed such scattering processes [29] and showed that the scattering time is around $0.1\mu s$, indeed significantly smaller than the photoluminescence decay time.

The radiative recombination rate R_R itself also features two contributions, one due to recombinations from the "bright" state R_1, the other one from recombinations from the "dark" state R_2, where the index denotes the angular momentum quantum number. Although the latter are in principle forbidden due to the conservation of angular momentum, in the solid state, phonon-assisted processes can make such transitions possible. Therefore, the ratio R_1/R_2 should be a large finite number, as R_2 should be significantly smaller than R_1. However, the occupation of the dark and bright states, which are split by the exchange interaction energy Δ is responsible for the temperature dependence of the radiative recombination rate, and one obtains:

$$R_R(T) = \frac{2R_2 + 2R_1 \exp\left(-\frac{\Delta}{K_B T}\right)}{2 + 2\exp\left(-\frac{\Delta}{K_B T}\right)}$$

Here, the factor 2 takes the twofold degeneracy of the dark and the bright state into account. This relation can be fitted to the experimental results (solid line in Fig. 9.9). From the fitting procedure, one obtains the exchange interaction energy $\Delta = 5.8\,\text{meV}$ and the ratio between the bright state and the dark state recombination rate $R_1/R_2 \approx 8$. The value for the exchange energy is comparably high, taking into consideration that the value for bulk silicon is as low as $140\,\mu\text{eV}$ [30]. In quantum confined systems such as the silicon nanoparticles studied here, however, the short range interaction is enhanced, leading to the observed large values for the exchange interaction [31]. The bright state/dark state recombination rate ratio is relatively low taking into account that the dark states can only recombine if angular momentum conservation is carried out by phonons. From the above model, one can determine the contribution from the dark and the bright state to the radiative recombinations. From this model, we obtain, as shown in Fig. 9.10, that for temperatures below $T = 35\,\text{K}$, the recombination rate from the dark state is higher than that from the bright state [20]. This elucidates the competition between the contributions from these two states to the total recombination and this is also the reason why the intensity exhibits such a non-monotic behavior.

In the next step, we study the radiative recombination times τ_{PL} for different emission energies—or different particle diameters—and different temperatures. Figure 9.11 gives an overview over the obtained results. For 80–300 K, we find an increase of the measured decay time with decreasing emission energy (increasing particle diameter). For temperatures below 40 K, however, the recombination rate is independent of the particle size. We attribute this size-independence of the recombination at low temperatures to nonradiative recombination processes with $\tau_{\text{NR}} = 200\,\mu\text{s}$. As the recombination rate is the sum of radiative and nonradiative recombination rates, the process with the smaller recombination time governs the observed behavior. While at high temperatures the radiative recombination rate is high, as discussed above, at low temperatures the nonradiative processes play the dominant role. Those do not exhibit any significant particle size dependence.

From the above discussions, we now have a reasonable estimate for the nonradiative recombination time—a quantity that is usually difficult to address directly. This allows us to assign quantitative values to the above derived temperature dependence of the radiative recombination time using the relation $\tau_R^{-1} = \tau_{\text{PL}}^{-1} - \tau_{\text{NR}}^{-1}$. This way, we can directly calculate the quantum efficiency of the silicon nanoparticles, which—due to the size dependence of the radiative recombination processes discussed above—also depends on the particles size. From $\eta = \tau_R^{-1}/(\tau_R^{-1} + \tau_{\text{NR}}^{-1})$ we obtain quantum efficiencies from 34 % for the largest particles considered here with $d = 6.1\,\text{nm}$ to 86 % for the smallest particles with $d = 2.7\,\text{nm}$. These quantum efficiencies are comparably high for an indirect semiconductor like silicon. However, similar values have also been reported by Mangolini et al. [32] and Jurbergs [33].

9 Optical Properties of Silicon Nanoparticles

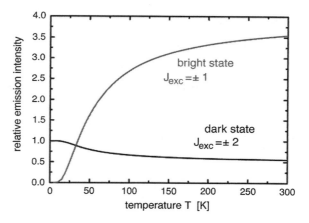

Fig. 9.10 Relative emission from *dark* and *bright* states, respectively

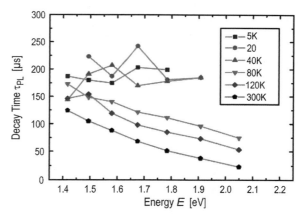

Fig. 9.11 Photoluminescence decay times for different temperatures as a function of the emission energy

Using the above derived information on the radiative lifetimes, we can also compute the oscillator strength of the observed transitions directly and are thus able to compare the recombination processes directly, e.g., to dipole transitions in atoms. The oscillator strength is related to the radiative lifetime of the transition via

$$f_{\text{OSC}}(\omega) = \frac{2\pi\varepsilon_0 m c_0^3}{e^2 n \omega^2}\frac{1}{\tau_R}.$$

Here, the n is the refractive index, $m = m_e + m_h$ is the exciton mass (in the weak confinement regime that applies for silicon nanoparticles of this size, with $m_e = 0.19\, m_0$ and $m_h = 0.286\, m_0$ [34, 35]) and ω is the frequency of the interband transition in consideration.

Figure 9.12 shows the results for the radiative lifetimes τ_R and the oscillator strength f_{OSC} for different emission energies, originating from particle diameters between 2.7 and 6.1 nm [36]. One can clearly recognize that the oscillator

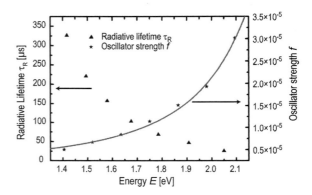

Fig. 9.12 Radiative lifetimes and oscillator strengths derived as discussed in the text. Increasing emission energy translates to decreasing particle size

strength increases monotonously with decreasing particle size, as expected due to the enhanced electron–phonon coupling and the increasing electron–hole wavefunction overlap. However, even for the smallest particles under consideration here, the oscillator strength is not very high with values around $f_{OSC} \approx 3 \times 10^{-5}$. Dipole transitions in atoms or ions typically exhibit values of $f_{OSC} > 10^{-1}$ for allowed transitions, while values of $f_{OSC} \approx 10^{-8}$ are found for forbidden transitions. When such values are considered, the silicon nanoparticles are emitters in an intermediate regime, because emission processes are typically second-order processes due to the involvement of at least one phonon as an additional particle. Combining the results shown in Fig. 9.12 with Ledoux' formula [23] given for the optical bandgap of silicon nanoparticles as a function of their size, one obtains an empirical expression for the size dependence of the oscillator strength [36]:

$$f_{OSC} = 1.4 \times 10^{-6} + 1.7 \times 10^{-6} \exp\left(\frac{11.24}{d(\text{nm})^{1.39}}\right)$$

This formula can be used for particle diameters down to $d \approx 2$ nm, as the experimental data used here do not cover smaller sizes. The agreement between the results of Trani et al. [16] obtained from theoretical calculations is excellent.

The indirect nature of the bandgap, which is expected to be responsible for the observed behavior, can directly be probed from the absorption spectra of the silicon nanoparticles. For a semiconductor with an indirect bandgap, the absorption coefficient α is expected to depend on the energy $\hbar\omega$ as given by

$$\alpha(\hbar\omega) = \omega^{-1}(\hbar\omega - E_g)^2.$$

Therefore, when the absorption data are plotted as $(\alpha\omega)^{1/2}$ versus ω, one expects a linear behavior for an indirect bandgap semiconductor, which can clearly be distinguished from direct bandgap semiconductors, for which the absorption coefficient is related to the square root of the energy. Figure 9.13 shows the result for two different particle sizes. It can be seen that—as expected—with decreasing particle size

9 Optical Properties of Silicon Nanoparticles

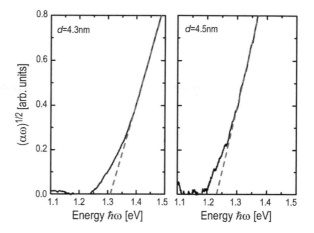

Fig. 9.13 Absorption spectra for silicon nanoparticles with mean particle diameters of $d = 4.3$ and $d = 4.8$ nm

also in the absorption spectra the optical band gap opens up. Moreover, the linear behavior of the absorption data plotted in the above discussed way clearly shows that the electronic band structure still has the character of an indirect gap. However, one can also see that for energies directly below the gap, already a soft increase of the absorption is observed. This onset of the absorption has two reasons. Firstly, absorption across the indirect bandgap requires contributions from phonons Thus, the energy conservation rule for this process is

$$E_f = E_i + \hbar\omega \pm \hbar\Omega,$$

where E_f and E_i are the final and initial state energies, is the energy of the photon and $\hbar\Omega$ is the energy of a phonon. Note that both, the creation and the annihilation of a phonon can contribute to this process and, thus, absorption is for indirect semiconductors already observed below the electronic bandgap.

In case of the particles studied here, however, this is not the only effect that needs to be taken into consideration. The samples used here contain nanoparticles with a lognormal size distribution. Typically, the geometrical standard deviation is about $\sigma = 1.3$. The lognormal distribution function for particles with a mean diameter of d_0 and a standard deviation σ is

$$f_{d_0,\sigma}(\tilde{d}) = \frac{1}{\sqrt{2\pi} \cdot \tilde{d} \cdot \sigma} \exp\left(-\frac{1}{2}\frac{\ln^2 \frac{\tilde{d}}{d_0}}{\ln^2 \sigma}\right)$$

Using this equation, we obtain for the absorption spectrum of an ensemble of nanoparticles with indirect bandgap:

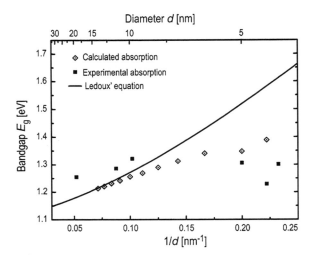

Fig. 9.14 Experimental and calculated optical bandgaps for a nanoparticle ensemble. The *solid line* is the relation for individual particles given in [23]

$$A(\hbar\omega) = \omega^{-1} \int_0^\infty f_{\mathrm{OSC}}(\omega, \tilde{d}) \cdot f_{d_0,\sigma}(\tilde{d}) \cdot (\hbar\omega - E_g(\tilde{d}))^2 \cdot d(\tilde{d})$$

For the size dependence of E_g, we use Ledoux' formula again, as discussed above. From this equation, absorption spectra for nanoparticle ensembles can be computed and the effect of the particle size distribution on the obtained optical bandgap of the ensemble sample can be derived. The results are shown in Fig. 9.14. One can clearly see that for small particles, the optical bandgap in the measured absorption spectra is significantly shifted toward lower energies. This is due to the fact that the power law dependence of the bandgap results in large gaps for the smallest particles, while the larger particles mostly contribute near the band gap energy of the bulk semiconductor. Therefore, for samples with a finite size distribution, the absorption band edge does not reflect the band edge of the particles with the mean particle diameter d_0.

9.4 Electroluminescence

Compared to the bulk material, silicon nanoparticles exhibit drastically improved optical properties. As an example, Fig. 9.1, above, demonstrates that the light emission after optical excitation can be increased by more than three orders of magnitude. A good photoluminescence efficiency in the visible spectral region may be of technological use for solid state lighting, as a phosphor for blue or UV light emitting diodes. The slow recombination rates shown in Figs. 9.7 and 9.8, on the other hand, make it questionable, whether Si nanoparticles are suitable for such color conversion. A more promising application of silicon nanoparticles in consumer products is the

9 Optical Properties of Silicon Nanoparticles

Fig. 9.15 Sample layout and schematic of the processing steps for the fabrication of an electroluminescent device. For details, see text

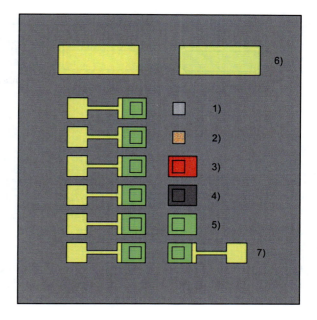

light emission from *electrically driven* devices. Here again, the slow recombination lifetimes may be disadvantageous, particularly for optical communication. For the fabrication of cost-effective light sources, which can be produced from an abundant and non-toxic element, however, silicon-based light emitting devices may be highly attractive.

Already 20 years ago, visible electroluminescence was observed from porous silicon [37]. In 2001, Valenta et al. demonstrated the fabrication of light emitting devices based on Si nanocrystals, embedded in a SiO_2 matrix [38]. These fabrication methods are in principle compatible with common CMOS technology, but require sophisticated process technologies. A first step toward devices based on nanoparticles, which can be produced in large quantities and may ultimately be processed by simple printing techniques, was demonstrated in Fojtik et al. [39] and by Cheong et al. [40]. They used Si nanocrystals, derived from pyrolysis of silane, as the optically active medium in electroluminescence devices.

In the following, we will discuss the fabrication of a simple light emitting device, based on silicon nanoparticles, which were produced from the gas phase as discussed in Chap. 1. Their optical properties were discussed in detail in the previous section.

The processing steps are schematically summarized in Fig. 9.15. The fabrication starts from a GaAs heterostructure, which consists of a 300 nm thick silicon-doped GaAs layer and a 100 nm thick undoped GaAs layer, grown by molecular beam epitaxy on a semi-insulating GaAs wafer. This starting material greatly facilitates the subsequent fabrication processes, as it provides well-defined and smooth contact and insulating layers. It is, however, not essential for the fabrication of electroluminescent

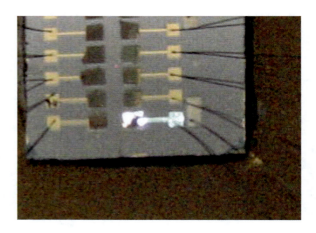

Fig. 9.16 Fully processed sample, containing 12 working devices. The device on the lower right is biased, and light emission from the Si nanoparticles is clearly observed

devices, and similar results as shown here have recently been achieved also on silicon substrates (Theis J (2011), unpublished). Using standard photolithography, recess areas of $300 \times 300\ \mu m^2$ size are etched into the capping layer, exposing the doped back contact (step 1 in Fig. 9.15). In a second photolithographic step, a larger window is defined on top of the contact region. The Si nanoparticles, dispersed in a solution of distilled water with 5 % ethanol, are drop-cast onto the sample and time is given for the solvents to evaporate. This results in a nanoparticle layer with an average thickness of about 150 nm (step 2). The sample is then covered by a 10 nm thick SiO_2 protection layer (step 3), and an 80 nm thick layer of indium tin oxide (ITO) is thermally evaporated (step 4). Up to this point, the photoresist has remained on the surface, so that in the subsequent lift-off process, all deposited layers are removed simultaneously, in a self-aligned fashion. In order to transform the evaporated ITO into a transparent conductive film, which serves as the top contact, the sample is annealed for 30 min at 380 °C under oxygen exposure (step 5). Electrical access to the conducting GaAs back contact and the ITO top layer is provided by annealed AuGe pads (step 6) and 50 nm thick gold electrodes (step 7), respectively.

Figure 9.16 shows a fully processed, mounted, and contacted sample, containing 12 working devices. As seen for the device on the lower right, when the structure is appropriately biased, light emission is easily observed even with the unaided eye. Closer inspection under a microscope reveals that the emission characteristic is very inhomogeneous, with a few bright spots contributing almost all the observed intensity. Furthermore, a weak emission is discernible, which seems to originate from the gold contact area. This spurious light may have different origins: Light from the Si particles, which is guided through the GaAs layers and scattered by the rough gold surface; GaAs electroluminescence in the infrared, which is picked up by the CCD camera; or surface plasmons in the gold layer. Note that the image in Fig. 9.16 is overexposed, so that the green emission is greatly enhanced (see also [41]). We will disregard this emission in the following and analyze the radiation from the nanoparticle-covered area.

9 Optical Properties of Silicon Nanoparticles

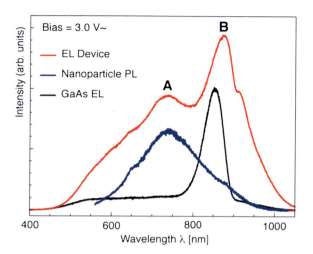

Fig. 9.17 Emission spectrum of the electroluminescence device at room temperature (*red line*). The observed emission peaks originate from the silicon nanoparticles (A) and from the GaAs substrate (B) as revealed from a comparison with the photoluminescence (PL) of the particles (*blue line*) and the electroluminescence (EL) from a reference sample without particles (*black line*)

Figure 9.17 displays the electroluminescence spectrum of our device at room temperature (red line). The sample is biased by a square AC voltage of 3.0 V amplitude. A broad emission, covering almost the entire visible spectrum is observed (red line). The spectral intensity exhibits two pronounced maxima, one (labeled A) at 740 nm and another (labeled B) at 870 nm, corresponding to 1.68 and 1.42 eV, respectively. Figure 9.17 also shows a photoluminescence spectrum of Si nanoparticles from the same batch as the ones used for the light emitting device (blue). Furthermore, an electroluminescence spectrum of a reference sample is shown (black), which was fabricated without the nanoparticle layer. A comparison of the three spectra allows us to identify the different peaks in the emission characteristics of the electroluminescence device. From the energetic position close to the GaAs bandgap and from the fact that it is clearly observed also for the reference sample without any particles, peak B can be identified as luminescence from the supporting GaAs layers. On the other hand, the similarity between the nanoparticle photoluminescence (blue) and the electroluminescence (red) on the short-wavelength side of the spectrum clearly shows that the light emission around peak A indeed originates from the electrically excited Si nanoparticles. Interestingly, the integrated intensity emitted from the nanoparticles is considerably higher than that from the GaAs. Since GaAs is an excellent optoelectronic material, this demonstrates the high optical quality of the Si particles, in agreement with the high quantum efficiency determined from photoluminescence studies (see above). For a recent, more in-depth study of GaAs-supported devices, see [41]. As mentioned above, electroluminescence can also be observed from similar structures, where the GaAs is replaced by a doped Si substrate.

Different models have been discussed in the literature to explain the electroluminescence from Si nanoparticles. Fojtik et al. [39] and Xu et al. [42] have attributed the light emission to bipolar injection, followed by radiative recombination of the simultaneously injected electrons and holes. Similarly, Walters et al. [43] have demonstrated Si nanoparticle electroluminescence from AC-driven samples, where the electrons and holes were injected successively during the half-cycles of different polarity. On the other hand, impact excitation of hot carriers was proposed by Irrera et al., Valenta et al., and Liu et al. as the mechanism that leads to the light emission [38, 44, 45]. In order to elucidate the dominant mechanism in the present devices, we have studied the bias dependence of the light emission. Under DC conditions, we do not observe any light emission for positive bias applied to the top contact. For negative voltages, electroluminescence of similar brightness is observed as for AC bias. No holes can be injected from the back contact, as it is n-doped and the bias is not sufficient to induce inversion conditions [41]. No holes can be injected from the (metallic) ITO top electrode either, because it is negatively biased. Therefore, we attribute the electroluminescence from our (unipolar) device to electron–hole generation by impact ionization of hot electrons, injected from the top electrode.

Finally, we would like to mention that in the present study, we have used particles, which can be produced in large quantities, using a continuous gas-phase based process. As demonstrated by Gupta et al. [46], these particles can be used in an ink-jet printing process to fabricate patterned layers of excellent optical quality. Thus, in principle, no high-temperature treatment, no vacuum process, and no elaborate lithography is necessary for the fabrication of the optically active layer. Similarly, the transparent top contact could be replaced by a printed layer of ITO nanoparticles [47], which are commercially available today. All this suggests that Si nanoparticles from the gas-phase constitute an excellent starting material for the fabrication of efficient, low-cost, printable light emitters.

References

1. L.T. Canham, Silicon quantum wire array fabrication by electrochemical and chemical dissolution of wafers. Appl. Phys. Lett. **57**, 1046 (1990)
2. S. Furukawa, T. Miyasato, Quantum size effects on the optical band gap of microcrystalline Si:H. Phys. Rev. B **38**, 5726 (1988)
3. H.Z. Song, X.M. Bao, Visible photoluminescence from silicon-ion-implanted SiO2 film and its multiple mechanisms. Phys. Rev. B **55**, 6988 (1997)
4. M. Zacharias, J. Heitmann, R. Scholz, U. Kahler, M. Schmidt, J. Blasing, Size-controlled highly luminescent silicon nanocrystals: a SiO/SiO[sub 2] superlattice approach. Appl. Phys. Lett. **80**, 661 (2002)
5. J. Knipping, H. Wiggers, B. Rellinghaus, P. Roth, D. Konjhodzic, C. Meier, Synthesis of high purity silicon nanoparticles in a low pressure microwave reactor. J. Nanosci. Nanotechnol. **4**, 1039 (2004)
6. H. Wiggers, R. Starke, P. Roth, Silicon particle formation by pyrolysis of silane in a hot wall gas-phase reactor. Chem. Eng. Technol. **24**, 261 (2001)
7. M.V. Wolkin, J. Jorne, P.M. Fauchet, G. Allan, C. Delerue, Electronic states and luminescence in porous silicon quantum dots: the role of oxygen. Phys. Rev. Lett. **82**, 197 (1999)
8. S. Godefroo, M. Hayne, M. Jivanescu, A. Stesmans, M. Zacharias, O.I. Lebedev, G. Van Tendeloo, V.V. Moshchalkov, Classification and control of the origin of photoluminescence from Si nanocrystals. Nat. Nano. **3**, 174 (2008)
9. V.G. Kravets, C. Meier, D. Konjhodzic, A. Lorke, H. Wiggers, Infrared properties of silicon nanoparticles. J. Appl. Phys. **97**, 084306 (2005)

9 Optical Properties of Silicon Nanoparticles

10. C. Meier, S. Lüttjohann, V.G. Kravets, H. Nienhaus, A. Lorke, H. Wiggers, Raman properties of silicon nanoparticles. Phys. E Low dimens. Sys. Nanostruct. **32**, 155 (2006)
11. M.H. Nayfeh, O. Akcakir, G. Belomoin, N. Barry, J. Therrien, E. Gratton, Second harmonic generation in microcrystallite films of ultrasmall Si nanoparticles. Appl. Phys. Lett. **77**, 4086 (2000)
12. H. Richter, Z.P. Wang, L. Ley, The one phonon Raman spectrum in microcrystalline silicon. Solid State Commun. **39**, 625 (1981)
13. I.H. Campbell, P.M. Fauchet, The effects of microcrystal size and shape on the one phonon Raman spectra of crystalline semiconductors. Solid State Commun. **58**, 739 (1986)
14. S. Brunauer, P.H. Emmett, E. Teller, Adsorption of gases in multimolecular layers. J. Am. Chem. Soc. **60**, 309 (1938)
15. R. Meyer, D. Comtesse, Vibrational density of states of silicon nanoparticles. Phys. Rev. B **83**, 014301 (2011)
16. F. Trani, G. Cantele, D. Ninno, G. Iadonisi, Tight-binding calculation of the optical absorption cross section of spherical and ellipsoidal silicon nanocrystals. Phys. Rev. B **72**, 075423 (2005)
17. J.C. Kim, H. Rho, L.M. Smith, H.E. Jackson, S. Lee, M. Dobrowolska, J.K. Furdyna, Temperature-dependent micro-photoluminescence of individual CdSe self-assembled quantum dots. Appl. Phys. Lett. **75**, 214 (1999)
18. Y.G. Kim, Y.S. Joh, J.H. Song, K.S. Baek, S.K. Chang, E.D. Sim, Temperature-dependent photoluminescence of ZnSe/ZnS quantum dots fabricated under the Stranski-Krastanov mode. Appl. Phys. Lett. **83**, 2656 (2003)
19. E.C. Le Ru, J. Fack, R. Murray, Temperature and excitation density dependence of the photoluminescence from annealed InAs/GaAs quantum dots. Phys. Rev. B **67**, 245318 (2003)
20. S. Lüttjohann, C. Meier, M. Offer, A. Lorke, H. Wiggers, Temperature-induced crossover between bright and dark exciton emission in silicon nanoparticles. Europhys. Lett. **79**, 37002 (2007)
21. M.L. Brongersma, P.G. Kik, A. Polman, K.S. Min, H.A. Atwater, Size-dependent electron-hole exchange interaction in Si nanocrystals. Appl. Phys. Lett. **76**, 351 (2000)
22. M. Dovrat, Y. Shalibo, N. Arad, I. Popov, S.T. Lee, A. Sa'ar, Fine structure and selection rules for excitonic transitions in silicon nanostructures. Phys. Rev. B **79**, 125306 (2009)
23. G. Ledoux, O. Guillois, D. Porterat, C. Reynaud, F. Huisken, B. Kohn, V. Paillard, Photoluminescence properties of silicon nanocrystals as a function of their size. Phys. Rev. B **62**, 15942 (2000)
24. L. Pavesi, M. Ceschini, Stretched-exponential decay of the luminescence in porous silicon. Phys. Rev. B **48**, 17625 (1993)
25. M. Paillard, X. Marie, E. Vanelle, T. Amand, V.K. Kalevich, A.R. Kovsh, A.E. Zhukov, V.M. Ustinov, Time-resolved photoluminescence in self-assembled InAs/GaAs quantum dots under strictly resonant excitation. Appl. Phys. Lett. **76**, 76 (2000)
26. S. Raymond, S. Fafard, S. Charbonneau, R. Leon, D. Leonard, P.M. Petroff, J.L. Merz, Photocarrier recombination in $Al_y In_{1-y} As/Al_x Ga_{1-x} As$ self-assembled quantum dots. Phys. Rev. B **52**, 17238 (1995)
27. D. Kovalev, H. Heckler, M. Ben-Chorin, G. Polisski, M. Schwartzkopff, F. Koch, Breakdown of the k-conservation rule in Si nanocrystals. Phys. Rev. Lett. **81**, 2803 (1998)
28. M.S. Hybertsen, Absorption and emission of light in nanoscale silicon structures. Phys. Rev. Lett. **72**, 1514 (1994)
29. B. Julsgaard, Y.-W. Lu, P. Balling, A.N. Larsen, Thermalization of exciton states in silicon nanocrystals. Appl. Phys. Lett. **95**, 183107 (2009)
30. J.C. Merle, M. Capizzi, P. Fiorini, A. Frova, Uniaxially stressed silicon: Fine structure of the exciton and deformation potentials. Phys. Rev. B **17**, 4821 (1978)
31. D.H. Feng, Z.Z. Xu, T.Q. Jia, X.X. Li, S.Q. Gong, Quantum size effects on exciton states in indirect-gap quantum dots. Phys. Rev. B **68**, 035334 (2003)
32. L. Mangolini, E. Thimsen, U. Kortshagen, High-Yield Plasma Synthesis of Luminescent Silicon Nanocrystals. Nano Lett. **5**, 655 (2005)

33. D. Jurbergs, E. Rogojina, L. Mangolini, U. Kortshagen, Silicon nanocrystals with ensemble quantum yields exceeding 60%. Appl. Phys. Lett. **88**, 233116 (2006)
34. J.-B. Xia, Electronic structures of zero-dimensional quantum wells. Phys. Rev. B **40**, 8500 (1989)
35. A.D. Yoffe, Low-dimensional systems: quantum size effects and electronic properties of semiconductor microcrystallites (zero-dimensional systems) and some quasi-two-dimensional systems. Adv. Phys. **42**, 173 (1993)
36. C. Meier, A. Gondorf, S. Lüttjohann, A. Lorke, H. Wiggers, Silicon nanoparticles: absorption, emission, and the nature of the electronic bandgap. J. Appl.Phys. **101**, 103112 (2007)
37. A. Richter, P. Steiner, F. Kozlowski, W. Lang, Current-induced light emission from a porous silicon device. IEEE Electron Device Lett. **12**, 691 (1991)
38. J. Valenta, N. Lalic, J. Linnros, Electroluminescence microscopy and spectroscopy of silicon nanocrystals in thin SiO2 layers. Opt. Mater. **17**, 45 (2001)
39. A. Fojtik, J. Valenta, T.H. Stuchlíková, J. Stuchlík, I. Pelant, J. Kocka, Electroluminescence of silicon nanocrystals in p-i-n diode structures. Thin Solid Films **515**, 775 (2006)
40. H.J. Cheong, A. Tanaka, D. Hippo, K. Usami, Y. Tsuchiya, H. Mizuta, S. Oda, Visible Electroluminescence from Spherical-Shaped Silicon Nanocrystals. Jpn. J. Appl. Phys. **47**, 8137 (2008)
41. J. Theis, M. Geller, A. Lorke, H. Wiggers, C. Meier, Electroluminescence from silicon nanoparticles fabricated from the gas phase. Nanotechnology **21**, 455201 (2010)
42. J. Xu, K. Makihara, H. Deki, S. Miyzazki, Electroluminescence from Si quantum dots/SiO2 multilayers with ultrathin oxide layers due to bipolar injection. Solid State Commun. **149**, 739 (2009)
43. R.J. Walters, G.I. Bourianoff, H.A. Atwater, Field-effect electroluminescence in silicon nanocrystals. Nat. Mater. **4**, 143 (2005)
44. A. Irrera, D. Pacifici, M. Miritello, G. Franzo, F. Priolo, F. Iacona, D. Sanfilippo, G. Di Stefano, P.G. Fallica, Excitation and de-excitation properties of silicon quantum dots under electrical pumping. Appl. Phys. Lett. **81**, 1866 (2002)
45. C.W. Liu, S.T. Chang, W.T. Liu, M.-J. Chen, C.-F. Lin, Hot carrier recombination model of visible electroluminescence from metal-oxide-silicon tunneling diodes. Appl. Phys. Lett. **77**, 4347 (2000)
46. A. Gupta, S.G. Khalil, M. Offer, M. Geller, M. Winterer, A. Lorke, H. Wiggers, Synthesis and ink-jet printing of highly luminescing silicon nanoparticles for printable electronics. J. Nanosci. Nanotechnol. **11**, 5028 (2011)
47. A. Gondorf, M. Geller, J. Weißbon, A. Lorke, M. Inhester, A. Prodi-Schwab, D. Adam, Mobility and carrier density in nanoporous indium tin oxide films. Phys. Rev. B **83**, 212201 (2011)

Chapter 10
Electrical Transport in Semiconductor Nanoparticle Arrays: Conductivity, Sensing and Modeling

Sonja Hartner, Dominik Schwesig, Ingo Plümel, Dietrich E. Wolf, Axel Lorke and Hartmut Wiggers

Abstract Electrical properties of nanoparticle ensembles are dominated by interparticle transport processes, mainly due to particle–particle and particle-contact interactions. This makes their electrical properties dependent on the network properties such as porosity and particle size and is a main prerequisite for solid- state gas sensors, as the surrounding gas atmosphere influences the depletion layer surrounding each particle. Different kinds of nanoparticle arrays such as pressed pellets, printed layer, and thin films prepared by molecular beam-assisted deposition are characterized with respect to their electrical transport properties. Experimental results are shown for the electrical and sensing properties of several metal oxide nanoparticle ensembles and the influence of porosity is investigated during compaction of nanoparticle powders exposed to an external force. A model describing these properties is developed and it is shown that for a given material only porosity, geometry, and particle size influence the overall electrical properties. The model developed for the description of current transport in particulate matter can also be utilized to describe current-assisted sintering.

10.1 Introduction

The electrical properties of nanoscale semiconductors differ in a number of ways from those of the corresponding bulk material. The most obvious reason for this fact is the missing translational invariance. Nanoparticulate materials are to an extreme

---

S. Hartner · D. Schwesig · I. Plümel · D. E. Wolf · A. Lorke · H. Wiggers (✉)
Institute for Combustion and Gasdynamics (IVG), Department of Physics,
CENIDE Center for NanoIntegration, University of Duisburg-Essen, 47057 Duisburg, Germany
e-mail: hartmut.wiggers@uni-due.de

H. Wiggers
Institute for Combustion and Gasdynamics, University of Duisburg-Essen,
Lotharstraße 1, 47057 Duisburg, Germany

A. Lorke et al. (eds.), *Nanoparticles from the Gas Phase*, NanoScience and Technology,
DOI: 10.1007/978-3-642-28546-2_10, © Springer-Verlag Berlin Heidelberg 2012

231

degree inhomogeneous, and their topology may not be simply connected (holey, sponge-like) or they may even have fractal dimension. All these properties will have immediate impact on the experimentally determined electrical properties of such materials. Furthermore, the complicated spatial arrangement makes it impossible to use the models, which have been long established for bulk semiconductors, to describe and model the electronic characteristics of nanoparticle-based materials.

While the above-mentioned structural properties can in principle be found in all particulate systems (macroscopic or microscopic), *nano*particulate materials bring in another aspect, i.e., size effects, which make them highly attractive for novel, tailored materials. When the typical structural dimensions of a system or a material are reduced below a characteristic length, the physical properties, which are associated with this length, may drastically change. Then, these properties are no longer solely given by the chemical composition of the material, but also by the size and shape of its nanoscale constituents. Examples for characteristic length scales, which affect the electrical properties, are the ballistic mean free path, the phase coherence length, and the deBroglie wavelength of the charge carriers.

When the size of the nanostructure is smaller than the *ballistic mean free path*, charge transport is no longer solely governed by random scatterers and reflection from the system boundaries becomes important. This 'guiding' of electron trajectories by surfaces and interfaces results in a number of intriguing transport phenomena, which have been studied in great detail in lithographically patterned semiconductor nanostructures, for an early review see [1]. One example for the impact that ballistic transport can have on the electrical properties is the breakdown of simple scaling laws: As an example, the resistance of a ballistic wire will no longer be proportional to its length as in a macroscopic conductor. The *phase coherence length* determines the scale on which quantum interference effects can be observed. In the phase coherent transport regime, the common (macroscopic) treatment in the framework of a Drude-type conductivity needs to be replaced by the Landauer-Büttiker approach [2], which derives the conductance of a multi-probe sample based on the transmission probabilities between the different quantum states in the leads. The smallest of the above-mentioned characteristic length scales is the *deBroglie wavelength* λ_{dB} of the charge carriers. Systems, which are smaller than λ_{dB} in 1, 2, or 3 dimensions are called quantum wells, quantum wires, or quantum dots, respectively. In this respect, semiconductor nanoparticles, which are considered in the framework of this review, can be regarded as quantum dots under ambient (room temperature) conditions when their diameters are below 10 nm. In nanoparticulate materials, they will form a network with—depending on the production process—weak or strong coupling between the individual dots.

A characteristic of nanoparticle arrays is their large surface-to-volume ratio, and charge carrier transport in such nanoparticulate networks is influenced by surface processes affecting their space charge region. This influence is described by the material-specific Debye length λ_D or depletion length of the (doped) nano-material. It also has a typical length scale in the nanometer range. As a result, the depletion region can extend over the entire volume of small particles in the 10 nm range. Such a system can be considered electronically to be 'surface only'. Thus, for weakly

10 Electrical Transport in Semiconductor Nanoparticle Arrays

compacted nanoparticle material, great care has to be taken to ensure a high enough carrier density to overcome this problem. On the other hand, the large surface-to-volume ratio offers great opportunities for using nanoparticulate layers in sensing devices. Physical processes such as molecular adsorption as well as chemical reactions between the nanoparticle's surface and the constituents of the surrounding gas lead to a change in surface potential and charge. This modification at the surface will translate into a change in carrier density, which in turn will affect the overall conductivity of the nanomaterial. The use of these effects for sensor applications will be discussed in Sect. 10.6.

In order to describe the complex dielectric response of a network of weakly coupled grains, the so-called "brick-layer-model" was established by van Dijk and Burggraaf [3]. An extension of this model will be introduced in Sect. 10.5, which takes into account the porosity of nanoparticulate material, produced by different common fabrication techniques as discussed in Sect. 10.4.

In applications where the above-mentioned surface effects and the low conductivity, caused by the weak interaction of the particles, is detrimental, the material needs to be sufficiently compacted. Two methods will be discussed in more detail within this review: Mechanical compaction by application of unidirectional force (Sect. 10.6.4) and a model describing particle compaction using a combination of pressure and electrical current-induced heat (Sect. 10.5.2). The latter method (employed, e.g., in spark plasma sintering) is of particular interest, because it can lead to material, which has the full density of bulk semiconductor, while maintaining the nanocrystalline local structure. Such materials are especially interesting for thermoelectric applications, where the thermal conductivity needs to be minimized without increasing the electrical resistance.

10.2 Principles of (Nano) Particle-Based Conduction Processes

In typical mesoscopic arrays of particulate matter such as solid-state sensor geometry, the elementary reactions of surface processes originating from adsorption and desorption of (gaseous) species are transduced into electrical signals that can be detected by an appropriate measurement technique. As a result, the initial sensing process itself takes place at the available surface of the respective material, usually consisting of semi-conducting metal oxides. Due to their high surface-to-volume ratio, nanosized materials provide a high surface area even at low mass. As long as a porous layer structure is obtained, its entire volume is accessible to the gases, and therefore the active surface is much larger than the area of the layer.

According to Barsan et al. [4], the electrical conductivity within a porous, three-dimensional network of interconnecting nanoparticles is complicated due to the presence of necks between the grains. A typical image of such an interconnecting network is shown in Fig. 10.1.

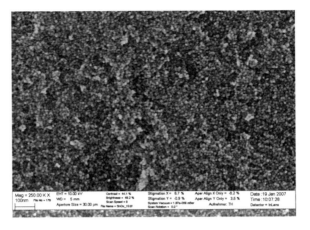

Fig. 10.1 Scanning electron microscopy image of a porous, interconnecting network of tin oxide nanoparticles

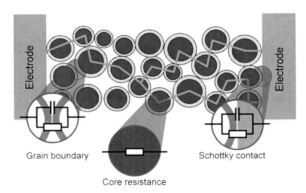

Fig. 10.2 Schematic representation of electrical transport processes within a mesoscopic network. The *gray* area surrounding the *blue* particles indicates the depletion layer. Equivalent circuits are commonly used to describe the different transport processes shown. Such an equivalent circuit model only applies, if the particles are not too small, which will be discussed in the introduction of Sect. 10.5

The overall conductivity in such a network can be separated into different types of conduction processes originating from grain boundaries, the (nano) particles themselves and the contact area between particles and connecting electrodes. A simple method to describe the electrical inter- and intraparticle transport processes uses equivalent circuits representing the different phenomena. The two main elements used within the equivalent circuits are capacitors (C) and resistors (R) and a lot of effort is made to describe the electrical behavior of (nano)particulate arrays by means of equivalent circuits [5]. We will denote the core resistance by R_{core}, the particle–particle-related impedance (grain-boundary) by Z_{part}, and the particle-electrode-related impedance (Schottky contact) by Z_{wall} (Fig. 10.2).

10.3 Electrical Measurement of (Nano) Particle Arrays

A typical, commonly used method in sensor technology, is the measurement of conductance or resistance by DC methods, either potentiostatic or galvanostatic. Besides the fact that these simple techniques deliver easily processible signals, much information with respect to the different electrical transport phenomena is lost. However, these methods usually deliver sufficient information in sensor application, but for detailed characterization dielectric or impedance spectroscopy (IS) has to be used, also accounting for capacity-related effects.

10.3.1 Impedance Spectroscopy

IS or dielectric spectroscopy is a very useful method for a much more detailed characterization of the electrical properties of (nano)particulate networks compared to DC methods. The measurement technique involves the application of an electrical stimulus (voltage or current) to materials connected to electrodes and investigates the corresponding response as a function of frequency. This process is performed with the basic assumption that the properties of the electrode-sample system do not vary within the measurement time. The objective of IS is to obtain information on the variation of the intrinsic properties of the sample materials based on parameters such as temperature, stoichiometry, and surrounding atmosphere and to relate these intrinsic properties to a suitable equivalent circuit. An excellent overview with respect to theory, application, and interpretation of impedance measurements is given by Barsoukov and Macdonald [5].

Several methods exist to induce stimuli for IS measurement. However, the most common approach, which is also used for the research presented here, involves the measurement of impedance directly in the frequency domain by applying a single-frequency voltage to the contacting electrodes. This is followed by measuring the phase shift as well as the amplitude of the resulting current at that frequency. The measured values can be converted into several complex values such as impedance (Z), admittance (Y), electric modulus (M), and relative permittivity (ε). The impedance measurement can be performed over a wide frequency range from mHz to MHz.

IS signals are often presented in frequency representation or in a Nyquist diagram, where the X-axis represents the real part Z' while the Y-axis represents the imaginary part Z'' of the complex impedance Z. The experiments are typically carried out within a specific temperature and frequency range. The analysis of the IS data is performed by fitting them to an equivalent circuit consisting of passive electronic devices [6]. This way, contributions from grain boundary, core conductivity, and Schottky contacts can be separated. In many cases, the specific conductivities follow an Arrhenius law:

$$\sigma = \sigma_0 \cdot e^{-E_a/kT} \tag{10.1}$$

with activation energies E_a depending on the transport process. This leads to admittances $Y = Z^{-1}$ of the equivalent circuit, which have the same temperature dependence.

10.3.2 Conductivity Measurements During Powder Compaction

One technique to manufacture mechanically stable, interconnecting (nano)particle arrays is to form stable pellets from the powder by applying a uniaxial force. Uniaxial pressing of thin disks leads to coplanar samples that can be directly mounted between connecting electrodes. This is a common technique to prepare samples for electrical testing.

When a powder is poured into a container, its particles settle into an initial arrangement which depends on, e.g., the shape, the size distribution, the surface properties, and the forces between and onto the particles the powder consists of. Also the shape of the container and the properties of its walls play a role. The arrangement of the particles has a big influence on the macroscopic mechanical and electrical properties of the powder since it determines the amount and shape of the particle contacts.

The porosity, meaning the ratio between the volume of void space in the material and the total volume of the container, is a measure to compare the arrangements of different powders. For large grain sizes, the gravitational force prevails over the other forces so that the powder settles into a random dense packing. Arc-like structures are formed due to cohesive forces and friction between the particles and between the particles and the walls, which lead to an inhomogeneous force distribution within the powder. Because of the void spaces underneath the arcs, this is a metastable state. Vibrations can break up the local particle network again and decrease the porosity of the powder slightly further.

Nanoparticle powders can also be called cohesive powders since the gravitational force is not dominant anymore on this size scale. For example, friction, van-der-Waals forces, and electrostatic interaction cause long-range particle networks leading to a high initial porosity of the particles. To be able to compare the properties of powders with different grain size, a similar porosity has to be reached. This can be done by applying an external force onto the powder. Using a cylindrical pressing tool the powders are compressed by means of a hydraulic press along the cylinder axis [7].

The experimental setup applied to prepare disks for electrical measurements was also used for conductivity measurements during powder compaction to investigate, whether size and morphology of our nanosized powders influence their electrical and mechanical behavior.

The powders are characterized by determining in situ the conductance, impedance, and density while a uniaxial mechanical force ranging from 10 to 100 kN is applied in a stepwise manner. The force is applied to the cylindrical sample cell (diameter 13 mm) shown in Fig. 10.3 along the cylinder axis using a hydraulic press. The force range corresponds to a pressure between 75 and 750 MPa for the given experimental setup. The load plates of the sample cell are used as electrodes for the electrical

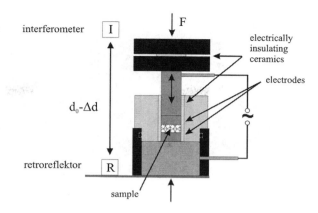

Fig. 10.3 Sample cell for powder compaction and characterization under a uniaxial force using a hydraulic press. The two electrodes are used for DC and impedance measurements while the laser interferometer determines the change in density [7]

measurements, and by measuring the I/V-characteristic it is verified that they can be used as ohmic contacts.

10.4 Formation of Nanoparticle Arrays

10.4.1 Compaction of Nanoparticle Powders

As already described above, thin disks of nanoparticles were prepared from nanoparticle powder by uniaxial pressing. The change in density during powder compaction was determined with high resolution by means of a laser interferometer. It is found that even after applying a force of 100 kN (750 MPa), which is the mechanical threshold of the setup, the minimum porosity of pressed nanoparticle powders with particle sizes between 4 and 50 nm in diameter never went below 40 % which is considerably more than the theoretical value for a dense packing of spheres. This behavior is attributed to the cohesive forces between the nanoparticles [8]. Only materials consisting of particles in the μm range could be compacted to disks with porosities of 30 % and less. Scanning electron microscopy of the pressed samples indicates that the surface is almost flat but porous, while the inner area consists of a highly interconnecting porous network of nanoparticles, see Fig. 10.4.

It can be noted that mechanical deformation and sometimes breaking of nanoparticles is found for particles that have been in direct contact with the pressing tool while no deformation is found for particles next to this first layer.

10.4.2 Printing of Nanoparticle Thin Films

Our second approach toward the formation of highly interconnecting nanoparticle networks is based on printing of dispersions made from nanoparticles as current semiconductor fabrication technologies such as photolithography are not feasible

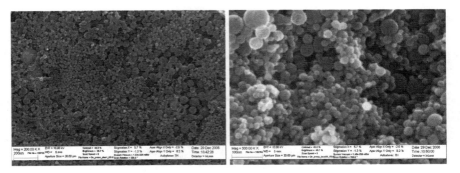

Fig. 10.4 Surface (*left*) and breaking edge (*right*) of a pressed nanoparticle disk. Please note: for better illustration of the materials properties, the images are shown with different resolution

for nanoparticle processing. If the particle concentration in the dispersion is very low due to stability requirements, technologies such as screen printing and spin coating are not useful. Additionally, these technologies need masks to write a well-defined pattern so that they are not suitable for the fabrication of a small number of samples. Using inkjet printing, the deposition of very small volume of suspensions in a well-defined pattern is possible and the direct structured deposition removes the need for masks leading to an efficient use of material. Moreover, inkjet printing offers the deposition of nanoparticle dispersions with very low concentration due to the possibility of multilayer printing. Therefore, it is not surprising that inkjet printing has become a quite important technology for optoelectronic and electronic devices such as diodes, light emitters, or organic electronics [9, 10].

10.4.2.1 Formation of Dispersions From Nanoparticles

Printing of any material requires stable dispersions of (nano) materials in a volatile solvent. Within the past years, water has become a very attractive solvent and as a result many substances were developed to stabilize particulate matter in water. We have used ZnO nanoparticles synthesized by chemical vapor synthesis (CVS) [11–13] to prepare stable dispersions of ZnO nanoparticles in water. The ZnO dispersions were produced by creating a mixture of 7 wt% ZnO nanoparticles and a polymeric stabilizer (in this case TEGO). Homogeneous and stable dispersions were achieved after sonication at a power of at least 30 W for at least 2.5 h. They were stable for several weeks as verified by dynamic light scattering (DLS).

10.4.2.2 Formation of Nanoparticle Thin Films on Interdigital Electrodes

Compared to common solid-state sensor geometry, the particles used for our devices are very small with a typical particle diameter of no more than a few 10 nm. Therefore,

Fig. 10.5 SEM picture of an interdigital electrode structure prepared on an Si/SiO$_2$ substrate

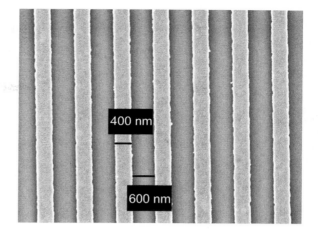

we developed small electrode structures with a very small interelectrode distance in the range of a few hundred nanometers [14, 15]. The main reason for utilizing the small structures is the following: In common sensor technology a thermal annealing step is applied [16] to strengthen the interparticle contacts and increase the overall conductivity substantially. In case of nanoparticles, this annealing step would lead to a significant particle coarsening effect (see for instance [17, 18]), destroying the specific nanoscopic features of the materials. Hence, we have minimized the distance of the interdigitating electrodes to receive suitable conductances even for as-prepared particles without any annealing.

Details of typical interdigital electrodes used for these experiments are shown in Fig. 10.5. It is prepared by electron beam lithography (EBL) on top of a silicon substrate covered by an electrically insulating SiO$_2$ layer. The electrodes consist of gold with an electrode–electrode distance of about 600 nm.

Preliminary to inkjet printing of nanoparticle thin films, the substrates with the electrodes were cleaned for 10 min in a "piranha" solution (3:1 sulfuric acid and hydrogen peroxide) and washed with de-ionized water.

Metal oxide thin films were produced from nanoparticle dispersion by means of a Dimatix 2800 inkjet printer. The printed films have a porosity of about 30% and show negligible roughness. They are smooth and crack free and were produced with thicknesses between 100 and 250 nm, see Fig. 10.6.

10.4.3 Molecular Beam-Assisted Deposition

The experimental facilities used for particle synthesis in gas phase reactors enabled us to deposit nanoparticles directly from a reaction chamber onto any substrate by molecular beam-assisted deposition of a nanoparticle layer (see Chap. 1). Usually,

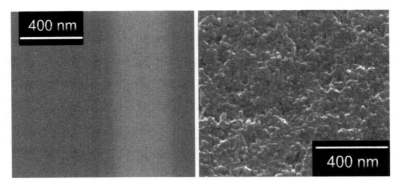

Fig. 10.6 *Left* SEM picture of a part of the inkjet printed film. One of the gold electrodes can be seen as a bright area. The film is very smooth and crack free. *Right* SEM picture of the surface of an inkjet printed film

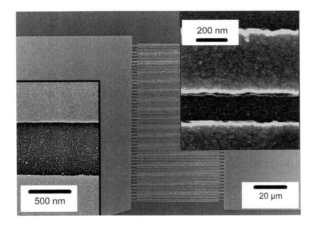

Fig. 10.7 SEM image of an interdigitated electrode structure and details of the deposited nanoparticle film

this technique is applied to TEM grids, but we used the same method to deposit multilayers of particles onto interdigitated electrode structures. The main advantage of this technique is the formation of a surfactant-free thin film because it does not require any stabilizers.

With this technique, porous thin films are prepared from individual single particles that are extracted from the reaction chamber by means of a supersonic molecular beam and subsequently impacted on the interdigitated electrode structure. The thin film shown in Fig. 10.7 consists of nanoparticles with a mean diameter of only 7 nm. Due to this fact, individual particles can hardly be observed. The film thickness was found to be about 50 nm. As shown in Chap. 1, a working sensor can be obtained

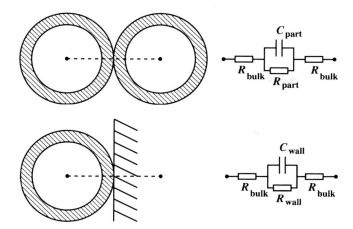

Fig. 10.8 A particle–particle contact is represented by the electrical circuit as shown at the *upper right*. *Below* Same for a particle-electrode contact

even with this small amount of particles, see also [14]. A very similar technique was used to prepare sensor structures from size-selected SnO_2 nanoparticles, see Chap. 4.

10.5 Modeling of Electrical Transport in (Nano) Particulate Networks

The current distribution in a nanoparticle agglomerate between electrodes does not only depend on the properties of the particles, but also on their geometrical arrangement. The impedance of such a system is a function of both, the particle and the geometrical properties. Computer simulations of particulate networks can help to disentangle these properties. Moreover they provide insight about the mechanisms behind current and pressure-induced structural changes.

As already implied by Fig. 10.2, it is assumed that the phase coherence length of electrons is much smaller than the particle diameter. Then only the transmission probability from one particle to the next has to be considered. Moreover, it is assumed that there is enough inelastic scattering inside a particle to guarantee local equilibrium, in other words: One can attribute to each particle a local electrochemical potential and a local temperature. Under these conditions one may replace the particulate network by a classical electrical circuit as shown in Fig. 10.8.

For simplicity let us assume that all particles are equal. They consist of a spherical conducting core covered by a thin, much less conducting surface layer. The particle centers are the nodes of the electrical circuit. Three parameters are needed to describe a particle–particle contact: the core resistance $R_{core} = 2R_{bulk}$, the tunnel resistance of the surface layers, R_{part}, and the capacitance C_{part} associated with the two conducting cores separated by the surface layers. The impedance of such a connection between two nodes k and l of the network to an alternating current of frequency ω is

$$Z_{kl}(\omega) = R_{\text{core}} + \frac{R_{\text{part}}}{1 + i\,\omega\,R_{\text{part}}C_{\text{part}}}. \tag{10.2}$$

Denoting the frequency-dependent admittance by $Y = Z^{-1}$, the current is given by

$$I_{kl} = Y_{kl}(\varphi_k - \varphi_l), \tag{10.3}$$

where φ_k is the electrochemical potential at site k.

The electrodes are represented by two external nodes of the network. In the following they will be labeled by the indices 0 and N. The connection between a particle and an electrode is modeled in the same way as a particle–particle contact, but with a different contact resistance R_{wall} and a different capacitance C_{wall} (note that half of R_{core} is then attributed to the electrode).

An important assumption made in this model is that the core capacitance C_{core} of the particles is negligible compared to the boundary capacitance, C_{part}. In the network model, C_{core} would be represented by a capacitor that is on one side attached to the node (the particle center) and grounded (or connected to a gate voltage) on the other. The core capacitance is proportional to the particle radius r, while the boundary capacitance is of the order r^2/w, where w is the thickness of the insulating layer between the two particles. The assumption is therefore justified as long as $w \ll r$.

In order to calculate the effective impedance Z_{eff} of the network as a whole, one must solve Kirchhoff's law, $\sum_k I_{kl} = 0$, for all nodes $l \neq 0, N$. For the external nodes one has $\sum_k I_{k,0} = -I_{\text{ext}}$ and $\sum_k I_{k,N} = I_{\text{ext}}$, because the external current I_{ext} enters the network at one electrode (node 0) and leaves it on the other side (node N). An efficient algorithm for solving this coupled set of linear equations was developed in [19]. As a result one gets the electrochemical potentials φ_k at all nodes and via Eq. (10.3) all currents $I_{k,l}$. The effective impedance of the network is then given by

$$Z_{\text{eff}} = \frac{\varphi_0 - \varphi_N}{I_{\text{ext}}}. \tag{10.4}$$

This calculation can be performed for any particle and electrode configuration and helps to interpret experimental impedance data. In the following, however, two key results will be discussed, where a particularly simple particle and electrode configuration has been chosen: The particles occupy a subset of sites of a simple cubic lattice of size $L_x \times L_y \times L_z$ with periodic boundary conditions in x- and y-direction. The electrodes are the boundaries perpendicular to the z-direction, i.e., they are planar and parallel, of size $L_x \times L_y$, and a distance L_z apart. All particle diameters are equal to the lattice constant d, so that only particles on nearest neighbor sites are in contact.

Fig. 10.9 Calculated currents through a particle configuration with parameter set 1 (Table 10.1) at an angular frequency $\omega = 10\,\text{Hz}$. Thickness of red lines indicates strength of local current. Electrodes are at the top and bottom

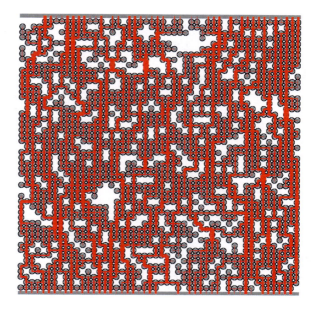

Table 10.1 Parameter sets $R_{core} = 100\,\Omega$

	1	2	3
R_{part}	5 MΩ	5 kΩ	5 kΩ
R_{wall}	5 kΩ	5 kΩ	5 MΩ
C_{part}	100 pF	10 pF	10 pF
C_{wall}	10 pF	10 pF	100 pF

10.5.1 How the Macroscopic Impedance Depends on Sample Geometry

In order to study the influence of powder porosity and electrode geometry on the macroscopic impedance, the lattice sites were randomly occupied by particles with a probability $(1 - p)$. Empty sites correspond to pores. Therefore, p may be called porosity. It must be small enough that there is a percolating path through particles, which connects the two electrodes [20]. Any isolated particles or particle clusters can be deleted. Figure 10.9 shows such a particle configuration with porosity $p = 0.2$ on a $50 \times 1 \times 50$ lattice.

The current sequentially passes through an electrode-sample contact, then through the sample, and finally again through a sample-electrode contact. Therefore, the effective impedance must be a sum of a sample impedance, which should only depend on Z_{part}, and 2 times a contact impedance, which should only depend on Z_{wall}. For dimensional reasons the effective impedance must therefore have the form

$$Z_{\text{eff}}(\omega) = g_{part} Z_{part}(\omega) + 2 g_{wall} Z_{wall}(\omega) \qquad (10.5)$$

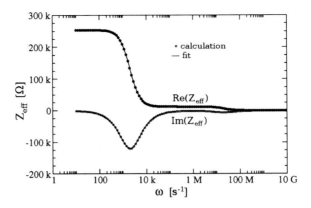

Fig. 10.10 Frequency-dependent real part (Z', upper data) and imaginary part (Z'', lower data) of the impedance; Eq. (10.4) can be fitted perfectly for all frequencies using geometry factors (Eq. 10.7) that do not depend on Ω. Data were calculated for parameter set 1 (Table 10.1). The impedance varies strongly only in the neighborhood of the characteristic frequencies $(R_{part}C_{part})^{-1} = 2\,\text{kHz}$ and $(R_{wall}C_{wall})^{-1} = 20\,\text{MHz}$

with dimensionless factors g_{part} and g_{wall}, which describe the geometrical properties of the sample, respectively, the sample-electrode contacts. Assuming that these geometry factors do not depend on frequency, they are determined by the low and high frequency limits of the impedance. For $\omega \to \infty$ the capacitances short circuit the contact resistances, so that

$$Z_{\text{eff}}(\omega \to \infty) = R_\infty = (g_{part} + 2g_{wall})R_{core}. \tag{10.6}$$

The DC-resistance is given by

$$Z_{\text{eff}}(0) = R_0 = g_{part}R_{part} + 2g_{wall}R_{wall} + R_\infty. \tag{10.7}$$

Solving for the geometry factors gives

$$\begin{aligned} g_{part} &= \frac{1}{R_{part} - R_{wall}}\left[R_0 - R_\infty\left(1 + \frac{R_{wall}}{R_{core}}\right)\right], \\ g_{wall} &= \frac{-1/2}{R_{part} - R_{wall}}\left[R_0 - R_\infty\left(1 + \frac{R_{part}}{R_{core}}\right)\right]. \end{aligned} \tag{10.8}$$

Figure 10.10 validates the assumption that the geometry factors do not depend on the frequency: Inserting Eq. (10.8) in (10.5) fits the real (Z') and the imaginary part (Z'') of the impedance perfectly in the whole frequency range.

Obviously, the geometry factors (Eq. 10.8) do not depend on the microscopic capacitances. Moreover, keeping the porosity and the electrode geometry fixed, it turns out that parameter sets 1 and 3 (Table 10.1) give the same values for g_{part} and

10 Electrical Transport in Semiconductor Nanoparticle Arrays 245

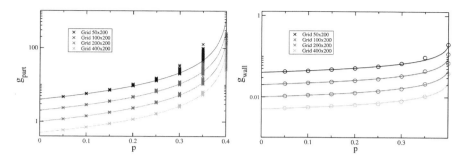

Fig. 10.11 Geometry factors versus porosity p. *Left* g_{part}, *right* g_{wall}. Data points were calculated with parameter set 3 (Table 10.1) and averaged over 100 particle configurations. Only data for $L_y = 1$ and $L_z = 200$ are shown. L_x equals 50 for the uppermost data, and doubles when going one curve down. The curves are fits with Eq. (10.9), where the following values were used as effective exponents $\nu_{\text{part}} = 1.25$, $\nu_{\text{wall}} = 0.38$; $p_c = 0.407$.

g_{wall}. For set 2 only the combination $g_{\text{part}} + 2 g_{\text{wall}}$ can be determined and agrees with the corresponding values obtained from the other two sets. This indicates that the geometry factors do not depend on the microscopic resistances either.

In conclusion, g_{part} and g_{wall} depend only on L_x, L_y, L_z, on the porosity p, and on the particle diameter d, which justifies to call them *geometry* factors. The impedance must diverge at the percolation threshold p_c, as no conducting path connects the electrodes any more for larger porosities. Figure 10.11 illustrates that

$$g_{\text{part}} = \frac{L_z d}{L_x L_y}\left(1 - \frac{p}{p_c}\right)^{-\nu_{\text{part}}}, \quad g_{\text{wall}} = \frac{d^2}{L_x L_y}\left(1 - \frac{p}{p_c}\right)^{-\nu_{\text{wall}}} \quad (10.9)$$

This result is very intuitive: The effective impedance depends on the sample dimensions as expected from Ohm's law. The effective frequency-dependent particle/particle and particle/wall contact resistivities are

$$\varsigma_{\text{part}}(\omega) = d\left(1 - \frac{p}{p_c}\right)^{-\nu_{\text{part}}} Z_{\text{part}}(\omega) \text{ and respectively}$$

$$\varsigma_{\text{wall}}(\omega) = d^2\left(1 - \frac{p}{p_c}\right)^{-\nu_{\text{wall}}} Z_{\text{wall}}(\omega). \quad (10.10)$$

They diverge as power laws, when the porosity approaches the percolation threshold. In the two-dimensional case ($L_y = 1$), $p_c = 0.407$ [21]. For larger L_y, percolation becomes easier, so that the porosity threshold p_c increases. The effective exponent $\nu_{\text{part}} \approx 1.25$ for larger systems should get closer to the literature value $\nu_{\text{part}} \approx 1.31$ [22, 23]. The effective exponent $\nu_{\text{wall}} \approx 0.38$ is significantly smaller.

10.5.2 A Simple Model for Current-Assisted Powder Compaction

Above it has been shown that electrical currents can be used to monitor the contact network in a powder and its connection to the electrodes. However, one also has to consider connectivity changes inflicted by the currents themselves. The contact parameters such as Rpart may change due to microscopic welding, or the network topology may evolve due to current-induced particle rearrangements. Both effects are well documented experimentally [24, 25].

In the following, results will be presented, which were obtained from a simple model for powder compaction assisted by a direct current [24].[1] Consider a sample geometry as shown in Fig. 10.9. The idea is that Joule heat is preferentially delivered in those cross-sections parallel to the electrodes, which contain a particularly low number of particles. These are the bottlenecks for the current, hence lead to a high current density. If the surface of a particle at such a hot spot becomes viscous, the particle may "flow" into a neighboring pore, where it cools down and sticks again. This can disrupt the current path and makes the cross-section even more suscep-tible for developing another hot spot. Ultimately, the cross-section may become so depleted that the structure yields to a pressure applied to the electrodes, which can be substantially lower than the consolidation pressure of the same sample in the absence of currents [8]. This is believed to be the mechanism behind current- assisted powder compaction.

This simple model is illustrated in Fig. 10.12. Particles are represented by occupied cells on a square lattice. Nearest neighbor particles are in contact and their connection has a resistance $R = R_{\mathrm{part}} = R_{\mathrm{wall}}$, which is assumed to be constant. The local currents are calculated, and a surface melting criterion is applied, which will be described in the next paragraph. Particles with molten surface are displaced into a randomly chosen empty neighbor cell, where they cool down and stick again. If no percolating path exists between the electrodes, the structure yields to the applied pressure, until percolation is reestablished.

Usually the current is applied in pulses of duration Δt_1. The time interval between two pulses $\Delta t_2 - \Delta t_1$, is assumed to be long enough to let all temperature differences within the sample relax. The simulation proceeds in time steps Δt_2. A current I_{ij} between particles i and j is assumed to deliver an equal amount of Joule heat to both sides, so that the total heat received by particle i in one pulse is

$$\Delta Q_i = \frac{1}{2} \sum_j R I_{ij}^2 \Delta t_1. \qquad (10.11)$$

The process can be regulated such that the average heat delivered in one pulse compensates for the heat losses between two pulses, so that the sample temperature is kept constant on average. The heat losses are proportional to the temperature difference between the sample and its environment, so that one can set

---

[1] In the literature the process is also called "spark plasma sintering" or "field assisted sinter tech-nique".

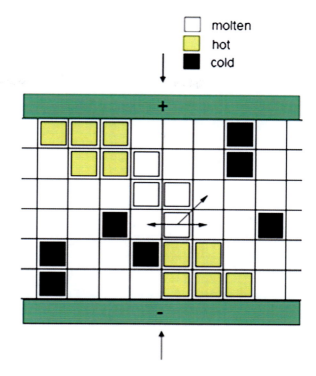

Fig. 10.12 Illustration of a lattice model for current-assisted powder compaction

$$\langle \Delta Q \rangle = \frac{1}{N} \sum_i \Delta Q_i = \kappa \left(T_{\text{sample}} - T_{\text{env}} \right) \Delta t_2. \tag{10.12}$$

Particle *i* begins to melt, if

$$\Delta Q_i \geq C \left(T_{\text{melt}} - T_{\text{sample}} \right), \tag{10.13}$$

where C denotes the heat capacity of the particle and T_{melt} its melting temperature. Using Eq. (10.12), the melting condition (Eq. 10.13) can be written in dimensionless form as

$$\frac{\Delta Q_i}{\langle \Delta Q \rangle} \geq \frac{C}{\kappa \Delta t_2} \left(\frac{T_{\text{melt}} - T_{\text{env}}}{T_{\text{sample}} - T_{\text{env}}} - 1 \right) \equiv m. \tag{10.14}$$

m is the only nontrivial parameter in the model. The smaller m, the closer the sample temperature T_{sample} to the melting point of the particles, T_{melt}. For large m, only extreme fluctuations above average heat delivery lead to melting, while for $m = 1$ any positive deviation from average suffices. The melting and hence the compaction process stops, when Joule heating becomes uniform enough.

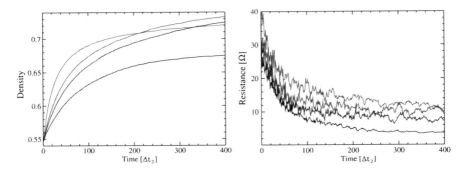

Fig. 10.13 Time evolution of the density (*left*) and the resistance (*right*) for the *m*-values 4, 12, 16, 24 (top down for small time). Data were averaged over 50 runs. Resistance fluctuations are much stronger than density fluctuations

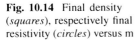

Fig. 10.14 Final density (*squares*), respectively final resistivity (*circles*) versus m

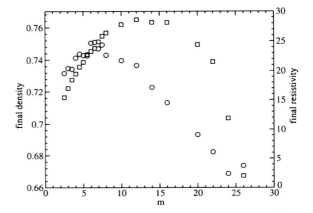

This model shows the expected current-assisted compaction, associated with a decrease of resistance, see Fig. 10.13. The product of density times resistance is proportional to the resistivity, which also decreases in all cases. In the long time limit the density and the resistivity approach stationary values, because the current pulses no longer lead to further compaction. These values depend sensitively on the sample temperature, i.e., on m, as shown in Fig. 10.14. Maximal compaction is reached near $m = m_0 = 12$. For larger m (lower sample temperature), isolated hot spots form very sporadically, and the chance that a cross-section of the sample becomes sufficiently depleted is too small. For smaller m (sample temperature closer to the melting point) current paths melt almost completely. Then the particle displacements do not deplete preferentially a low density cross-section further.

Surprisingly, the final density and the final resistivity vary in a correlated way, with the exception of a small interval $6 < m < 12$. Optimal compaction does not imply optimal conductivity. At first sight this result seems counterintuitive, because one expects that moving the electrodes toward each other creates more parallel current

10 Electrical Transport in Semiconductor Nanoparticle Arrays 249

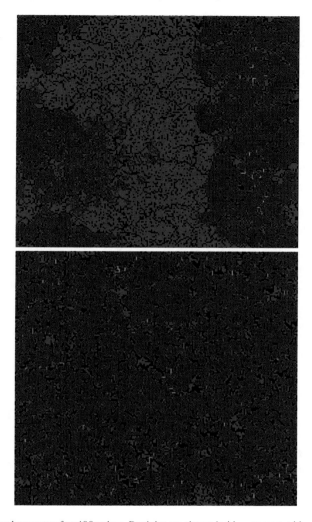

Fig. 10.15 Local currents after 400 pulses. Particles are shown in *blue*, pores are *black*. *Top* Optimal conditions for compaction, $m = m_0$. Density is 0.727 particles per cell, and density × resistance [R] is 17.27. This value (proportional to resistivity) is high because of little parallelism in the network. *Bottom* $m = 2m_0$. Density is 0.676, and density × resistance [R] is 2.78. Resistivity is much lower, because of high parallelism in the network. Color coding of the local currents uses the same scale in both pictures: maximal currents are shown as *white*, medium ones as *red*, and weak ones as *black* connections between neighboring particles. Currents weaker than the maximal current/1000 are not shown

paths and hence reduces the resistivity of the sample. This intuition is correct for compaction at fixed m. However, if one compares samples compacted at different temperatures, the intuition is misleading. The reason is a subtle, temperature-dependent self-organization of pores in the sample. For compaction it is favorable, if cross-sectional depletion zones get amplified, i.e., if pores grow preferentially parallel to the electrodes. This, however, cuts off parallel current paths. This explains, why optimizing the final density may increase the final resistivity of the sample. This picture is confirmed by Fig. 10.15

10.6 Examples of Nanoparticle Array Conductivity and Sensitivity

10.6.1 Tin Dioxide

A systematic investigation of impedance properties of particle networks based on almost monodisperse nanoparticles has been carried out by electrical characterization of size-selected SnO_2 nanoparticles. As has been described in detail before, the overall impedance depends on geometrical factors, particle size, and porosity (see Eqs. 10.7–10.9). As long as the geometry and porosity are kept constant, materials with different size can be directly compared with respect to information available from impedance measurements.

10.6.1.1 Electrical Characterization

Tin dioxide nanoparticles with 10, 15, and 20 nm in diameter and a geometric standard deviation of less than 1.2 (lognormal size distribution) were deposited by means of impaction on interdigitated structures similar to those shown in Fig. 10.7. Details concerning the preparation of the samples can be found in Chap. 4 and [15, 26]. For all particle sizes, the geometry of the interdigitated structure was identical and from electron microscopy investigations it was found that the porosity was low and did not vary significantly with particle size. Because the porosity is far away from the percolation threshold, it can be assumed (see Eq. 10.9) that the influence of small changes in porosity is negligible and that therefore changes in the impedance mostly originate from changes in particle size.

Temperature-dependent impedance measurements of SnO_2 were carried out with a Solatron SI 1255 frequency response analyzer (FRA) in combination with a dielectric interface SI 1296. Data were sampled in the frequency range between 1 Hz and 1 MHz with 17 impedance measurement points/decade and in the temperature range between 100 and 250 °C in steps of 10 °C. Before starting an impedance measurement, the system was thermally stabilized by waiting 30 min at the desired temperature. Due to this procedure, the temperature deviation during the measurements was less than ±0.3 °C.

Figure 10.16 shows a typical impedance measurement received from 10 nm SnO_2 particles. For comparison it was plotted according to Fig. 10.10. As can be seen,

10 Electrical Transport in Semiconductor Nanoparticle Arrays 251

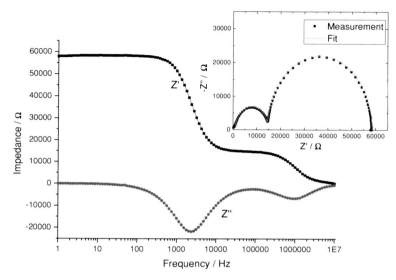

Fig. 10.16 Impedance measurement of 10 nm SnO$_2$ nanoparticles deposited on an interdigitated structure. The inset shows the respective Nyquist plot and the fitted data (see text)

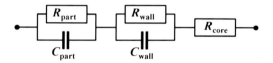

Fig. 10.17 Equivalent circuit used to fit the experimental results of the impedance measurements on tin dioxide nanoparticle arrays. Note that the denotation used here is associated with the effective values while the denotation used in Sect. 10.5 describes the microscopic values of each contact

the material exhibits two resonance frequencies around 2.5 kHz and 1 Mhz and the consistency of both, the theoretical description based on the simple model shown in Fig. 10.2 and the experiment is obvoius. The inset in Fig. 10.16 displays the same measurement in the Nyquist representation.

10.6.1.2 Data Fit and Transport Processes

The data were fitted by means of a nonlinear least square fit to an equivalent circuit representing all contributions to the total impedance as shown in Fig. 10.17. An almost perfect agreement between measured data and fit is found (correlation coefficient $= 0.99$) supporting the possibility to describe the overall impedance of the SnO$_2$ nanoparticle network by the chosen combination of passive electrical elements. (Note that not only the position and diameter of each semicircle is fitted, but also that every measured point in the complex plane $(Z'(\omega), Z''(\omega))$ is accurately reproduced.)

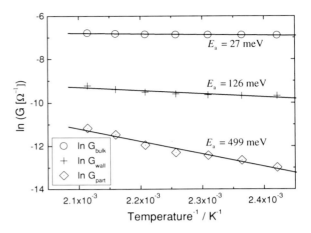

Fig. 10.18 Arrhenius diagram of the different charge carrier transport processes. The activation energies E_a calculated from the slope are given in the plot

Due to the fact that the resistance of metal-oxide semiconductors such as tin dioxide can be affected by temperature [27], the different contributions of bulk, particle–particle, and particle-electrode impedance should be detectable by a change in resistance of R_{bulk}, R_{part} and R_{wall}. This assumption holds as long as no changes in the nanoparticle network occur that lead to a change in capacity. Temperature-dependent measurements were fitted to the equivalent circuit shown in Fig. 10.17 and the parameters for the different resistances and capacities were extracted from the fits. The determined values for C_{part} and C_{wall} were constant within the accuracy of the fits (confirming the above assumption that no change in the network topology occured), while the values for the different resistances show a clear trend. The fit results received for the different resistances were used to calculate the activation energies of the different conduction processes using

$$G = G_0 \cdot e^{-E_a/kT} \qquad (10.15)$$

with $G = 1/R$. The values received for G_{bulk}, G_{part} and G_{wall} are summarized in Fig. 10.18.

The Arrhenius plots show the conductances $G = 1/R$ calculated from the non-linear least square fit. The smallest activation energy of 27 meV is the fit result obtained for G_{bulk}. The assignments of G_{wall} and G_{part} are based on the simple model that the distance—which is the main parameter for charge carrier transport—between two conducting particle cores is bigger than the distance between one core and the electrode. Therefore, it can be assumed that the particle-electrode (G_{wall}) and particle–particle contribution (G_{part}) are related to activation energies of $E_a(G_{wall}) = 126$ meV and $E_a(G_{part})$ 499 meV, respectively.

The impedance spectra of 15 and 20 nm are shown in Fig. 10.19. In contrast to the measurements on 10 nm particles no clearly separated semicircles are observed

10 Electrical Transport in Semiconductor Nanoparticle Arrays

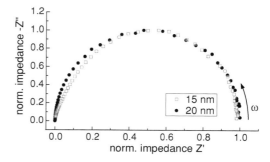

Fig. 10.19 Normalized Nyquist plots of 15 and 20 nm SnO$_2$ nanoparticles measured at 150 °C

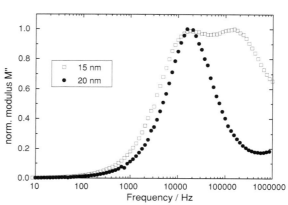

Fig. 10.20 Normalized imaginary part of the modulus function against frequency for 15 and 20 nm SnO$_2$ nanoparticles measured at 150 °C

as can be seen from the normalized Nyquist plots, only a slight shoulder at high frequencies of the graph representing the 15 nm particles measurement can be seen.

A closer insight can be obtained by plotting the imaginary part of the modulus versus the measuring frequency. M'' has been calculated from

$$M'' = \omega \cdot C_0 \cdot Z' \qquad (10.16)$$

with the angular velocity $\omega = 2\pi f$ and the background capacity C_0 of the interdigitated structure. This kind of representation enhances the high frequency regime and enables a more detailed visualization of structures which are almost hidden in the Nyquist plot. Figure 10.20 shows the normalized imaginary part of the modulus for the two samples. In the case of 20 nm particles only one signal at about 10^4 Hz can be observed. Its maximum corresponds to the maximum of the semicircle in the Nyquist representation. The spectrum of the 15 nm particles exhibits two signals that are only slightly separated.

From the graphs shown in Figs. 10.16, 10.19 and 10.20 it is obvious that the distance between the two maxima indicating two different charge transport processes decreases with increasing particle size until only one signal is left for particles with 20 nm in size. This is due to the fact that in such cases the parameters determining the electrical properties of the different transport processes lead to similar resonance frequencies (see also parameter set two in Table 10.1). In such cases, different mechanisms cannot be separated, neither visually nor with support of fit programs. As a result, one of the RC elements shown in Fig. 10.17 has to be neglected because otherwise the fit of the measurement data is overdetermined. In such cases, the respective fit result of the residual RC element contains the sum of both transport processes (particle–particle and particle-electrode).

Impedance spectra showing almost ideal semicircles as it is found for the size-selected particles discussed in this section are rarely found for impedance measurements on particulate systems. This is due to the fact that the interparticle transport in these systems is almost identical due to the small particle size distribution. In many cases, so-called "depressed arcs" are observed originating from broad distributions and therefore varying interparticle transport properties.

10.6.2 Tungsten Oxide

Semiconductor metal oxides such as TiO_2, SnO_2, WO_3 and In_2O_3 are suitable materials for gas-sensing applications [28–31]. However, some of these materials have the disadvantage of poor selectivity, cross sensitivity, and instability against humidity. Among the above-mentioned metal oxides, tungsten oxide is known to be a very promising candidate with a good selectivity for sensing of the air pollutants NO/NO_2, CO/CO_2, and ethanol [32–34]. The sensing behavior of semiconducting metal oxides such as WO_3 is known to be a function of their chemical and physical properties such as surface area, crystallinity, stoichiometry, and phase composition and is mainly affected by physicochemical reactions on the surface of the sensing material.

Tungsten oxide nanoparticles with a mean particle size of about 7 nm were produced by premixed flame synthesis (for details see Chap. 1 and [14]) from dilute mixtures of tungsten hexafluoride (WF_6) in argon as precursor material. One percentage of WF_6 in Argon was burnt together with hydrogen and oxygen resulting in the formation of tungsten oxide nanoparticles. By means of molecular beam sampling, sensor devices were realized by depositing the tungsten oxide nanoparticles on top of a sensor substrate based on an interdigital capacitor (see Figs. 10.7 and 10.21).

The device was achieved by placing the sensor substrate along the path of the particle-laden molecular beam, which arises from the pressure difference in the various segments of the synthesis reactor. Assuming a close arrangement of particles with a filling factor of 87 %, a minimum film thickness of about 50 nm is calculated.

10.6.2.1 Sensing Mechanism

The sensing mechanism of nanoparticles with respect to CO and NO measurements is attributed to interactions between different species on the particle surface. It has been

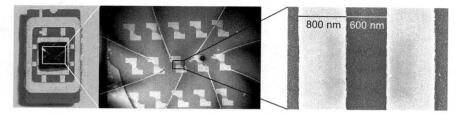

Fig. 10.21 Images of the WO₃ sensor device; the *left* image shows the complete sensor chip, the *right* image shows a scanning electron microscopy image of two contacts (*bright*) with a spacing of 600 nm covered with the WO₃ film

established that both gases react with preadsorbed oxygen or oxygen from the lattice, and release electrons into the conduction band. The dependence of conductivity on the CO or NO ratio in the atmosphere is described in a model by Morrison [35] and also related to temperature by Reyes et al. [36]. The reducing chemical surface reactions of NO and CO with the WO₃ particles are displayed in Eq. (10.17) and (10.18) in Kröger-Vink notation [37]:

$$W_W^x + O_O^X + NO \Rightarrow W_W'' + V_O^{\bullet\bullet} + NO_2 \qquad (10.17)$$

$$W_W^x + O_O^X + CO \Rightarrow W_W'' + V_O^{\bullet\bullet} + CO_2 \qquad (10.18)$$

Additionally, a disproportionation of NO at the particle surface takes place as shown in Eq. (10.19):

$$3NO \Rightarrow N_2O + NO_2 \qquad (10.19)$$

Therefore, further oxidizing reactions can occur:

$$W_W'' + V_0^{\bullet\bullet} + N_2O \Rightarrow W_W^X + O_O^X + N_2 \qquad (10.20)$$

10.6.2.2 Electrical Characterization

For a detailed characterization of the sensing process itself, AC measurements were performed under the influence of reactive gases. Therefore, the sensor device was kept in a glass tube with a gas inlet and outlet at the ends of the tube. The inlet was connected to a gas mixing system as previously described by Rakesh et al. [38]. The complete measurement setup was encased by a temperature-controlled tube furnace to vary the temperature of the sample in the range between 553 and 583 K, after annealing the sample at 583 K for several hours. Starting from 583 K, impedance spectroscopy (IS) measurements were performed under synthetic air, synthetic air with 1000 ppm CO, and synthetic air with 1000 ppm NO in steps of 10 down to

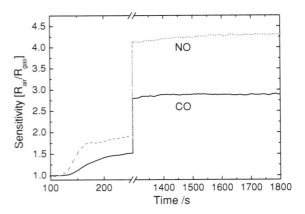

Fig. 10.22 Sensitivity against time of a WO$_3$ sensor device under synthetic air with 1000 ppm NO and CO, respectively

553 K. After each measurement series the sample was annealed again at 583 K under synthetic air.

10.6.2.3 Sensitivity

To investigate the sensitivity of the particulate tungsten oxide thin films, the change of the sum of the different resistances, R_{bulk}, R_{part} and R_{wall} under synthetic air and with addition of the reactive gases CO and NO, respectively, were measured at 583 K. For this purpose the impedance data were recorded at a fixed frequency of 100 Hz (which is still in the DC range) as a function of time.

Figure 10.22 shows the normalized sensitivity $S = R_{air}/R_{gas}$ for 1000 ppm NO and 1000 ppm CO in synthetic air. In both cases the reactive gas has been added after 120 s. Both, the CO and NO data exhibit a fast increase in sensitivity during the first ten seconds followed by a creeping increase and a leveling off at about $t = 1400$ s. At this time, a sensitivity $S = 4.3$ for NO and $S = 2.9$ for CO is reached. For both gases sensitivity related to an increase in conductivity can be explained by the injection of electrons due to the release of oxygen as shown in Eq. (10.17) and (10.18). However, the data reveal a faster increase of sensitivity for NO as well as a better sensitivity. The fast change in sensitivity during the first few seconds is usually not seen with common WO$_3$ sensors and can be attributed to the high surface-to-volume ratio of the nanosized oxide. To further substantiate this, impedance spectra have been measured to investigate core- and grain-boundary sensitivity in detail.

10.6.2.4 Impedance Measurements

The results received from the sensitivity measurements reveal stable conditions before the reactive gases are added and also about 1400 s after the reactive gases were added. Therefore, it is possible to perform time-consuming IS under these

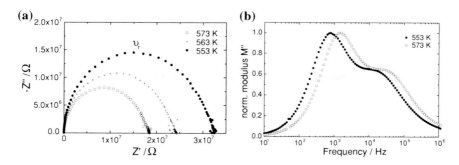

Fig. 10.23 a Impedance data in Nyquist representation for the sensor device measured under 1000 ppm CO. **b** Normalized modulus spectra measured at 553 and 573 K

stable conditions. After annealing and stabilizing with 1000 ppm NO and 1000 ppm CO, impedance data were recorded from 583 to 553 K in steps of 10 K. Impedance measurements were performed with the same setup as described above.

Figure 10.23a shows the Nyquist diagram of the impedance measurements in the temperature range between 553 and 573 K under CO exposure. The typical decrease of resistance with increasing temperature as it is expected for semiconducting materials is observed. All measurements exhibit an almost perfect quarter of a circle for frequencies higher than the resonance frequency υ_r (vertex of the semicircles) while for frequencies lower than the resonance frequency (right hand side of the spectra) the measured spectra show a widening toward higher impedances. Similar results are obtained for granular tungsten oxide by Ling et al. [39], and for solid thin films by Labidi et al. [40]. This behavior indicates a second scattering process at low frequencies and has been described in detail by Barsan et al. [41]. The Modulus representation clearly indicates the presence of two processes as can be seen from Fig. 10.23b.

For a quantitative analysis of the results, the data were fitted using an equivalent circuit similar to that shown in Fig. 10.17 by means of a nonlinear least square fit algorithm. In contrast to the known equivalent circuit, we replaced one capacitor by a constant phase element (CPE, for details see [5]), as the quality of the fit in the low frequency range could be significantly improved by using the CPE instead of a capacitor. The utilization of a CPE considers the typical broadening of impedance signals originating from the polydispersity of particulate materials.

10.6.2.5 Itemized Sensitivity

The low frequency contribution was identified to originate from the particle–particle contacts and R_{part} as well as R_{bulk} could be successfully received from the fit results. The resistances obtained from the fit of measurements under CO and NO, respectively, were compared with those received from the measurements under synthetic

Fig. 10.24 Core (S_{bulk}) and grain boundary (S_{part}) sensitivities on CO and NO against temperature ($S = R_{air}/R_{gas}$). The open symbols indicate the sensitivity of CO and the filled ones those of NO

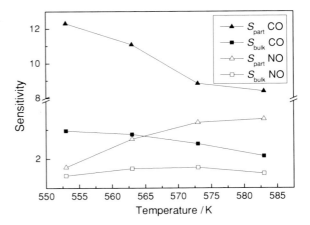

air to calculate the specific sensitivities on NO and CO. Figure 10.24 shows the grain boundary sensitivity S_{GB} originating from the particle–particle contacts and the bulk sensitivity S_{bulk} for both gases.

The data of the separated sensitivities are in good agreement with the overall sensing results (see Fig. 10.22). From Fig. 10.24 it is obvious that for both gases the sensitivity related to the grain boundary process is significantly enhanced for NO as well as CO due to the fact that the resistance of the particle–particle contacts changes much more than that of the core. Moreover, it is obvious that the sensitivity for CO is enhanced at lower temperature while that for NO increases with temperature. Infrared studies on NO adsorption on W/Al$_2$O$_3$ catalysis have shown that NO contrary to CO favors adsorption in twin form on catalysis surface, and different nitrogen oxides such as N$_2$O, N$_2$O$_3$ and NO$_2$ are observed [42, 43]. Weingand et al. reported that the stable species produced after coadsorption of NO and O$_2$ on WO$_3$/ZrO$_2$ surfaces are NO$^+$ and nitrates (NO$_3^-$) [44]. Thus, it is concluded that adsorption/desorption and surface chemistry of NO is quite complex compared to CO. Moreover, it is generally accepted that NO interacts more strongly with transition metal oxides than CO. As a result, the adsorption of CO and desorption of CO$_2$ is expected to occur at lower temperatures compared to NO/NO$_2$.

The results of the AC measurements at 583 K are in good agreement with those derived from the DC measurements at the same temperature. Based on the results shown we attribute the increase in sensitivity within the first ten seconds (see Fig. 10.22) to a fast surface reaction at the grain boundaries followed by a slowly rising sensitivity due to a change in bulk conductivity. This can explain, why a fast change in sensitivity during the first seconds is usually not seen with common WO$_3$ sensors. Compared to common sensors, our device is based on a material with a much higher surface-to-volume ratio and high specific surface area boosting the initial sensing process.

10.6.3 Zinc Oxide

Despite their application as gas sensors, semiconducting metal oxides are increasingly used as transparent conducting oxides, TCOs. Some of the already used electronic devices such as flat-panel displays, touch-panel controls, or even solar cells mainly utilize indium tin oxide (ITO) as transparent conducting layer [45, 46]. However, a couple of disadvantages such as price and availability promote material with optical and electrical properties similar to the established ITO. Zinc oxide and especially aluminum-doped ZnO (AZO) nanomaterials seem to have the most promising properties for replacing the ITO. They show very good optical properties compared to the ITO [27, 47–50], but too little is known with respect to their electrical properties. Therefore, ZnO nanoparticles synthesized from diethylzinc by CVS were investigated with respect to their electrical properties (cf. also Chap. 2). The gas-phase materials show high crystallinity, low defect density, and also a high amount of electrically active dopants. The synthesis route itself is very cheap and a high production rate can be achieved. Within this section the electrical properties of ZnO nanoparticles are discussed. The influence of different gases as well as aluminum doping of the ZnO is investigated. Additionally, the influence of moisture and the sensing properties of inkjet printed ZnO films are shown.

10.6.3.1 Sample Preparation

Thin pellets with 5 mm in diameter were prepared from about 10 mg of ZnO particles using the standard procedure described in Sect. 10.4.1. The green density of the compacted ZnO nanoparticles was found to be usually 80 % of the ZnO bulk density. The produced pellets were sandwiched between two spring-loaded platinum disks in the measurements cell situated in a gastight ceramic tube with gas in- and outlet and equipped with a temperature-controlled tube furnace.

For electrical characterization, a Hewlett Packard HP4192A impedance spectrometer was used. All samples were measured between 10 Hz and 10 MHz with 20 frequency points/decade and in the temperature range between 50 and 400 °C in steps of 25 °C. Before starting an impedance measurement, the system was thermally stabilized for 5 min at the desired temperature ensuring a temperature stability of ± 0.5 °C during measurement. After reaching 400 °C, the samples were cooled down and measured again. For all samples discussed in this section, the temperature-dependent measurements were repeated 5 times, three times under hydrogen atmosphere and after cooling down in hydrogen three times under dry synthetic air (20.5 % O_2, 5.0 in N_2 5.0).

10.6.3.2 Measurements on Undoped ZnO Nanoparticles

Figure 10.25 shows a typical Nyquist plot of undoped ZnO in hydrogen atmosphere. Only data for the real part of the impedance are measured and no frequency dispersion

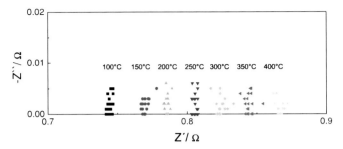

Fig. 10.25 Typical Nyquist plot for the measurement of undoped zinc oxide in hydrogen atmosphere for a selection of temperatures. A pure ohmic transport behavior is observed

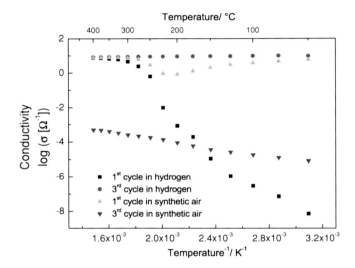

Fig. 10.26 Specific conductivity of undoped ZnO for the first and third measurement cycles in hydrogen and synthetic air, respectively. The measurements in hydrogen show a much higher conductivity than the measurement in synthetic air

was observed. The conductivity for all measured samples was very high and due to the fact that—within the accuracy of the measurement—all measurement values for one temperature are assembled in one point, it can be stated that the sample shows pure ohmic behavior. It has to be mentioned that the values shown in Fig. 10.25 include the impedance of the sample itself as well as parasitic resistances within the measurement cell such as contact and transfer resistances.

A more detailed analysis indicates that the first cycle in hydrogen atmosphere shows a strong increase in conductivity with increasing temperature, see the Arrhenius plot shown in Fig. 10.26. The temperature cycles two and three keep very high conductivity that was reached after the first cycle and show identical values; therefore, only cycle 3 is shown.

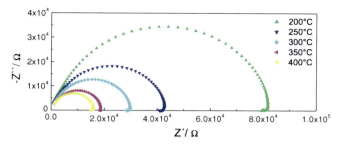

Fig. 10.27 Nyquist plot of zinc oxide nanoparticle samples measured under synthetic air. A decrease in resistance with increasing temperature is observed indicating semiconducting behavior

The very high conductivity in hydrogen atmosphere results from a generation of free charge carriers due to a population of shallow donor sites [51], as hydrogen acts as a dopant in zinc oxide. The incorporation of hydrogen and the subsequent release of electrons can be described as follows:

$$2Zn_{Zn}^x + 2O_O^x + H_2(g) \Leftrightarrow 2Zn_{Zn}^x + 2OH_O^\bullet + 2e^- \qquad (10.21)$$

In contrast to the measurements under hydrogen, the Nyquist plots of the same samples measured under synthetic air show the typical semicircles as they have also been observed for tin oxide and tungsten oxide, see Fig. 10.27.

The fact that the semicircles become smaller with increasing temperature is characteristic for semiconducting materials. Nevertheless, the type of conductivity observable for zinc oxide can be varied from metallic (under hydrogen) to semiconducting (synthetic air) behavior. As has been shown, the conductivity in air is about six orders of magnitude lower compared to the measurements in hydrogen atmosphere. A similar behavior has also been described in the literature [52] and is related to the release of the hydrogen during heating in synthetic air as already mentioned by Thomas et al. [53]. The graphs shown in Fig. 10.27 also indicate the well-known presence of two transport processes, as the data does not exhibit a perfect semicircular shape.

10.6.3.3 Influence of Aluminum Doping on the Electrical Properties of ZnO

From multiple temperature-dependent impedance measurements we ensured that after the first heating cycle the sample has stabilized leading to almost identical results for the second and third cycles. Therefore, all measurements shown are taken from the third cycle measured under the respective atmosphere if not indicated otherwise.

Aluminum-doped zinc oxide nanoparticles were prepared similar to the process described before by addition of triethylaluminum to the precursor mixture. Samples containing between 5.4 and 37.6 % of aluminum were characterized with respect to their electrical properties. Figure 10.28 shows the conductivities as a function of temperature for the differently doped ZnO nanoparticles in hydrogen atmosphere.

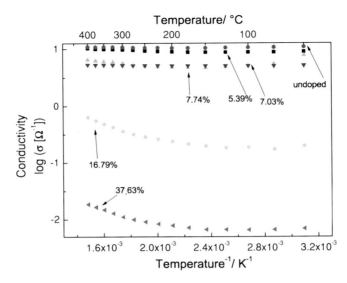

Fig. 10.28 Arrhenius representation of aluminum-doped zinc oxide nanoparticles measured under hydrogen

Surprisingly, the conductivity decreases with rising aluminum content and the samples doped with 16.79 and 37.63 % of aluminum show conductivity values orders of magnitude lower compared to the remaining materials. This behavior was not expected, as aluminum, like hydrogen, is responsible for the formation of free charge carriers. Additionally, it was observed that similar to undoped ZnO, doped materials of up to 7 % of Al show a slightly positive temperature coefficient, while for concentrations higher than 7 % a negative temperature coefficient is observed.

The decrease in conductivity with increasing aluminum content can be explained as follows: ZnO nanoparticles consist of Zn and oxygen atoms occupying different lattice sites. Typically, a substitutional replacement of the Zn ions in the lattice by Al ions as well as the incorporation of aluminum on interstitial positions is observed, both generating free charge carriers. This can best be described starting from aluminum oxide inserted into a zinc oxide lattice as follows:

$$2Al_2O_3 \overset{ZnO}{\Rightarrow} 4Al^{\bullet}_{Zn} + 4O^x_O + O_2\,(g) + 4e^- \tag{10.22}$$

$$2Al_2O_3 \overset{ZnO}{\Rightarrow} 4Al^{\bullet\bullet\bullet}_i + 3O_2\,(g) + 12e^- \tag{10.23}$$

Each crystalline material possesses lattice defects, for example point or surface defects. Due to the addition of dopants such as aluminum, the defect concentration increases, increasingly leading to charge carrier scattering processes. As a result, the overall conductivity decreases due to increased disorder. The negative temperature

10 Electrical Transport in Semiconductor Nanoparticle Arrays

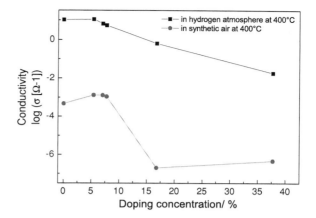

Fig. 10.29 Conductivity of differently doped zinc oxide nanoparticles measured at 400 °C under hydrogen and synthetic air, respectively

coefficient can be explained by progressively thermally activated charge carriers originating from the (high) aluminum doping. The strong increase in disorder is also responsible for the poor conductivity of the highly doped materials.

In synthetic air, the conductance is lower than under hydrogen as also found for pure ZnO. Nevertheless, it increases with higher aluminum concentration reaching a maximum at 7.03 % aluminum. Similar to the results observed under hydrogen, the two highest doped ZnO nanoparticles also show very poor conductivity. A comparison of the specific conductivities under hydrogen and synthetic air, respectively, is shown in Fig. 10.29. In all cases, the conductivity is calculated from the extrapolated DC resistance (intersection of the low-frequency range of the semicircle in Nyquist representation with the real axis Z'). The increase in conductivity for concentrations of up to 7.03 % of aluminum can be explained easily by the increase of charge carriers due to the doping, while the poor conductivity found for the high dopant concentrations is again owing to the poor crystallinity (note that conductivities measured around 10^{-7} S/cm are at the limit of the impedance spectrometer). Most importantly it can be stated that—despite the fact that charge carrier transport within particulate matter is almost determined by grain boundary processes—aluminum doping is able to enhance the conductivity by about one order of magnitude. The highest increase is found for a dopant concentration of about 7 % of aluminum.

10.6.3.4 ZnO Sensor Devices

Zinc oxide nanoparticles like many other materials are quite interesting for gas-sensing applications as already mentioned above [54, 55], and thin films consisting of ZnO nanoparticles have nice opportunities for electrical and optoelectronic devices. ZnO sensor devices were produced from undoped ZnO nanoparticle dispersions by inkjet printing [56]. As determined by X-ray diffraction, the particles have a mean crystallite size of 20 nm while the particle size within the aqueous dispersion is about

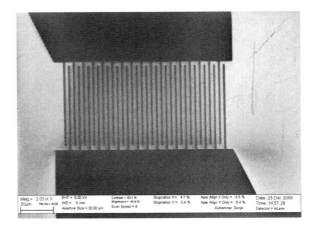

Fig. 10.30 SEM picture of the interdigital structure consisting of evaporated gold contacts on a silicon substrate covered with an insulating SiO$_2$ layer

44 nm. Zeta-potential measurements and DLS indicated a stability of the dispersions for several weeks. Inkjet printing was done on interdigital structures prepared by EBL. The structure consists of interdigitated gold fingers, see Figs. 10.6 and 10.30.

After printing, the zinc oxide films had an inhomogeneous thickness between 100 and 250 nm due to the "coffee ring" effect originating from the drying process and well known from the inkjet printing of particle dispersions [57]. Nevertheless, it was possible to produce dense and crack-free films, see Fig. 10.6. After printing, the resulting film was characterized by XRD indicating a mean crystallite size of 36 nm. This increase compared to the initial crystallite size is due to a particle growth process and will be explained later. In contrast to thin disks prepared by uniaxial pressing, the inkjet printed ZnO films exhibit a high contact area to the environment, which is required for electronic applications [58].

Similar to the procedure described for tungsten oxide (see Sect. 10.6.2), the substrate with the inkjet printed film was placed in a chip carrier and bonded to its contacts. Measurements were performed in a heated, gastight ceramic tube with gas outlet and the inlet connected to mass flow controllers. Impedance characterization was done with a Solatron SI1255 impedance analyzer between 25 and 200 °C in the frequency range between 10 Hz and 10 MHz in ambient and in hydrogen atmosphere, respectively.

Typical semicircles as also observed for the other materials were measured as can be seen from Fig. 10.31. An increase in conductivity with increasing temperature was found as expected and can be explained with the increasing amount of thermally activated charge carriers.

For a detailed characterization of the sensor properties, the conductivity is plotted as a function of the temperature under ambient conditions and in hydrogen, respectively (Fig. 10.32).

A five times higher conductivity is observed for measurements in hydrogen compared to the measurements under ambient atmosphere. This sensitivity of five is very promising for sensing application due to the fact that no annealing of the thin film at

10 Electrical Transport in Semiconductor Nanoparticle Arrays 265

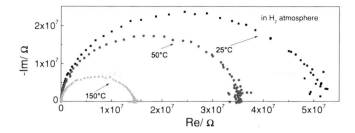

Fig. 10.31 Nyquist plot of the impedance measurements on inkjet printed zinc oxide layer. The measurements shown were performed in hydrogen atmosphere at different temperatures

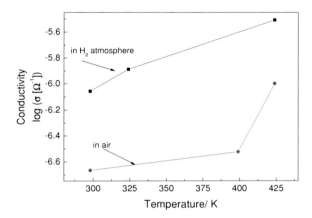

Fig. 10.32 Conductivity of inkjet printed ZnO thin films measured under ambient conditions and hydrogen, resp., as a function of temperature

400 °C was done as required for the tungsten oxide sensor devices. Multiple measurements with changing atmosphere showed that the sensing process is reversible, except some very small differences after the first heating cycle. Therefore, the sensor made from printed zinc oxide nanoparticles can also be realized on temperature-sensitive substrates such as polymers.

10.6.3.5 Influence of Moisture on the Electrical Properties of ZnO Nanoparticles

There are a lot of investigations dealing with humidity sensors based on ZnO, and sensitivities according to an increase in conductivity are reported [59, 60]. Some groups paid attention to ZnO nanorods and nanotubes, which seem to have higher sensitivity for different humidity levels compared to zinc oxide nanoparticles [61, 62]. However, a report concerning the influence of moisture on the properties of as-prepared sensor devices and the evolution of conductivity during and after exposure to moisture atmosphere is still missing.

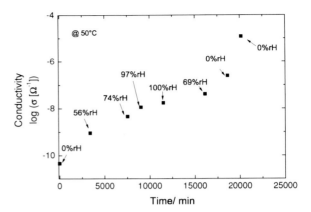

Fig. 10.33 Plot of the specific conductivity of a ZnO pellet at 50 °C as a function of time. The specific conductivity was calculated from the interpolated DC resistance as described above. Measurements with different levels of humidity were performed consecutively as indicated in the graph

For the characterization of electrical properties during and after exposure to humidity, IS of a compacted 5 mm pellet was used, see Sect. 10.6.3.1. First, the sample was measured in argon atmosphere between 50 and 400 °C for three times. After cooling down to 50 °C, a controllable amount of humidity was added to the argon gas flow using a temperature-controlled water bubbler. To monitor the settings, the relative humidity (rH) was also measured using a commercial humidity sensor. Starting from 0 % rH, temperature-controlled measurements (three ramps from 50 to 400 °C) at 56, 74, 97, 100, 67, and again 0 % rH were performed (Fig. 10.33).

The measured impedance spectra in moisture atmosphere and also in dry argon exhibit the characteristic semicircles. A typical Nyquist plot for several measurements from 473 to 573 K at a moisture level of 74 % is shown in Fig. 10.34. From the asymmetric shape of the spectra it is obvious that the conduction is based on two different charge carrier transport processes and consequently, the data of all measurements could be fitted to the known equivalent circuit shown in Fig. 10.17. Unfortunately, the data did not show a clear systematic, temperature-dependent trend at a given humidity and it was also observed that in contrast to the measurements on tungsten oxide and tin oxide the fit results for the capacities (C_{part} and C_{wall}) were not constant indicating a change in morphology. Therefore, it was concluded that the used ZnO nanoparticles are not stable under the given conditions.

However, as a first result it was found that the conductivity increased by raising the moisture level as expected, see Fig. 10.33. The difference in conductivity between the measurements in dry argon (0 rH %, first data point) and 56 % rH (second data point) is more than one order of magnitude, which is attributed to the adsorption of water molecules and the formation of $Zn(OH)_2$, as can be described by Eq. (10.24) [13].

$$V_O^{\bullet\bullet} + O_O^X + H_2O \rightarrow OH_O^{\bullet} + OH_O^{\bullet} + 2e^- \qquad (10.24)$$

Up to the measurements with 100 % rH the transport behavior meets the expectations. The formation of hydroxyl groups on the surface of metal oxides is commonly

10 Electrical Transport in Semiconductor Nanoparticle Arrays

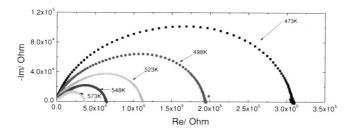

Fig. 10.34 Nyquist plot of impedance measurements on nanocrystalline zinc oxide measured at 74 % rH and at different temperatures

known and was also described for sensor systems by Henrich et al. [63]. Water molecules that are absorbed on the surface of the nanoparticles generate hydroxyl groups which lead to a higher charge carrier concentration leading to an increased conductivity. As expected, an increase in conductivity with increasing humidity was observed.

After being exposed to 100 % rH and reducing the relative humidity, a decrease in conductivity was expected. Surprisingly, the ZnO pellets show a high conductivity which still increases with time even after setting the moisture level to 0 % rH. Referring to the observation of increasing crystallite size after inkjet printing (see Sect. 10.6.3.4) and to results published by Ali et al. [13], we expect that the small ZnO nanoparticles grow by exposing them to moisture. This particle growth leads to a strong decrease in grain boundaries and will also change the morphology of the initially porous structure, also changing the surface-to-volume ratio. Even after removing all moisture from the gas atmosphere, the ZnO nanoparticles still grow due to physisorbed water molecules at the surface of the ZnO nanoparticles. In combination with Zinc interstitials (Zn_i) originating from initial defects formed during particle formation in the gas phase, new ZnO layer can be produced leading to bigger crystals. The increase in conductivity will reach saturation after the physisorbed water is depleted and a particle size limit is expected depending on temperature and humidity [13]. However, the high surface activity of pure, nanocrystalline materials must always be taken into account when using porous, high surface materials for any application.

10.6.4 Electrical Properties of Nanoscale Powders During Compaction

The mechanical and electrical properties of a powder strongly depend on its porosity. This effect can be directly observed by performing in situ IS while applying an external force onto the powder as described in Sect. 10.3.2. Impedance spectra were measured from nanocrystalline silicon powders prepared from hot-wall reactor

Fig. 10.35 Capacity of a pressed silicon nanoparticle sample depending on applied pressure. After each pressure step, the applied force was kept constant for 35 min. The diagram shows the capacities fitted from four impedance measurements taken during this time

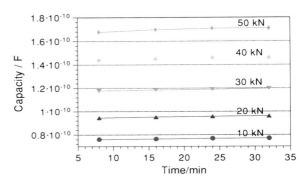

synthesis [64] with a mean crystallite diameter of about 40 nm. The nominal applied external force was increased in steps of 10 kN between 10 and 50 kN. After each pressure step, the force was kept constant for 30 min and several impedance spectra were recorded. The received spectra were fitted according to the equivalent circuit shown in Fig. 10.17. Similar to the measurements on bigger SnO_2 nanoparticles, the contribution of the particle-electrode contact was neglected as it could not be identified from the spectra (see also discussion on SnO_2, Sect. 10.6.1.2).

The capacitances received from fit are shown in Fig. 10.35. Each graph represents four subsequently measured impedance spectra. It can be observed that the capacitance of the sample cell increases slightly over time on each force step, while the capacity change between the steps is much bigger. As described before (see Sect. 10.5.1), this behavior is easy to understand as the capacity is determined by the geometry of the measurement setup. Considering the load plates of the sample cell as plate capacitor, its capacitance is according to

$$C = \varepsilon_0 \varepsilon_{r,Si} \frac{A}{d} \quad (10.25)$$

where A is the area of the plates, d the distance between them, ε_0 the electric constant, and ε_r, S_i the relative permittivity of the silicon powder. Both the distance between the plates and the relative permittivity depend on the porosity and increase while pressing the powder together. The electrical behavior of the powder can be explained by a slow creeping compaction over time and a large effect when the force is increased. To investigate this behavior in more detail, a long-time measurement was performed at a pressure of 20 kN. Figure 10.36 summarizes the results received from the fit of C_{part} and R_{part}.

It is obvious that during this long-term measurement a slow creeping process takes place, indicated by the increasing capacity with time. Even after more than 3 h, the densification of the powder has not finished. In parallel, the resistance decreased by more than 50% indicating an increasing amount of conduction paths.

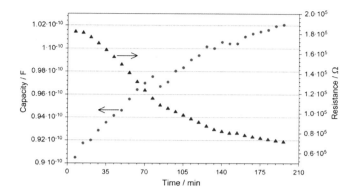

Fig. 10.36 R_{part} and C_{part} fitted to the time-dependent impedance measurements during compaction of silicon nanoparticles at a force of 20 kN. The discontinuities observable from the capacity data are due to the fact that during the long-term measurement the applied force of 20 kN had to be readjusted several times

10.7 Conclusions

Electrical properties of mesoporous and nanoscale materials are an important feature with respect to gas sensors, transparent conducting layers, and optoelectronic properties. Many devices made by established physical or chemical techniques such as physical vapor deposition, chemical vapor deposition, epitaxial growth, or sputtering lead to nanocrystalline but almost dense structures with negligible or no porosity. These devices usually exhibit very good electrical performance and show properties that are only related to the small surface area that is in contact with the ambient. However, the production of such devices is costly as it is mostly performed under (high) vacuum conditions and often requires elevated temperatures constraining the range of suitable substrates.

Sol–gel-based synthesis is a well-known way to produce materials that exhibit nanoscale and porous structures. These particles from the sol–gel route are always covered by a functional surface layer such as ligands or spacer, stabilizing particle size and morphology, and ensuring a good dispersibility in liquid media. Therefore, cost-efficient techniques such as printing can be applied to produce functional surfaces and layer. Nevertheless, the purity of the as-prepared materials is limited and surface chemistry, i.e., with respect to sensor applications is influenced by their surface coating.

Pure materials from gas-phase synthesis have the ability to combine both, high purity and dispersibility enabling a cost-efficient formation of functional structures. As has been shown in this contribution, the electrical properties of porous structures based on such particles is always dominated by charge carrier transport processes between individual particles as well as between particles and contacts. It is shown that the density of the porous structure plays an important role and that the sustainability of the structure can be determined by IS. We found a very good accordance between modeling and experiment of impedance data taking three different charge carrier transport processes into account. With these results it is possible to separate between processes originating from surface-related particle–particle and particle-electrode contacts and those originating from the bulk material itself.

Acknowledgments The financial support of this work through the German research foundation (DFG) within SFB445 is gratefully acknowledged. The authors are also grateful to Lothar Brendel and Gabi Schierning for their productive and rewarding joint research.

References

1. C.W.J. Beenakker, H. Vanhouten, Solid State Phys. Adv. Res. Appl. **44**, 1–228 (1991)
2. M. Büttiker, Y. Imry, R. Landauer, S. Pinhas, Phys. Rev. B **31**(10), 6207–6215 (1985)
3. T. vanDijk, A.J. Burggraaf, Phys. Status Solidi A Appl. Res. **63**(1), 229–240 (1981)
4. N. Barsan, U. Weimar, J. Electroceram. **7**(3), 143–167 (2001)
5. E. Barsoukov, J.R. Macdonald, *Impedance Spectroscopy: Theory, Experiment, and Applications* (Wiley, New York, 2005)
6. B.A. Boukamp, Solid State Ion. **169**(1–4), 65–73 (2004). doi:10.1016/j.ssi.2003.07.002
7. I. Plümel, H. Wiggers in *In-situ Investigation of the Mechanical and Electrical Properties of Nanosized Silicon Powders*, MRS Spring Meeting, San Francisco, 2008, Materials Research Society: San Francisco, 2008, pp. 1083-R05-06
8. M. Morgeneyer, M.Röck, J. Schwedes, L. Brendel, K. Johnson, D. Kadau, D.E. Wolf, L.-O. Heim, in *Behavior of Granular Media*, eds. by P. Walzel, S. Linz, C. Krülle, R. Grochowski (Shaker, Aachen, 2006), pp. 107–136
9. P. Calvert, Chem. Mater. **13**(10), 3299–3305 (2001)
10. E. Tekin, P.J. Smith, U.S. Schubert, Soft Matter **4**(4), 703–713 (2008). doi:10.1039/B711984D
11. M. Ali, N. Friedenberger, M. Spasova, M. Winterer, Chem. Vapor Depos. **15**(7–9), 192–198 (2009). doi:10.1002/cvde.200806722
12. S. Hartner, M. Ali, C. Schulz, M. Winterer, H. Wiggers, Nanotechnology **20**(44), 445701 (2009). doi:10.1088/0957-4484/20/44/445701
13. M. Ali, M. Winterer, Chemistry of Materials **22**(1), 85–91 (2010). doi:10.1021/cm902240c
14. T.P.Hülser, A. Lorke, P. Ifeacho, H. Wiggers, C. Schulz, J. Appl. Phys. **102**(12), 124305 (2007)
15. T.P. Hülser, H. Wiggers, F.E. Kruis, A. Lorke, Sens. Actuators B Chem. **109**(1), 13–18 (2005)
16. N. Barsan, U. Weimar, J. Phys. Condes. Matter **15**(20), R813–R839 (2003)
17. Z.Z. Fang, H. Wang, Int. Mater. Rev. **53**(6), 326–352 (2008). doi:10.1179/174328008x353538
18. J.R. Groza, Nanostruct. Mater. **12**(5–8), 987–992 (1999)
19. H.A. Knudsen, S. Fazekas, J. Comput. Phys. **211**(2), 700–718 (2006)
20. S. Kirkpatrick, Rev. Mod. Phys. **45**(4), 574–588 (1973)
21. D. Stauffer, Phys. Rep. Rev. Sec. Phys. Lett. **54**(1), 1–74 (1979)
22. P. Grassberger, Phys. A **262**(3–4), 251–263 (1999)
23. T. Kiefer, G. Villanueva, J. Brugger, Phys. Rev. E **80**(2), 021104 (2009)
24. D. Schwesig, G. Schierning, R. Theissmann, N. Stein, N. Petermann, H. Wiggers, R. Schmechel, D.E. Wolf, Nanotechnology **22**(13), 135601 (2011)
25. E. Falcon, B. Castaing, Am. J. Phys. **73**(4), 302–307 (2005)
26. M.K. Kennedy, F.E. Kruis, H. Fissan, B.R. Mehta, S. Stappert, G. Dumpich, J. Appl. Phys. **93**(1), 551–560 (2003)
27. A. Dieguez, A. Romano-Rodriguez, J.R. Morante, J. Kappler, N. Barsan, W. Gopel, Sens. Actuators B Chem. **60**(2–3), 125–137 (1999)

10 Electrical Transport in Semiconductor Nanoparticle Arrays

28. J. Tamaki, T. Hayashi, Y. Yamamoto, M. Matsuoka, Electrochemistry (Tokyo, Japan) **71**(6), 468–474 (2003)
29. Y. Shimizu, A. Kawasoe, Y. Takao, E. Makoto, in *Proceedings—Electrochemical Society 96-27 (Ceramic Sensors)* (1997), pp. 117–122
30. C.A. Papadopoulos, D.S. Vlachos, J.N. Avaritsiotis, Sens. Actuators B Chem. **B32**(1), 61–69 (1996)
31. G. Lu, N. Miura, N. Yamazoe, Sens. Actuators B Chem. **B35**(1–3), 130–135 (1996)
32. A. Ponzoni, E. Comini, M. Ferroni, G. Sberveglieri, Thin Solid Films **490**(1), 81–85 (2005)
33. G. Sberveglieri, L. Depero, S. Groppelli, P. Nelli, Sens. Actuators B Chem. **B26**(1–3), 89–92 (1995)
34. F. Ahmed, S. Nicoletti, S. Zampolli, I. Elmi, A. Parisini, L. Dori, A. Mezzi, S. Kaciulis, in *Sensors and Microsystems, Proceedings of the 7th Italian Conference*, Bologna, Italy, February 4–6, 2002, pp. 197–204
35. S.R. Morrison, Sens. Actuators **11**, 283–287 (1987)
36. L.F. Reyes, A. Hoel, S. Saukko, P. Heszler, V. Lantto, C.G. Granquist, Sens. Actuators B **117**, 128–134 (2006)
37. F.A. Kröger, H.J. Vink, Solid State Phys. Adv. Res. Appl. **3**, 307–435 (1956)
38. R.K. Joshi, F.E. Kruis, Appl. Phys. Lett. **89**(15), 153116-1-3 (2006)
39. Z. Ling, C. Leach, R. Freer, J. Eur. Ceram. Soc. **23**(11), 1881–1891 (2003)
40. A. Labidi, C. Jacolin, M. Bendahan, A. Abdelghani, J. Guerin, K. Aguir, M. Maaref, Sens. Actuators B **106**, 713–718 (2005)
41. N. Barsan, U. Weimar, J. Phys. Condens. Matter **15**, R813–R839 (2003)
42. K. Hadjiivanov, P. Lukinska, H. Knötzinger, Catal. Lett. **82**(1–2), 73–77 (2002)
43. Y. Yan, Q. Xin, S. Jiang, X. Guo, J. Catal. **131**, 234–242 (1991)
44. T. Weingand, S. Kuba, K. Hadjiivanov, H. Knötzinger, J. Catal. **209**, 539–546 (2002)
45. Y. Furubayashi, T. Hitosugi, Y. Yamamoto, K. Inaba, G. Kinoda, Y. Hirose, T. Shimada, T. Hasegawa, Appl. Phys. Lett. **86**(25), 252101 (2005)
46. T. Minami, Semicond. Sci. Technol. **20**(4), S35–S44 (2005)
47. N.S. Baik, G. Sakai, K. Shimanoe, N. Miura, N. Yamazoe, Sens. Actuators B Chem. **65**(1–3), 97–100 (2000)
48. V. Bhosle, A. Tiwari, J. Narayan, Appl. Phys. Lett. **88**(3), 032106 (2006)
49. E. Fortunato, D. Ginley, H. Hosono, D.C. Paine, MRS Bull. **32**(3), 242–247 (2007)
50. R.G. Gordon, MRS Bull. **25**(8), 52–57 (2000)
51. C.G. van de Walle, Phys. Rev. Lett. **85**(5), 1012–1015 (2000)
52. M. Arita, H. Konishi, M. Masuda, Y. Hayashi, Mater. Trans. **43**(11), 2670–2672 (2002)
53. D.G. Thomas, J.J. Lander, J. Chem. Phys. **25**(6), 1136–1142 (1956)
54. G. Kwak, K.J. Yong, J. Phys. Chem. C **112**(8), 3036–3041 (2008)
55. L. Liao, H.B. Lu, J.C. Li, H. He, D.F. Wang, D.J. Fu, C. Liu, W.F. Zhang, J. Phys. Chem. C **111**(5), 1900–1903 (2007)
56. A.S.G. Khalil, S. Hartner, M. Ali, H. Wiggers, M. Winterer, J. Nanosci. Nanotechnol. **11**(12), 10839–10843 (2011). doi:10.1166/jnn.2011.4043
57. R.D. Deegan, O. Bakajin, T.F. Dupont, G. Huber, S.R. Nagel, T.A. Witten, Nature **389**(6653), 827–829 (1997)
58. K. Okamura, N. Mechau, D. Nikolova, H. Hahn, Appl. Phys. Lett. **93**(8), 083105 (2008)
59. B.M. Kulwicki, J. Am. Ceram. Soc. **74**(4), 697–708 (1991)
60. W.P. Tai, J.H. Oh, J. Mater. Sci. Mater. Electron. **13**(7), 391–394 (2002)
61. F. Fang, J. Futter, A. Markwitz, J. Kennedy, Nanotechnology **20**(24), 245502 (2009)
62. Y.S. Zhang, K. Yu, D.S. Jiang, Z.Q. Zhu, H.R. Geng, L.Q. Luo, Appl. Surf. Sci. **242**(1–2), 212–217 (2005)
63. V.E. Henrich, P.A. Cox, Appl. Surf. Sci. **72**(4), 277–284 (1993)
64. H. Wiggers, R. Starke, P. Roth, Chem. Eng. Technol. **24**(3), 261–264 (2001)

Chapter 11
Intrinsic Magnetism and Collective Magnetic Properties of Size-Selected Nanoparticles

C. Antoniak, N. Friedenberger, A. Trunova, R. Meckenstock, F. Kronast, K. Fauth, M. Farle and H. Wende

Abstract Using size-selected spherical FePt nanoparticles and cubic Fe/Fe-oxide nanoparticles as examples, we discuss the recent progress in the determination of static and dynamic properties of nanomagnets. Synchroton radiation-based characterisation techniques in combination with detailed structural, chemical and morphological investigations by transmission and scanning electron microscopy allow the quantitative correlation between element-specific magnetic response and spin structure on the one hand and shape, crystal and electronic structure of the particles on the other hand. Examples of measurements of element-specific hysteresis loops of single 18 nm sized nanocubes are discussed. Magnetic anisotropy of superparamagnetic ensembles and their dynamic magnetic response are investigated by ferromagnetic resonance as a function of temperature at different microwave frequencies. Such investigations allow the determination of the magnetic relaxation and the extraction of the average magnetic anisotropy energy density of the individual particles.

11.1 Introduction

Fe containing nanoparticles gained a lot of interest over the last decades due to their various possible applications ranging from future high-density storage media [1] to biomedical applications [2]. Here, we focus on two examples: hard-magnetic FePt

---

C. Antoniak (✉) · N. Friedenberger · A. Trunova · M. Farle · H. Wende · R. Meckenstock
Faculty of Physics and Center for Nanointegration Duisburg-Essen (CENIDE),
University of Duisburg-Essen, Lotharstraße 1, 47057 Duisburg, Germany
e-mail: carolin.antoniak@uni-due.de

F. Kronast
Helmholtz-Zentrum Berlin für Materialien und Energie (HZB), Albert-Einstein-Str. 15,
12489 Berlin, Germany

K. Fauth
Experimentelle Physik IV, Universität Würzburg, Am Hubland, 97074 Würzburg, Germany

---

A. Lorke et al. (eds.), *Nanoparticles from the Gas Phase*, NanoScience and Technology,
DOI: 10.1007/978-3-642-28546-2_11, © Springer-Verlag Berlin Heidelberg 2012

nanoparticles which are one of the prime candidates for technological interests and Fe/Fe-oxide nanoparticles which are soft-magnetic and biocompatible. It is the aim of this work to characterise the nanoparticles regarding their collective magnetic behaviour and magnetisation dynamics as well as to study the intrinsinc magnetic properties like spin and orbital magnetic moments and the switching behaviour of individual nanoparticles.

The FePt system is one of the materials with the highest magnetocrystalline anisotropy energy (MAE) density in the bulk material, i.e. $K = E_A/V \approx 6 \times 10^6$ J/m^3 [3–6] in the chemically ordered state with L1$_0$ symmetry forcing the magnetisation direction parallel to one distinguished crystallographic axis. From the technological perspective it is desirable to align the equilibrium magnetisation directions of the individual particles parallel, up to now they are usually randomly oriented for spherically shaped particles. Changing their shape to cubic gives the possibility to achieve one distinguished axis perpendicular to the substrate on which the particles are deposited. For the case of Fe-nanoparticles, the cubic shape could already be realised [7, 8] within this work.

However, in all cases, when reducing the dimensionality of the system towards nanoparticles with a small volume V, a high MAE density is essential to overcome the so-called superparamagnetic limit. Otherwise, the anisotropy energy E_A which stabilises the magnetisation direction in the equilibrium state can be overcome by the thermal energy yielding fluctuations of the magnetisation direction and thus, inhibiting long-term data storage. The temperature limit, above which the magnetisation direction is fluctuating in a given time window, is called blocking temperature, T_B. However, in ensembles of nanoparticles the interactions between nanoparticles—mainly exchange interaction and magnetic dipole interaction—may also strongly influence the stabilisation of the magnetisation direction in the single particles. Experiments have shown for example that different particle configurations in macroscopic ensembles can cause opposite shifts of T_B [9] which has been modelled in terms of different magnetically coupled particle configurations [10]. Since a quantification of all interactions among the particles in an ensemble and their inclusion in the data interpretation may be a delicate task, a temperature-dependent effective anisotropy density K_{eff} is introduced including MAE and all interaction energies in the system. In this example, the importance of investigating both individual and collective properties of nanoparticles becomes already evident: usually, the determination of K_{eff} is carried out using integral measurement techniques—very often on more than millions of nanoparticles—due to the unavailability of single particle detection. Such techniques can give averaged results for the magnetic properties of individual particles and offer only limited clues about the effects of locally varying interactions between the nanoparticles which is crucially important as previously mentioned.

Within the last years, several experimental approaches have been developed to address the determination of individual magnetic properties of nanomagnets. For example, first experiments to measure the switching fields at low temperatures ($35\,$mK $< T < 30\,$K) of 20 nm Co-nanoparticles embedded in Nb have been reported by Thirion et al. using a micro-SQUID technique [11]. Also, a bolometer detection scheme to record the static and high-frequency dynamic magnetic response of

11 Intrinsic Magnetism and Collective Magnetic Properties

individual colloidal nanoparticles with diameters below 10 nm and a magnetic moment of about $10^5 \mu_B$ per particle has been proposed [12]. Furthermore, techniques as ballistic Hall micro-magnetometry [13], differential phase contrast microscopy, holography and energy-loss magnetic chiral dichroism in the transmission electron microscope (TEM) [14–17] have been used to determine magnetic hystereses and domain configurations for nanomagnets in the range between 30 nm and 1 μm. Recent developments in magnetic imaging by soft X-ray spectroscopies [18] like X-ray holography [19], (scanning) transmission X-ray microscopy [20–23] offer new possibilities in this field with a claimed lateral resolution of about 25 nm. In pioneering work the size-dependent spin structure of ensembles of Fe-clusters prepared by a gas-phase cluster source and exchange coupled to a ferromagnetic Co-substrate has been analysed by X-ray photoemission electron microscopy (XPEEM) in combination with atomic force microscopy [24–26].

In this work, we used XPEEM in combination with scanning electron microscopy (SEM) to measure the magnetic switching behaviour of individual 18 nm Fe-nanocubes and study the influence of interactions among the particles in few particle configurations (dimers and trimers). Since the question remains, if the particle under investigation is representative for all particles from a macroscopic batch, on this system also the magnetisation dynamics like relaxation behaviour of the whole particle ensemble has been studied.

Besides the influence of interactions among nanoparticles, also intrinsic properties like crystal structure and coordination number may change the magnetic properties significantly with respect to the corresponding bulk material. Therefore, our experimental results on the magnetism of nanoparticles is complemented by a detailed structural characterisation. In the case of bimetallic nanoparticles, also the local chemical composition may strongly influence the magnetic moments and MAE making a local probe of the composition indispensable. In contrast to averaging methods like energy-dispersive X-ray spectroscopy, the analysis of the extended X-ray absorption fine structure (EXAFS) has turned out to be a useful tool for structural investigations of nanoparticle systems and several examples can be found in the literature Co[27], CdS [28], CdSe [29], SnO_2 [30], Au [31]-nanoparticles, and also Ag-nanoparticles embedded in glass [32]. In this work, the standard EXAFS analysis based on Fourier transform is compared to the quite new field of wavelet transforms (WT) that have the potential to outperform traditional analysis especially in bimetallic alloys.

The organisation of this chapter is as follows: in the first section, structural results on Fe-nanocubes and FePt nanoparticles are presented including 3D tomography and EXAFS analysis in addition to the well-established electron microscopy imaging and diffraction methods like X-ray diffraction. In Sect. 11.3 we present the influence of local chemical environment and crystal symmetry on the element-specific magnetic properties of FePt nanoparticles before we turn to the discussion of the magnetism of Fe ions in oxides and the possibility to measure magnetic hystereses separately for Fe ions on inequivalent lattice sites. In Sect. 11.4 results of X-ray spectroscopies on monomers, dimers, and trimers of Fe-nanocubes are presented focussing on their magnetic switching behaviour. Magnetisation dynamics of ensembles of Fe/FeO_x and Fe-nanocubes are presented in Sect. 11.5 before a short summary is given.

11.2 Structural Characterisation

As will be shown in this work, a detailed knowledge about the structure of the nanoparticles is essential for data interpretation on the magnetic properties. Therefore, we used not only standard characterisation methods like (high-resolution) transmission electron microscopy, electron diffraction, and X-ray diffraction (XRD), but also tomography measurements to obtain 3D images of the Fe/Fe-oxide nanocubes. In the case of bimetallic systems like FePt, detailed EXAFS analyses are presented as a useful tool to determine the local structure and composition.

11.2.1 Fe/Fe-Oxide Nanocubes

Nearly perfect monodisperse Fe/Fe-oxide nanoparticles (standard deviation of diameter distribution: $\sigma = 9\%$) with a cubic shape and an average edge length of 13.6 nm were synthesised in an organic solvent by the decomposition of a metal–organic precursor $[FeN(SiMe_3)_2]_2$ [7, 8, 33] under H_2 atmosphere. A combination of hexadecylamine (HAD) and a long-chain acid (oleic acid) was used as a surfactant that covers the growing particle, controls its size and shape, and prevents the agglomeration of the particles. Figure 11.1a shows a typical bright-field TEM image of an assembly of the Fe/FeO_x cubes deposited on a thin carbon film. A representative high-resolution HR-TEM image of a single nanocube is shown in Fig. 11.1b. It reveals a core–shell contrast typical for air exposed particles. The metallic Fe core appears darker in contrast because of the higher electron density compared to the FeO_x shell with a thickness of about 3 nm. From HR-TEM studies we observed that the nanocubes are single crystalline and oriented in (001) direction with a maximum tilt of $1°$ parallel to the normal of the substrate surface. From the perspective of X-ray absorption spectroscopy, such core–shell structures have been analysed and discussed e.g. in Ref. [35].

Larger Fe/FeO_x nanocubes with a side length of about 18 nm were synthesised by Shavel [36]. Beside standard TEM imaging of these nanoparticles, also tomography measurements in the TEM on the as-prepared Fe/FeO_x nanocubes were performed recently allowing the determination of the particles morphology. For the acquisition of the tomography data the scanning mode (STEM) was used and a corresponding image of a six nanocubes configuration is shown in Fig. 11.2. The resulting reconstructed image of a tomography data series on this particle configuration is displayed in Fig. 11.2b. It is easy to see, that the particles are not perfect cubes but have rough surfaces and curved edges. Furthermore, there can be variations in the side lengths of more than 10%. From the reconstructed data, the dimensions of the 2nd particle (from the left) in the row were measured yielding $x = (19 \pm 2)$ nm, $y = (22 \pm 2)$ nm and $z = (20.5 \pm 2)$ nm.

11 Intrinsic Magnetism and Collective Magnetic Properties

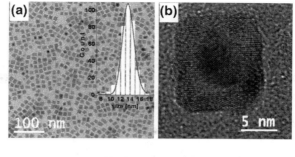

Fig. 11.1 a Bright-field TEM image of Fe-nanocubes assembled on a carbon-coated Cu grid and their size distribution. b Typical high-resolution TEM image of an individual as-prepared Fe/Fe-oxide cube with core–shell contrast [34]

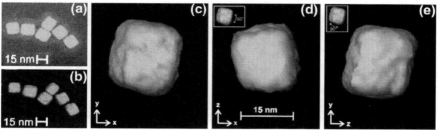

Fig. 11.2 a STEM-micrograph of six Fe-oxide nanocubes. b Reconstructed 3D image of the particle configuration shown in a. Top views for different viewing directions of the *red* coloured nanocube are displayed in (c)–(e)

11.2.2 FePt Nanoparticles

The FePt nanoparticles [37] exhibit a spherical shape with a log-normal distribution of diameters around the mean value of 4.4 nm and a standard deviation of $\sigma(d) = 0.12$. The composition was measured using energy-dispersive X-ray spectroscopy (EDX) in the TEM and found to be (56 ± 6) at% Fe and (44 ± 6) at% Pt. HR-TEM analyses indicate a lattice expansion [38] with respect to the corresponding bulk material in agreement to selected area electron diffraction (SAED) and XRD results shown in Fig. 11.3. For larger FePt nanoparticles with a mean diameter of 6.3 nm, no lattice expansion was found within experimental error bars. In order to distinguish between changes of the lattice parameters caused by oxidation at the surface and caused by the reduced dimensionality of the system itself, the structural characterisation should be done on oxide-free nanoparticles. Thus, we employed the analysis of EXAFS measurements in ultra-high vacuum on hydrogen plasma cleaned FePt nanoparticles. The efficiency of the plasma treatment was checked by XANES at both the Fe $L_{3,2}$ and carbon K absorption edges [39]. After the sample was exposed to a soft hydrogen plasma at room temperature and a pressure of 5 Pa, a pure metallic X-ray absorption near-edge structure (XANES) at the Fe $L_{3,2}$ edges was obtained indicating that all oxides have been reduced. At the carbon K edge no absorption peaks were observed after the cleaning procedure, since the hydrogen plasma also

Fig. 11.3 Room-temperature X-ray diffraction data for chemically disordered FePt nanoparticles with a mean diameter of 6.3 nm and 4.4 nm, respectively [42]

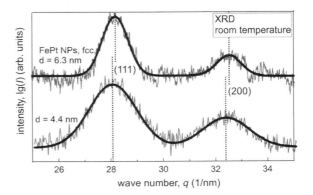

removes organic impurities at the surface as is also known for reactive oxygen plasma [40, 41].

EXAFS measurements were performed at room temperature in fluorescence yield (FY) at the undulator beamline ID12 at the ESRF both at the Pt L_3 absorption edge and Fe K absorption edge. In Fig. 11.4, the EXAFS signals at the Pt L_3 absorption edge of the plasma cleaned FePt nanoparticles and of bulk material with the same composition are shown as well as their Fourier transform. It can clearly be seen that the frequency of EXAFS oscillations as a function of photoelectron wavenumber—which is proportional to the nearest-neighbour distance—is higher for the case of nanoparticles than for the bulk material indicating a larger lattice constant in the particles [42]. In the Fourier transform it is evident by the shifted main peak that is connected to the geometric nearest-neighbour distance. Note that, the data are not corrected for the EXAFS phase shift, therefore r is not the geometric distance. A quantification of the lattice constant is possible by comparison of the experimental data to calculated ones, e.g. by using the FEFF programme [43, 44] for ab initio multiple scattering calculations of X-ray absorption fine structures. It is commonly used also for the determination of the coordination number of the probe atoms and gives information on their chemical environment. A detailed discussion of the results can be found in Refs. [42, 45, 46], a summary is given in Table 11.1. It shows that there is not only a lattice expansion of 1–2 % in the 4.4 nm particles with respect to the corresponding bulk material, but also a compositional inhomogeneity in the nanoparticles: while the Pt probe atoms are located in a Pt-rich environment with respect to the composition determined by EDX, the Fe probe atoms are in an Fe-rich environment. The different compositions found by EDX on the one hand and EXAFS on the other hand reflects the fact that EDX is an averaging method whereas EXAFS means a local probe around the absorbing atoms.

A visualisation of Fe and Pt seggregation, respectively, offers the method of WT [47]. It overcomes the obvious disadvantage of Fourier transform which has only a resolution in Fourier space and not in the space of original data. That means the Fourier transform of experimental EXAFS data provides some kind of radial

11 Intrinsic Magnetism and Collective Magnetic Properties 279

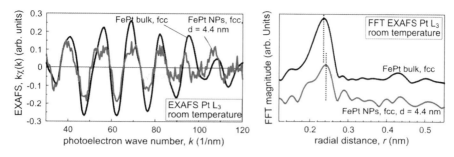

Fig. 11.4 Oscillations of the X-ray absorption in the EXAFS regime as a function of the wave number for a bulk Fe$_{56}$Pt$_{44}$ sample (*black line*) and 4.4 nm FePt nanoparticles, respectively (*red line*). The Fourier transform of the data is shown on the *right* [42]

Table 11.1 Results of structural and compositional characterisation of FePt bulk material and nanoparticles of two different sizes employing different methods, where a denotes the lattice constant and x the Fe content

System	Method	a/nm	x/at%
FePt bulk, fcc	XRD	0.384 ± 0.002[a,b]	
	Fe EXAFS	0.383 ± 0.004[a,b]	55 ± 6[a]
	Pt EXAFS	0.383 ± 0.003[a,b]	59 ± 4[a]
	EDX		56 ± 3[a]
FePt NPs, fcc, d = 6.3 nm	XRD	0.386 ± 0.002	
	EDX		50 ± 6
FePt NPs, fcc, d = 4.4 nm	XRD	0.388 ± 0.002[a,b]	
	Fe EXAFS	0.387 ± 0.008[a,b]	70 ± 12[a]
	Pt EXAFS	0.387 ± 0.004[a,b]	40 ± 8[a]
	EDX		56 ± 6[a]

Note that the EXAFS results reflect the local environment of the absorbing atoms as discussed in the text [a]from Ref. [42], [b]from Ref. [45]

distribution function of backscattering atoms but the information is lost at which wave number k the scatterer contributes. Since the position in k space is related to the atomic species of the backscattering atom, important information is lost in the magnitude of the transformed signal. As a rule of thumb one may keep in mind that the heavier the element, the larger the k value at which the backscattering amplitude is maximum. The WT now gives the possibility to achieve high resolution both in real space and in k space [45, 48–52]. Figure 11.5 shows the WT of the EXAFS signal at the Pt L$_3$ absorption edge of as-prepared FePt nanoparticles and the k-dependence of the backscattering amplitude of carbon, Fe and Pt. Thus, it can easily be distinguished between EXAFS contributions of light backscatters like oxygen or carbon (at k values between 10 and 20/nm), contributions of Fe (around $k \approx 60$/nm) and of Pt ($k \geq 120$/nm). The WTs of the EXAFS signals of cleaned nanoparticles and bulk material both measured at the Pt L$_3$ absorption edge are shown in Fig. 11.6. The colour scale is the same for both wavelet transforms. It can already be

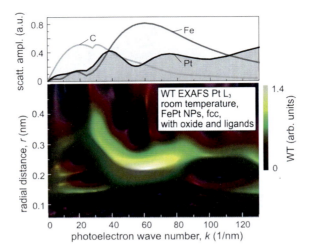

Fig. 11.5 Calculated backscattering amplitude of carbon, Fe, and Pt (*upper panel*) and wavelet transform (WT) of experimental EXAFS data of chemically disordered 4.4 nm FePt nanoparticles in the as-prepared state, i.e. with surface oxides and organic impurities at the surface [45]

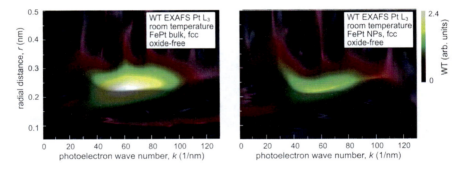

Fig. 11.6 Wavelet transform (WT) of experimental EXAFS data of FePt bulk material (*left*) and 4.4 nm cleaned nanoparticles (*right*) [45]. The colour scale is for both graphics the same

seen that (i) the contributions of light elements at low k values almost vanished in both cases, (ii) the overall amplitude of the wave transform is reduced for the nanoparticles indicating a lower coordination number of the absorber or a modified chemical environment, and (iii) the relative contribution of Pt backscattering atoms at high k values is increased for the case of nanoparticles with respect to the bulk sample. The latter is even better visible in the WT of nanoparticles' EXAFS signal after subtraction of the signal of the bulk material as shown in Fig. 11.7: while at the k position of Fe maximum backscattering amplitude (around 60/nm) the difference has a minimum, at large values of k there is a maximum that can be assigned to a higher number of Pt backscatterers in the vicinity of Pt absorbing atoms in the nanoparticles. Note that, also the fine structure in the k-dependent backscattering amplitude of Pt

11 Intrinsic Magnetism and Collective Magnetic Properties

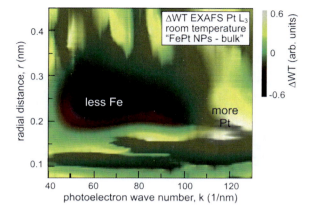

Fig. 11.7 Wavelet transform of experimental EXAFS data of cleaned 4.4 nm FePt nanoparticles after subtraction of the wavelet transform of experimental EXAFS data of the corresponding bulk material [45]. The data before subtraction are shown in Fig. 11.6

can be identified in the difference of the wavelet transforms. At the Fe K absorption e.g. it is exactly the other way round (not shown here): in the nanoparticles, we found less Pt backscatters around the Fe absorbers and more Fe backscatters than in the bulk material. The EXAFS spectra obtained at the Fe K absorption e.g. Fourier and WTs can be found elsewhere [45].

11.3 Element-Specific, Site-Selective Magnetism

The element-specific electronic structure and correlated magnetic moments can be analysed by means of the X-ray magnetic circular dichroism (XMCD). In addition, this method offers the possibility to extract also site-specific information for the case that the same atomic species can be found in different crystal fields. As examples, we present our results on FePt nanoparticles as well as on Fe/FeO$_x$ nanocubes, in which the Fe ions are either tetrahedrally or octahedrally surrounded by oxygen anions.

11.3.1 Influence of Local Composition and Crystal Symmetry on the Magnetic Moments

The inhomogeneous composition found in the chemically disordered FePt nanoparticles as presented in the previous section may also have an influence on the electronic structure and magnetic moments of the nanoparticles. In this section, we discuss the element-specific magnetic moments determined from XMCD analyses of FePt nanoparticles and bulk-like films as references as well as from band structure calculations for bulk materials in this context before we present the results on the influence of changing the crystal symmetry by introducing chemical order.

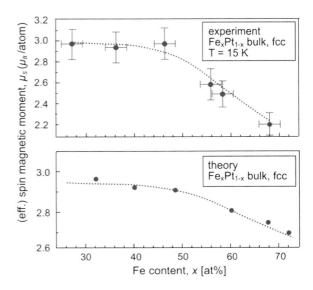

Fig. 11.8 Effective spin magnetic moment of Fe in chemically disordered bulk-like Fe_xPt_{1-x} films as determined from XMCD (*upper panel*) and calculated spin magnetic moments using the SPR-KKR methods (*lower panel*) [45]

One should note, that all spin magnetic moments determined from XMCD by the standard sum rules based [53–55] analysis are *effective* spin magnetic moments including also an intra-atomic dipole term T_z accounting for a possible asphericity of the spin density distribution which might not cancel out especially in the case of nanostructured materials [56]. Examples of spectra can be found in Fig. 11.10, details on the analysis and corrections for saturation and self-absorption effects [57, 58] can be found elsewhere [45]. XANES and XMCD measurements at the Fe $L_{3,2}$ absorption edges were performed in total electron yield (TEY) mode at the PM-3 bending magnet beamline at the HZB—BESSY II synchrotron radiation source in external magnetic fields of up to ± 3 T. To analyse the influence of the composition of chemically disordered Fe_xPt_{1-x} alloys on the magnetic moments at the Fe sites, bulk-like films of (46 ± 5) nm thickness (determined from Rutherford backscattering measurements) have been investigated with $28 \leq x \leq 68$ at% using the XMCD technique. It was found that the higher the Fe content, the lower the magnetic moment per Fe atom [45] in agreement to our theoretical studies using the spin-polarised relativistic Korringa–Kohn–Rostoker method as implemented in the Munich SPR-KKR package [59]. Both experimentally obtained and calculated values are shown in Fig. 11.8. Starting from this result, the seggregation of Fe (and Pt, respectively) in FePt nanoparticles is expected to reduce the magnetic moments at the Fe sites with respect to the case of a random occupation of lattice sites with Fe and Pt atoms. In fact, the magnetic moments at the Fe sites in FePt nanoparticles are reduced by 20–30% compared to a chemically disordered FePt alloy and match the values obtained experimentally for Fe-rich Fe_xPt_{1-x} with $x \approx 70$ at%. Note that spin canting effects that may also reduce the averaged spin magnetic moments were found to be unlikely due to the large exchange length of about 40 nm in Fe_xPt_{1-x} alloys [60]. Reducing

11 Intrinsic Magnetism and Collective Magnetic Properties

Table 11.2 Size dependence of effective spin magnetic moment and orbital magnetic moment of Fe in FePt nanoparticles as determined by XMCD

System	$\mu_s^{\text{eff}}(\text{Fe})/\mu_B$	$\mu_l(\text{Fe})/\mu_B$	$\mu_l/\mu_s^{\text{eff}}(\text{Fe})/\%$
FePt NPs, fcc, d = 6.3 nm[a]	2.28 ± 0.25	0.048 ± 0.010	2.1 ± 0.5
FePt NPs, fcc, d = 4.4 nm[a]	2.13 ± 0.21	0.062 ± 0.014	2.9 ± 0.5
FePt NPs, fcc, d = 3.4 nm[a]	2.01 ± 0.16	0.068 ± 0.015	3.4 ± 0.5

[a] from Ref. [39]

the particles' size increases this effect of reduced magnetisation as can be concluded from the experimental values of magnetic moments in Table 11.2. In addition, these values show that decreasing the particles size, i.e. increasing the surface fraction, yields enhanced orbital magnetic moments.

If chemical order is induced in the FePt system, the crystal structure changes from fcc to fct with an $L1_0$ symmetry. The tetragonal distortion in combination with the large spin–orbit coupling in Pt is discussed as a reason for the large magnetocrystalline anisotropy in this system. The phase transformation towards the $L1_0$ phase is driven by volume diffusion and can be achieved by thermal treatment of the nanoparticles at about $600\,^{\circ}\text{C}$, while special care has to be taken to avoid sintering of the nanoparticles [37, 38, 62]. As can be seen in Fig. 11.9, a distinct increase of the coercive field after annealing the FePt nanoparticles for 30 min at $600\,^{\circ}\text{C}$ indicates the (partial) formation of the high-anisotropic $L1_0$ phase. The magnetic hystereses have been recorded by measuring the XMCD at the L_3 absorption edge of Fe at $T \approx 11\,\text{K}$ as a function of the external magnetic field [63] applied under an angle $\theta = 75^{\circ}$ with respect to the sample normal. Before annealing, the coercive field is quite small, $\mu_0 H_c = (36 \pm 5)\,\text{mT}$ and the ratio of remanence-to-saturation magnetisation $m_r/m_s \approx 0.2$, indicating strong magnetic dipole interactions among the nanoparticles forcing the easy direction(s) of magnetisation in the sample plane. After annealing, the coercive field increased by a factor of about eight to $\mu_0 H_c = (228 \pm 8)\,\text{mT}$ and $m_r/m_s \approx 0.5$ as expected in the Stoner–Wohlfarth model for non-interacting nanoparticles with randomly oriented axes of uniaxial intrinsic anisotropy. Since the total magnetic moments of the nanoparticles do not significantly change after annealing—as known from the bulk material [61] and will be evidenced later in this work for the case of nanoparticles—and their arrangement remains the same, the dipolar interactions should be in the same order of magnitude before and after annealing. However, this simply reflects the fact that the magnetic anisotropy of the individual FePt nanoparticles is increased significantly due to the thermal treatment and hence, the inter-particle interactions become negligible. From temperature-dependent measurements of the coercive field as reported in Ref. [39], the anisotropy can be estimated according to the equation derived by Sharrock [64]:

$$H_c(T) \approx \frac{K_{\text{eff}}}{M}\left[1 - \left(\frac{25k_B T}{K_{\text{eff}} V}\right)^{2/3}\right] \qquad (11.1)$$

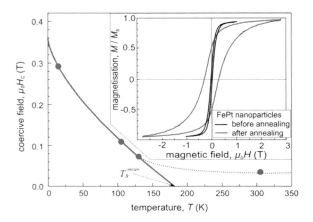

Fig. 11.9 Temperature-dependent coercive field of FePt nanoparticles with (partial) L1$_0$ order. *Inset* normalised field-dependent magnetisation measured at the Fe L$_3$ absorption edge of FePt nanoparticles before and after annealing, i.e. in the chemically disordered fcc state and the (partial) L1$_0$ phase

Note that, in this calculation the temperature dependence of K_{eff} [68–70] and the volume distribution are not taken into account. The magnetisation M can be calculated by the magnetic moments and lattice constants determined experimentally. With this method, we obtained $K_{\text{eff}} = (4.7 \times 10^5)$ J/m^3 which is large compared to the case of chemically disordered FePt [60], but still one order of magnitude smaller than the magnetocrystalline anisotropy in chemically ordered L1$_0$ bulk material of FePt. This reduction of the anisotropy in nanoparticles with respect to the bulk material is also reported in the literature for other FePt nanoparticles [65, 66]. A possible explanation in accordance to our structural investigations presented above is a lower degree of chemical order in the nanoparticles compared to thin films or bulk FePt, due to the competing tendency towards a seggregation of Fe and Pt, respectively, that hinders the formation of the L1$_0$ order. In the literature, it is also discussed that other types of atomic structures may be energetically favourable over the L1$_0$ symmetry in small nanoparticles [67].

The change in the crystal structure induced by the annealing process is also expected to modify the magnetic moments, especially the orbital magnetic moment that is a sensitive monitor to changes in the crystal symmetry. Therefore, we determined spin and orbital magnetic moments at the Fe and Pt sites by XMCD analyses at the L$_{3,2}$ absorption edges. Examples of the spectra can be found in Fig. 11.10, again, details on the analysis and corrections for saturation and self-absorption effects can be found elsewhere [45]. The magnetic moments are summarised in Table 11.3. It can clearly be seen that the magnetic moments at the Pt sites remain largely constant as well as the spin magnetic moment at the Fe sites. The orbital magnetic moment of Fe increased due to the annealing by almost a factor of four from $\mu_l(\text{Fe}) = (0.048 \pm 0.010)\,\mu_B$ to $\mu_l(\text{Fe}) = (0.204 \pm 0.020)\,\mu_B$. This can be understood in terms of unquenching the orbital moment when changing the crystal structure from cubic (in the chemically disordered, non-annealed state) to tetragonally distorted (in the partial L1$_0$ state).

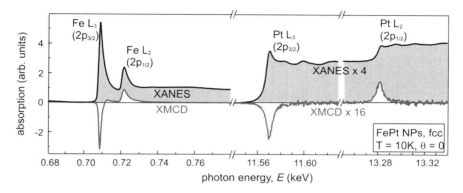

Fig. 11.10 XANES and XMCD of magnetically saturated chemically disordered FePt nanoparticles measured at $T \approx 10\,\text{K}$ at the $L_{3,2}$ absorption edges of Fe and Pt, respectively. Note the different scaling factors of XANES and XMCD for the case of Pt spectra

Table 11.3 Element-specific magnetic moments as determined by XCMD at for chemically disordered bulk material, nanoparticles and chemically ordered nanoparticles

System	$\mu_s^{\text{eff}}(\text{Fe})/\mu_B$	$\mu_l(\text{Fe})/\mu_B$	$\mu_s^{\text{eff}}(\text{Pt})/\mu_B$	$\mu_l(\text{Pt})/\mu_B$
FePt bulk, fcc[b]	2.92 ± 0.29	0.083 ± 0.012	0.47 ± 0.02	0.045 ± 0.006
FePt NPs, fcc, d = 6.3 nm[a,b]	2.28 ± 0.25	0.048 ± 0.010	0.41 ± 0.02	0.054 ± 0.006
FePt NPs, L1$_0$, d = 6.3 nm[a,b]	2.38 ± 0.26	0.204 ± 0.020	0.41 ± 0.04	0.042 ± 0.008

Note that the values taken from Ref. [62] have been recalculated for the case of Fe using a number of unoccupied d-states $n_h = 3.41$, [a]from Ref. [62], [b]from Ref. [39]

11.3.2 Magnetic Response of Fe on Different Lattice Sites in Fe/FeO$_x$ Nanocubes

In the bulk material, Fe may form four types of oxides: FeO, Fe$_3$O$_4$, α-Fe$_2$O$_3$, and γ-Fe$_2$O$_3$. FeO (wuestite) is metastable at temperatures below 850 K and thus will not be discussed here. Fe$_3$O$_4$ (magnetite) crystallises in an inverse spinel structure and Fe in this type of oxide exhibits a mixed valence state of Fe^{2+} on octahedral lattice sites surrounded by O^{2-} ions and Fe^{3+} equally distributed on octahedral and tetrahedral lattice sites. The highest degree of oxidation exhibits Fe in Fe$_2$O$_3$ which exists in two different states: ferrimagnetic γ-Fe$_2$O$_3$ (maghemite) includes Fe^{3+} ions only and has also an inverse spinel structure like Fe$_3$O$_4$ but with additional vacancies on octahedral sites, whereas α-Fe$_2$O$_3$ (hematite) has a Corundum structure and is an antiferromagnet below its Néel temperature of 263 K.

In order to analyse the contributions of Fe^{2+} and Fe^{3+} at different lattice sites, i.e. either in tetrahedral or octahedral environment, calculations of the XANES and XMCD spectra were performed using the CTM4XAS Charge Transfer Multiplet Program [71]. This semi-empirical programme is based on a Hartree–Fock method corrected for correlation effects to solve the atomic Hamiltonian [72]. It includes the core and valence spin–orbit coupling, the core-valence two-electron integrals

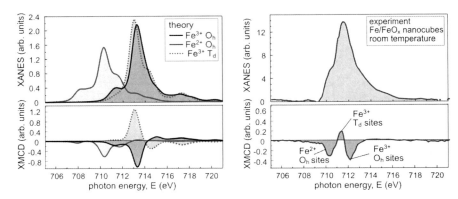

Fig. 11.11 Calculated XANES and XMCD at the Fe L₃ absorption edge in Fe-oxide (*left*) and experimental spectra of naturally oxidised Fe-nanocubes (*right*) measured at room temperature in a magnetic field of 2.8 T applied perpendicular to the sample plane

(multiplet effects), and the effects of strong correlations within the charge transfer model. More details about the programme and some examples of applications can be found e.g. in Refs. [73, 74]. Calculated spectra are shown in Fig. 11.11 for Fe^{3+} in either octahedral (O_h) or tetrahedral (T_d) symmetry and Fe^{2+} in octahedral symmetry as it occurs in Fe_2O_3 or Fe_3O_4. The t_{2g}-e_g splitting was set to 10Dq = 1.5 eV for Fe ions in O_h symmetry and 10Dq = 0.7 eV for Fe ions in tetrahedral environment as, suggested in the literature [75, 76]. The spectra were calculated with a Lorentzian broadening of 0.25 eV and a Gaussian broadening of 0.3 eV to account for lifetime effects and finite energy resolution in experiments. Although shifted in energy, it can clearly be seen that the spectral shape for Fe^{2+} and Fe^{3+} in the same octahedral environment is similar. The maximum absorption is shifted by about 3 eV to lower energies in the case of Fe^{2+}, the maximum dichroism is shifted accordingly. The smaller intensity is due to a smaller number of unoccupied final d states in the case of Fe^{2+} ($3d^6$) with respect to Fe^{3+} ($3d^5$). Compared to Fe^{3+} in octahedral symmetry, the tetragonally coordinated Fe^{3+} shows a maximum absorption slightly shifted to lower energies, i.e. by less than 0.5 eV as the maximum dichroism is shifted as well. The reversed sign of the dichroic signal indicates the antiparallel alignment of the spins of Fe^{3+} in tetrahedral symmetry with respect to Fe ions on octahedral lattice sites.

The XMCD signals of both Fe_3O_4 and γ-Fe_2O_3 exhibit a W-like spectral shape at the L₃ edge as can be found by summing up the contributions of non-equivalent Fe cations to the total XMCD signal. An example of experimental data obtained on naturally oxidised Fe-nanocubes is shown in Fig. 11.11 where the spectral shape is referred to the contributions of Fe cations in different crystal fields. By recording the field-dependent XMCD as a measure of magnetisation [63], magnetic hystereses were measured for the different lattice sites as shown in Fig. 11.12. Having a closer look on the hystereses, one may realise that the Fe ions at the tetrahedrally coordinated lattice sites are already magnetically saturated for magnetic fields of about ±2 T.

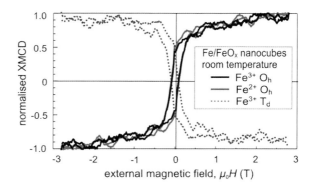

Fig. 11.12 Site-selective magnetic hystereses of Fe in naturally oxidised Fe-nanocubes measured at room temperature with the external magnetic field applied perpendicular to the sample plane

In contrast, there is still a slope in the hysteresis loops for the case of Fe ions on octahedral lattice sites. This may reflect different anisotropies due to vacancies and/or defects that are preferentially located at octahedral lattice sites [77].

11.4 Spectro-Microscopy of Individual Nanoparticles

In the previous chapter, the magnetic properties of nanoparticles were presented as deduced from measurements of ensembles. Such measurements can only yield average values for the MAE or magnetisation per atom. To be able to determine the MAE of individual particles, detection schemes allowing the magnetic characterisation of single particles are required. Here, we discuss measurements of the element-specific electronic structure and magnetic response as a function of external magnetic field amplitude and orientation for single Fe-nanocubes with 18 nm edge length. Magnetic states and interactions of monomers, dimers and trimers are analysed by XPEEM for different particle arrangements. Furthermore, this individual local approach allows for the detailed investigation of dipolar interactions within different configurations and their influence on the blocking temperature which is a measure for a stable magnetisation during the time window of the measurement.

Here, XPEEM is employed in combination with SEM and the magnetisation reversal behaviour is studied, i.e. the magnetic hysteresis of size-selected individual Fe-nanocubes with sizes around 18 nm in different local configurations at room temperature [78]. The magnetic characterisation is correlated with a detailed imaging of the orientation and crystalline structure of the individual nanoparticles. The coercive field and the shape of the single particle hysteresis loops recorded for different orientations of the magnetic field provide a measure of the MAE which is analysed within the framework of the Stoner–Wohlfarth theory of coherent reversal of the magnetisation in single domain nanomagnets. Summing over all the locally resolved hysteresis loops allows for developing an understanding of the macroscopic collective magnetic response. In the next section, magnetic hysteresis measurements for different nanoparticle configurations are presented and discussed.

11.4.1 Magnetic Hysteresis and Spectroscopy of Monomers, Dimers, Trimers and Many Particle Configurations

XPEEM in combination with SEM was employed to determine the magnetic hysteresis and electronic structure of individual nanoparticles. Experiments were performed at the HZB-BESSY II synchrotron radiation source using the experimental XPEEM setup schematically shown in Fig. 11.13. The XPEEM instrument is based on a commercial Elmitec photoemission electron microscope attached to a soft X-ray micro-focus beamline (UE49) with $10 \times 15\,\mu m$ spot size and full polarisation control providing a spatial resolution of 25 nm. The XPEEM image in Fig. 11.13 demonstrates that this resolution is sufficient to image individual Fe-nanocubes but not to resolve configurations. Therefore, the XPEEM data has to be matched to images with a higher lateral resolution allowing for the discrimination of particle configurations. A lithographically designed grid of Au-markers on the sample substrates allows for the identification of the same sample position in PEEM and SEM. As a result of the colloidal preparation technique of the nanocubes, the metallic core of the as-grown nanoparticles is surrounded by an oxide shell and organic ligands. The specimen were prepared by drying the nanoparticle suspension on the naturally oxidised Si substrates provided with lithography markers. Acting as a spacer, the ligand shell keeps neighbouring particles at a minimum distance of about 2 nm. Different configurations are obtained ranging from single nanocubes to dimers, trimers and more complex configurations as visible in the SEM and TEM images shown in Fig. 11.13a. To obtain metallic, oxide-free Fe-nanocubes for our magnetic characterisation, the organic ligands were removed and Fe-oxides were reduced by a plasma etching technique [40, 41] prior to the magnetic measurements. The position of the particles remains unchanged by this treatment. The samples were protected against re-oxidation by an Al layer grown subsequently in situ.

The local coordination and orientation of the cubes were determined from SEM images, the corresponding chemical and magnetic contrast image series (Fig. 11.13b, d) were determined by means of spectromicroscopy at the Fe L_3 absorption edge. Background normalisation was performed by using a surrounding region containing no particle. In Fig. 11.13c, the resulting spectra for a dimer configuration are shown, which are extracted from image stacks where each image was recorded at a different photon energy ($\Delta E = 0.2\,eV$) in an external magnetic field of 33 mT applied during imaging. Recording a stack of images during a magnetic field sweep allows us to extract the hysteresis of almost any nanoparticle configuration within our field of view: hysteresis loops of different particle configurations were extracted by plotting the magnetic (XMCD) signal at Fe L_3 edge for each pixel or collection of pixels (Fig. 11.13d) as a function of the magnetic field. An example for a

Fig. 11.13 Magnetic imaging and spectro-microscopy of individual Fe nano-cubes in an applied magnetic field of up to ±33 mT. **a** *left* schematics of the photoemission electron microscope, *right* SEM image of a sample with Au-markers and Fe-nanocubes. **b** XPEEM image stacks of the same area at different photon energies and **d** magnetic contrast at the Fe L_3 edge as a function of external magnetic field applied along the indicated direction. *Blue* and *red colour* indicate opposite magnetisation directions in the particles. **c** XANES of one dimer marked by the *circles* in **b** and **d**, *green diamonds* correspond to positive, *red circles* to negative helicity of the photons. **e** A hysteresis loop for a dimer aligned with its easy axis parallel to the applied field recorded at the Fe L_3 edge

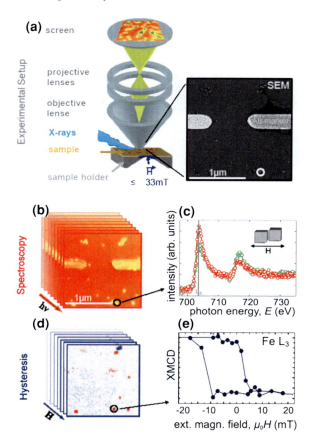

spatially resolved hysteresis loop of a typical dimer configuration consisting of two Fe-nanocubes oriented with the {100} facets facing each other (marked with a circle in Fig. 11.13b, d and the SEM image of Fig. 11.13a is shown in Fig. 11.13e. To improve statistics, we took about 600 images (one second exposure time each) at each field step. Images were corrected for drift and summed up for each magnetic field and helicity. Finally, the intensity from a particular region of interest containing an individual particle or cluster to calculate the corresponding XMCD signal was integrated and a local background subtraction was performed.

The spectral data show no evidence of oxidation of the Fe-nanocubes. In Fig. 11.13d, the sequence of images of the same sample area as for the spectroscopy which show the magnitude of the XMCD signal is displayed in false colour where blue and red colour indicate opposite magnetisation directions. The dichroic signal yields the magnetic moment per atom which was determined for the dimer configuration with the magnetic field $\mu_0 H = -20$ mT applied parallel to the dimer axis (Fig. 11.13c, e). Utilizing the sum rules [53–55] and assuming the

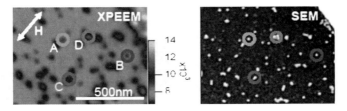

Fig. 11.14 Chemical contrast XPEEM image (*left*) at the Fe L$_3$ absorption edge and SEM image of the same area (*right*). The magnetisation behaviour of the particles denoted A–D are analysed in more details (cf. Fig. 11.15)

bulk-like number of Fe d-holes $n_{3d} = 3.4$ a bulk-like ratio of orbital-to-spin magnetic moment $\mu_l/\mu_S = 0.04 \pm 0.02$ and a spin magnetic moment per Fe atom of $\mu_S = 1.05 \pm 0.2\,\mu_B$ is calculated which is about half the value of bcc Fe in the bulk. This reduction is most likely due to thermal fluctuations of the magnetisation over the acquisition time of the ten averaged spectra (about 200 min) at room temperature, i.e. the spectra were not recorded in magnetic saturation. Spin canting effects at the surface could also reduce the measured magnetisation but are unlikely the reason for such a significant reduction of the averaged magnetisation. Alloying between Fe and Al of the capping layer may also reduce the magnetic moment of Fe, but for the large size of nanocubes, these interface effects are negligible. However, small magnetic moments were also reported for an ensemble of similar Fe-nanocubes [36] in magnetic saturation. An appropriate explanation may be the occurrence of interactions between the nanocubes by complex stray fields as discussed at the end of this paragraph that may yield a reduced magnetisation on average.

Magnetic hysteresis loops with respect to different orientations of the applied magnetic field for individual nanoparticles in either a dimer or a trimer configuration are presented and compared to micromagnetic simulations in Fig. 11.15 using the object oriented micromagnetic framework (OOMMF) code [79]. The positions of the corresponding particles are indicated by the respective colours in the XPEEM and SEM images of Fig. 11.14. For individual Fe-nanocubes we find the expected dependence of the hysteresis loops on the orientation of the applied magnetic field (Fig. 11.15, configurations A and B) indicating an easy axis of the magnetisation along a $\langle 100 \rangle$ direction (cube edge) and a magnetically hard direction along the $\langle 110 \rangle$ direction. The coercive fields of about 2 mT are much smaller than the ones obtained by our simulations using bcc bulk Fe parameters and $T = 0$ K. As mentioned above, this is most likely due to instabilities of the magnetisation over the time scale of the measurement and the dependence of the blocking temperature on the particle's volume which is not taken into account in the simulation. In addition, the morphology of the nanoparticles, in particular a deviation from the cubic shape, may change the effective anisotropy and influence the switching behaviour [80, 81]. For an Fe cube of about 18 nm side length with bulk-like MAE and reduced magnetisation, the transition into a superparamagnetic state for these kind of measurements takes place close to the measuring temperature of 300 K. Qualitative differences in the shape and

11 Intrinsic Magnetism and Collective Magnetic Properties

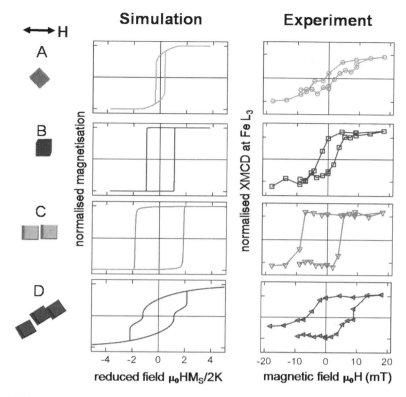

Fig. 11.15 Field dependent magnetisation of differently coordinated Fe-nanocubes marked in Fig. 11.14: micromagnetic simulations of the configurations depicted in A–D with respect to the direction of the indicated magnetic field H. The magnetic field is normalised with respect to the anisotropy field $2K/M_S$ (M_S saturation magnetisation). Experimental data obtained at the Fe L_3 absorption edge are shown on the *right* panel

remanent magnetisation between the hysteresis loops correspond rather nicely to the simulated ones. A much better quantitative agreement can be achieved when one considers the atomically rough topography of the Fe cubes and the small deviations from perfect cubic symmetry as discussed in the next section. To generally confirm the orientation of the easy axis of magnetisation in the nanocubes, the orientations using high resolution SEM images and hysteresis loops of nearly 100 nanocubes were analysed. Half of the hysteresis loops were obtained for particles with the edge axis parallel to the applied magnetic field H, the other half for H applied at an angle of $45°$. In the averaged hysteresis loops shown in Fig. 11.16 blue squares represent the loop for cubes oriented with their $\langle 100 \rangle$ axis along the magnetic field direction while circles represent the one for cubes oriented with the $\langle 110 \rangle$ axis along the magnetic field direction. The rounding in comparison to ideal simulated Stoner–Wohlfarth loops is due to the fact that loops of cubes which were oriented within $\pm 22°$ with

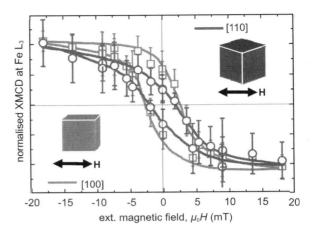

Fig. 11.16 Averaged hysteresis loops of about 100 single nanocubes recorded with the magnetic field oriented parallel to the cube edge (⟨100⟩) or at an angle of 45° (⟨110⟩)

respect to the magnetic field direction were also averaged. The strongly enhanced remanence and higher coercive field of the hysteresis recorded with the magnetic field along the ⟨100⟩ axis clearly evidences the presence of magnetocrystalline anisotropy with a preferred magnetisation along this axis as confirmed by our micromagnetic simulations shown in Fig. 11.15 for configurations A and B. The two average loops shown in Fig. 11.16 resemble the loops of a macroscopic measurement which masks the influence of variations in particle size and orientation and averages over small changes of MAE from particle to particle. For the dimer and trimer (Fig. 11.15, configurations C and D) and also for other more complex configurations (not shown here) we find large variations in the coercive field and the shape of the hysteresis loops which can be qualitatively understood by micromagnetic simulations at T = 0 K again assuming the parameters of bulk Fe.

The dimer configuration (C) consists of two nanocubes arranged face to face with adjacent {100} planes and the long axis of the dimer was oriented along the axis of the magnetic field. Its magnetic response in Fig. 11.15 shows a wide open and almost rectangular hysteresis with an enhanced coercivity of about 7.5 mT. The 100% remanence magnetisation at zero field demonstrates the contribution of magnetic dipolar coupling between the neighbouring Fe-nanocubes. Obviously, the magnetic dipolar coupling enhances the effective anisotropy and stabilises the magnetisation of the dimer resulting in suppressed magnetisation fluctuations, i.e. a higher blocking temperature. Whereas this explanation is confirmed by micromagnetic calculations showing an increased coercivity, the experimental hysteresis shows an unexpected horizontal shift to negative fields. The latter one is reminiscent of the exchange bias effect [82, 83] usually due to the unidirectional coupling between an antiferromagnet and a ferromagnet. This horizontal shift is observed in this work only for dimer or linear trimer configurations, not for single particles. In configurations of several closed packed nanocubes this shift is very small or not detectable. The origin of this peculiar shift for some particles configurations is not clear, yet. An antiferromagnetic material is not apparent in our sample. Fe-Al alloys have shown antiferromagnetic

11 Intrinsic Magnetism and Collective Magnetic Properties

correlations which might be a possible route for explanation. However, the existence of an intermixed antiferromagnetic Fe-Al interface layer cannot be the explanation, since in this case the shift should also be present for single particles. This argument holds also for the exclusion of a possible antiferromagnetic Fe-oxide layer (α-Fe_2O_3) at the bottom of the nanocubes. Thus, classical exchange bias effects cannot be the reason for the observed shift of hysteresis curves. As discussed in the literature [84], there may be a shift of the hysteresis loop of nanoparticles if a minor loop is recorded. However, since the hysteresis e.g. shown in Fig. 11.15 C is rectangular, clearly saturated and shifted, it is not an appropriate explanation. Also, the trivial effect of an apparent shift due to the contributions of magnetic stray fields from other nanocubes can be ruled out for the dimers discussed here. According to our micromagnetic simulations the stray field of the Fe-nanocubes (assuming full bulk-like magnetisation) is smaller than 2 mT at a distance of approximately 60 nm. As a result, we have to consider a complex unidirectional coupling due to diverging stray fields at the cube edges that could be present only in specially arranged dimers or more complex configurations. However, at the current status of our micromagnetic simulations at T = 0 K this complex situation involving a strongly inhomogeneous magnetisation per cube cannot be interpreted and requires further investigations.

In Fig. 11.15 the magnetisation curve of a trimer (D) with the magnetic field oriented as indicated in the figure is shown. This configuration shows an interesting two-step magnetization reversal which is also seen for other, slightly different configurations of trimers. In a corresponding micromagnetic simulation we find a similar behaviour. Compared to the dimer discussed above (config. C) this configuration has a significantly reduced saturation magnetisation pointing towards the presence of magnetic fluctuations or frustration, caused by the larger distance between the Fe-nanocubes or their non collinear alignment, respectively. Unfortunately, the individual nanocubes in this particular configuration cannot be resolved in the experiment. Nevertheless the measured hysteresis demonstrates the complexity of magnetic coupling in configurations of more than two Fe-nanocubes which can only be addressed with the appropriate spatial resolution and specific contrast. Instrumental developments such as aberration corrected PEEM [85] promise a much more detailed insight into the magnetic properties of individual nanomagnets in the near future, making additional SEM images dispensable. Furthermore, element-specific studies on individual bimetallic, complex nanomagnets, e.g. core–shell nanoobjects with unconventional, non-aligned spin configuration will become feasible.

For a quantitative comparison of the simulated and measured magnetic hysteresis, the temperature has to be taken into account that yields smaller coercive fields in the experiment due to thermal fluctuations which are not negligible for the room-temperature experiments. In addition, deviations from a perfect cubic shape will strongly influence the effective anisotropy of the individual nanoparticles while for the simulations the anisotropy of Fe in the bulk with its four-fold symmetry was assumed. However, the elongation along one direction that was found by tomography as presented in Sect. 11.2, will yield an uniaxial contribution to the anisotropy. In particular, the magnetisation direction is favoured parallel to the distinguished elongated axis exhibiting maximum coercivity for the magnetic field applied also parallel to this axis. Along all other axis, the coercivity will be reduced. A detailed discussion on these effects will be presented elsewhere [80].

11.5 Magnetisation Dynamics of Nanoparticle Ensembles

A useful method to study the magnetisation dynamics of nanoparticle ensembles is the ferromagnetic resonance (FMR) technique. By microwave absorption in a quasi-static magnetic field, not only the spectroscopic splitting factor (g-factor) can be determined, but also the damping of the stimulated magnetisation precession can be studied as presented below. For the magnetic characterisation by means of FMR, the Fe/FeO$_x$-nanocubes were deposited onto GaAs substrates. Experiments were performed at different microwave frequencies and temperatures in conventional microwave cavities (see e.g. [86]). Recently, FMR techniques have been suggested which offer the possibility to even measure single nanoparticles [87].

In general, microwave absorption of the sample can be detected if the precession frequency of magnetisation equals the frequency of the irradiated microwave $\omega = 2\pi \nu$. In the ground state of the system, all spins of a ferromagnet are aligned parallel due to the exchange interaction while precessing around the effective magnetic field $\mathbf{H}_{\text{eff}}$ consisting of the external magnetic field, anisotropy fields, exchange field and the magnetic component of the microwave. In the specific case of the nanoparticles ensemble, the dipolar coupling between the particles is also included. Thus, the relation between microwave frequency and resonance field is no longer linear and the resonance condition can be written as

$$(\omega/\gamma)^2 = \mu_0^2(H_{\text{res}}^2 - H_{\text{eff}} H_{\text{res}}) \tag{11.2}$$

where $\gamma = g\mu_B/\hbar$ is the magnetogyric ratio with the spectroscopic splitting factor g and H_{res} is the resonance field, i.e. the externally applied magnetic field at which the resonant microwave absorption is maximum.

Relaxation processes like e.g. energy dissipation into the lattice yield to a damping of the magnetisation precession that can be phenomenologically described for small damping factors α by the Landau–Lifshitz–Gilbert equation:

$$\dot{\mathbf{M}} = -\gamma \mu_0 \mathbf{M} \times \mathbf{H}_{\text{eff}} + \frac{\alpha}{M_S} \mathbf{M} \times \dot{\mathbf{M}} \tag{11.3}$$

Note that, for this ansatz a constant absolute value of the magnetisation vector $\mathbf{M}$ is assumed. The microscopic mechanism behind this damping behaviour can be traced back to the spin–orbit coupling: in a simple model, the orbital motion is coupled to the spin precession and may be disturbed by phonons yielding a phase shift which implies damping. An illustration of the damped magnetisation precession is given in Fig. 11.17. The time evolution after the excitation time t_0 for the magnetisation component that is parallel to the quantisation axis, i.e. the axis of the effective magnetic

Fig. 11.17 Illustration of the precession of the magnetisation vector in an effective magnetic field including damping (*left*) and the related change of the magnetisation component along the direction of the effective magnetic field (*right*)

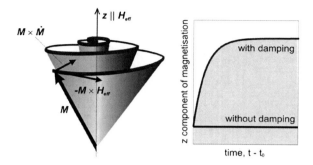

field, is shown for this case and additionally for a fictive case without any damping mechanisms. It turned out, that the Gilbert-type damping is sufficient for the case of nanocubes investigated here. Details on additional contributions beyond the Gilbert damping like two-magnon or four-magnon scattering can be found elsewhere (e.g. [88]).

Experimental room-temperature FMR spectra at different frequencies are shown in Fig. 11.18. Note that the first derivative of the absorbed microwave power is plotted as a function of external magnetic field, i.e. maximum power in resonance is absorbed at the zero-crossings of the spectra. The solid lines represent fits of the experimental data using the first derivative of Lorentzian lines necessary for a reliable extraction of the resonance fields and linewidths. The latter is connected to the intrinsinsic damping of the magnetisation precession while the resonance field gives the possibility to determine the spectroscopic splitting factor as discussed above. From the experimental data on the naturally oxidised Fe-nanocubes, we obtained $g = 2.07 \pm 0.03$ and a damping parameter of $\alpha = 0.032 \pm 0.008$. While the g-factor is the same within the error bars as for bulk Fe ($g = 2.09$), the damping is enhanced by about one order of magnitude. In order to exclude effects of the Fe-oxide shell on these values, FMR measurements were also performed on the sample after a plasma cleaning procedure [] and subsequent capping by Ag and Pt.

The corresponding FMR signal of the cleaned Fe-nanocubes is shown in Fig. 11.19 for two different configurations: with the external magnetic field applied either perpendicular or parallel to the sample plane. In the latter case, two different contributions to the overall signal can clearly be seen by eye which were not observed for the as-prepared (oxidised) sample. To interpret the observed spectra, the imaginary part of the high-frequency susceptibility as a function of the external magnetic field has been calculated which is proportional to the absorbed microwave power. To simulate the experimental FMR spectra, the Landau–Lifshitz equation for the magnetisation precession was solved. Details on the calculations are published elsewhere [89]. The best fit to the experimental data that is shown in Fig. 11.19 was obtained using the following parameters: cubic anisotropy energy density $K_4 = 4.8 \times 10^4 \, \text{J/m}^3$, $g = 2.09$, a damping constant $\alpha = 0.03$, and a magnetisation $M(H_{\text{res}}) = 6.7 \times 10^5 \, \text{A/m}$, which was determined independently by

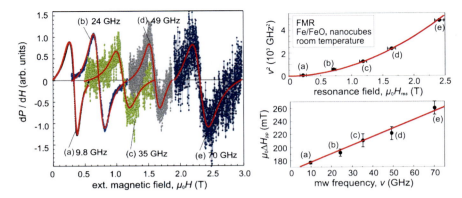

Fig. 11.18 Ferromagnetic resonance measurements of Fe/FeO$_x$ nanocubes at room temperature (*left*) and extracted relation between the squared microwave frequency and resonance field (*upper right graphic*) and frequency dependence of the FMR linewidth (*lower right graphic*)

Fig. 11.19 Experimental FMR spectra (*symbols*) of plasma cleaned and capped Fe-nanocubes at room temperature for different measurement geometries and simulations (*lines*)

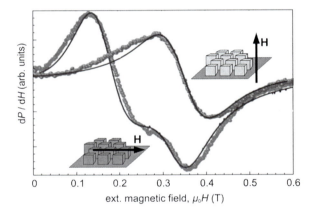

superconducting quantum interference device (SQUID) magnetometry. We assumed a random distribution of crystallographic axes in the substrate plane, while the [001] direction is oriented perpendicular to the substrate plane as known from HR-TEM analyses. Thus, the FMR spectrum measured with the external magnetic field applied in the sample plane is given by the sum of the resonance spectra obtained for the particles with all possible orientations. This average spectrum is shown in Fig. 11.19 and is in good agreement to the experimental data. The contribution at low fields can be assigned to Fe-nanocubes with the easy in-plane axis [100] or [010] oriented parallel to the external magnetic field. The spectrum at the highest field values results from the nanocubes with their hard axis ($\langle 110 \rangle$) aligned parallel to the magnetic field direction. All other orientations of the nanocubes contribute to the spectrum at intermediate field values. With the same set of parameters the FMR spectrum measured with the magnetic field applied perpendicular to the sample plane was fitted.

11 Intrinsic Magnetism and Collective Magnetic Properties

Table 11.4 Room-temperature FMR results for the spectroscopic splitting factor (g-factor), Gilbert damping parameter α and cubic anisotropy energy density K_4

System	g-factor	Damping α	K_4 / (kJm^{-3})
Fe/Fe-oxide nanocubes	2.07 ± 0.03	0.032 ± 0.008	
Fe-nanocubes	2.09 ± 0.01	0.03 ± 0.005	48 ± 5
Fe 3 nm granular film		0.0081 [a]	
Fe 8 nm continuous film		0.0046[a]	
Fe bulk	2.09[b]	0.0043[c]	45[b]

[a]from Ref. [90] [b]from Ref. [91] [c]from Ref. [92]

Again, the experimental data could be well reproduced. Note that an improved fit was obtained if one takes into account a slight variation of 0.2° for the [001] direction with respect to the substrate plane. This small misalignment is in agreement with HR-TEM results.

For the plasma cleaned Fe-nanocubes, the anisotropy and g-factor are in agreement to the values of the corresponding bulk material, whereas the damping parameter is increased by about one order of magnitude (Table 11.4). It seems to be a general trend that with decreasing dimensions the damping will be increased: for example, while epitaxial Fe films of 8 nm thickness exhibit a bulk-like damping parameter [92], it is significantly larger for 3 nm Fe films. For the latter case, the film is not continuous, but granular and can be understood as epitaxial Fe nanoislands with a thickness of 3 nm. For the case of the nanocubes investigated in the work, the lateral confinement is more distinct and the damping parameter is further increased.

11.6 Summary

In conclusion, we present an experimental work allowing for the detailed study of the magnetic properties of nanoparticle ensembles, individual nanoparticles and few particles configurations. Intrinsic properties like local structure, coordination number and chemical environment have been shown to modify the static magnetisation with respect to the corresponding bulk material as well as the magnetisation dynamics: besides changes in the spin and orbital magnetic moments, the lower coordination number at the surface is discussed to increase the magnetisation damping in Fe-nanocubes. In addition, the influence of dipolar coupling on the magnetic response turned out to be significant since it stabilises the magnetisation direction. In addition, the crucial influence of slight variations in shape and morphology towards the switching behaviour of individual nanocubes is discussed pointing out the need of individual particle magnetic probing techniques. Thus, our experimental approaches open up new possibilities for a more detailed understanding of the single and collective behaviour of magnetic nanoparticles, a knowledge that is also of importance for modern engineering and biomedical applications.

Acknowledgments We would like to thank the HZB—BESSY II staff, in particular T. Kachel, R. Schulz, and H. Pfau for their kind support during beamtimes. For their great help in the measurements and technical support, F. Wilhelm, A. Rogalev, P. Voisin, and S. Feite (ESRF) are gratefully acknowledged. For nanoparticle synthesis we would like to thank S. Sun (Brown U.) and O. Margeat (U. de la Méditerranée Marseille), and J.-U. Thiele (Seagate) for preparation of the epitaxial films. For help with the SPR-KKR package J. Minár, M. Košuth, S. Mankovsky and H. Ebert (LMU Munich) are acknowledged. We thank all other collaborators and members of the SFB 445—a number too large to be mentioned here in particular—for their help and for fruitful discussions. This work was financially supported by the DFG (SFB445), EU (MRTN-CT-2004-005567), ESRF and BMBF (05 ES3XBA/5).

References

1. J.L. Dorman, D. Fiorani, E. Tronc, Magnetic relaxation in fine-particle systems. Adv. Chem. Phys. **98**, 283 (1997)
2. C.K. Kim, D, Kan, T, Veres, F. Normadin, J.K. Liao, H.H. Kim, S.-H. Lee, M. Zahn, M. Muhammed, Monodispersed FePt nanoparticles for biomedical applications. J. Appl. Phys. 97, 10Q918 (2005).
3. O.A. Ivanov, L.V. Solina, V.A. Demshina, L.M. Magat, Determination of the anisotropy constant and saturation magnetization, and magnetic properties of powders of an iron-platinum alloy. Phys. Met. Metall. **35**, 81 (1973)
4. M.R. Visokay, R. Sinclair, Direct formation of ordered CoPt and FePt compound thin films by sputtering. Appl. Phys. Lett. **66**, 1692 (1995)
5. J.-U. Thiele, L. Folks, M.F. Toney, D. Weller, Perpendicular magnetic anisotropy and magnetic domain structure in sputtered epitaxial FePt (001) L1$_0$ films. J. Appl. Phys. **84**, 5686 (1998)
6. T. Shima, K. Takanashi, Y.K. Takahashi, K. Hono, Coercivity exceeding 100 kOe in epitaxially grown FePt sputtered films. Appl. Phys. Lett. **80**, 2571 (2004)
7. F. Dumestre, B. Chaudret, C. Amiens, P. Renaud, P. Fejes, Superlattices of iron nanocubes from Fe[N(SiME$_3$)$_2$]$_2$. Science **303**, 821 (2004)
8. O. Margeat, F. Dumestre, C. Amiens, B. Chaudret, P. Lecante, M. Respaud, Synthesis of iron nanoparticles: size effects, shape control and organisation. Prog. Solid. State Chem. **33**, 71 (2005)
9. C. Frandsen et al., Interparticle interactions in composites of nanoparticles of ferrimagnetic γ-Fe$_2$O$_3$ and antiferromagnetic (CoO, NiO) materials. Phys. Rev. B **70**, 134416 (2004)
10. D.V. Berkov, Density of energy barriers in fine magnetic particle systems. IEEE Trans. Magn. **38**, 2637–2639 (2002)
11. C. Thirion, W. Wernsdorfer, D. Mailly, Switching of magnetization by nonlinear resonance studied in single nanoparticles. Nat. Mater. **2**, 524–527 (2003)
12. I. Rod, O. Kazakova, D.C. Cox, M. Spasova, M. Farle, The route to single magnetic particle detection: a carbon nanotube decorated with a finite number of nanocubes. Nanotechnology **20**, 335301 (2009)
13. L.T. Kuhn et al., Magnetisation of isolated single crystalline Fe-nanoparticles measured by a ballistic Hall micro-magnetometer. Europ. Phys. J. D **10**, 259–263 (2000)
14. T. Uhlig, J. Zweck, Recording of single-particle hysteresis loops with differential phase contrast microscopy. Ultramicroscopy **99**, 137–142 (2004)
15. E. Snoeck et al., Magnetic configurations of 30 nm iron nanocubes studied by electron holography. Nano Lett. **8**, 4293–4298 (2008)
16. P. Schattschneider et al., Detection of magnetic circular dichroism on the two-nanometer scale. Phys. Rev. B **78**, 104413 (2008)
17. P. Schattschneider et al., Detection of magnetic circular dichroism using a transmission electron microscope. Nature **441**, 486–488 (2006)

11 Intrinsic Magnetism and Collective Magnetic Properties

18. H.A. Dürr et al., A closer look into magnetism: opportunities with synchrotron radiation. IEEE Trans. Magn. **45**, 15–57 (2009)
19. S. Eisebitt et al., Lensless imaging of magnetic nanostructures by X-ray spectro-holography. Nature **432**, 885–888 (2004)
20. E. Amaladass, B. Ludescher, G. Schütz, T. Tyliszczak, T. Eimüller, Size dependence in the magnetization reversal of Fe/Gd multilayers on self-assembled arrays of nanospheres. Appl. Phys. Lett. **91**, 172514 (2007)
21. P. Fischer et al., Element-specific imaging of magnetic domains at 25 nm spatial resolution using soft x-ray microscopy. Rev. Sci. Instrum. **72**, 2322–2324 (2001)
22. M.-Y. Im et al., Direct real-space observation of stochastic behavior in domain nucleation process on a nanoscale. Adv. Mater. **20**, 1750–1754 (2008)
23. D.Y. Kim, Magnetic soft x-ray microscopy at 15 nm resolution probing nanoscale local magnetic hysteresis. J. Appl. Phys. 99, 08H303 (2006)
24. A. Fraile Rodríguez, F. Nolting, J. Bansmann, A. Kleibert, L.J. Heyderman, X-ray imaging and spectroscopy of individual cobalt nanoparticles using photoemission electron microscopy. J. Magn. Magn. Mater. **316**, 426–428 (2007)
25. J. Bansmann et al., Magnetism of 3d transition metal nanoparticles on surfaces probed with synchrotron radiation–from ensembles towards individual objects. Phys. Stat. Sol. B **247**, 1152–1160 (2010)
26. A. Fraile Rodríguez et al., Size-dependent spin structures in iron nanoparticles. Phys. Rev. Lett. **104**, 127201 (2010)
27. G. Cheng, J.D. Carter, T. Guo, Investigation of Co nanoparticles with EXAFS and XANES. Chem. Phys. Lett. **400**, 122–127 (2002)
28. J. Rockenberger, L. Trger, A. Kornowski, T. Vossmeyer, A. Eychmüller, J. Feldhaus, H.J. Weller, EXAFS studies on the size dependence of structural and dynamic properties of CdS nanoparticles. Phys. Chem. B **101**, 2691–2701 (1997)
29. M.A. Marcus, L.E. Brus, C. Murray, M.G. Bawendi, A. Prasad, A.P. Alivisatos, EXAFS studies of cadmium chalcogenide nanocrystals. Nanostruct. Mater. **1**, 323–335 (1992)
30. S. Davis, A. Chadwick, J.J. Wright, A combined EXAFS and diffraction study of pure and doped nanocrystalline tin oxide. Phys. Chem. B **101**, 9901–9908 (1997)
31. D. Zanchet, H. Tolentino, M.C. Alves, O.L. Alves, D. Ugarte, Inter-atomic distance contraction in thiol-passivated gold nanoparticles. Chem. Phys. Lett. **323**, 167–172 (2000)
32. M, Dubiel, J. Haug, H. Kruth, H. Hofmeister, W. Seifert, Temperature dependence of EXAFS cumulants of Ag nanoparticles in glass. J. Phys. Conf. Ser. **190**, 012123-1–012123-6 (2009)
33. R.A. Andersen, K. Faegri, J.C. Green, A. Haaland, W.-P. Leung, K. Rypdal, Synthesis of bis[bis(trimethylsilyl)amido]iron(II). Structure and bonding in $M[N(SiMe_3)_2]_2$ (M = manganese, iron, cobalt): two-coordinate transition-metal amides. Inorg. Chem. **27**, 1782–1786 (1988)
34. A.V. Trunova, R. Meckenstock, I. Barsukov, C. Hassel, O. Margeat, M. Spasova, J. Lindner, M. Farle, Magnetic characterization of iron nanocubes. J. Appl. Phys. **104**, 093904-1–093904-5 (2008)
35. K. Fauth, E. Goering, G. Schütz, L.T. Kuhn, Probing composition and interfacial interaction in oxide passivated core-shell iron nanoparticles by combining X-ray absorption and magnetic circular dichroism. J. Appl. Phys. **96**, 399 (2004)
36. A. Shavel, B. Rodríguez-González, M. Spasova, M. Farle, L.M. Liz-Marzán, Synthesis and characterization of iron/iron oxide core/shell nanocubes. Adv. Funct. Mat. **17**, 3870–3876 (2007)
37. S. Sun, C.B. Murray, D. Weller, L. Folks, A. Moser, Monodisperse FePt nanoparticles and ferromagnetic FePt nanocrystal superlattices. Science **287**, 1989–1992 (2000)
38. R.M. Wang, O. Dmitrieva, M. Farle, G. Dumpich, H.Q. Ye, H. Poppa, R. Kilaas, C. Kisielowski, Layer resolved structural relaxation at the surface of magnetic FePt icosahedral nanoparticles. Phys. Rev. Lett. **100**, 017205-1–017205-4 (2008)
39. C. Antoniak, M. Farle, Magnetism at the nanoscale: the case of FePt. Mod. Phys. Lett. B **21**, 1111–1131 (2007)

40. H.-G. Boyen, K. Fauth, B. Stahl, P. Ziemann, G. Kästle, F. Weigl, F. Banhart, M. Heßler, G. Schütz, N.S. Gajbhije, J. Ellrich, H. Hahn, M. Büttner, M.G. Garnier, P. Oelhafen, Electronic and magnetic properties of ligand-free FePt nanoparticles. Adv. Mater. **17**, 574–578 (2005)
41. U. Wiedwald, K. Fauth, M. Heßler, H.-G. Boyen, F. Weigl, M. Hilgendorff, M. Giersig, G. Schütz, P. Ziemann, M. Farle, From colloidal Co/CoO core/shell nanoparticles to arrays of metallic nanomagnets: surface modification and magnetic properties. Chem. Phys. Chem. **6**, 2522–2526 (2005)
42. C. Antoniak, A. Trunova, M. Spasova, M. Farle, H. Wende, F. Wilhelm, A. Rogalev, Lattice expansion in nonoxidized FePt nanoparticles: an x-ray absorption study. Phys. Rev. B **78**, 041406(R)-1–041406(R)-4 (2008)
43. A.L. Ankudinov, B. Ravel, J.J. Rehr, S.D. Conradson, Real-space multiple-scattering calculation and interpretation of x-ray-absorption near-edge structure. Phys. Rev. B **58**, 7565–7576 (1998)
44. S.I. Zabinsky, J.J. Rehr, A. Ankudinov, R.C. Albers, M.J. Eller, Multiple-scattering calculations of X-ray-absorption spectra. Phys. Rev. B **52**, 2995–3009 (1995)
45. C. Antoniak, M. Spasova, A. Trunova, K. Fauth, F. Wilhelm, A. Rogalev, J. Minár, H. Ebert, M. Farle, H. Wande, Inhomogeneous alloying in FePt nanoparticles as a reason for reduced magnetic moments. J. Phys. Cond. Mat. **21**, 336002 (2009)
46. C. Antoniak, A. Warland, M. Darbandi, M. Spasova, A. Trunova, K. Fauth, E.F. Aziz, M. Farle, H. Wende, X-ray absorption measurements on nanoparticles: self-assembled arrays and dispersions. J. Phys. D Appl. Phys. **43**, 474007 (2010)
47. A. Grossmann, J. Morlet, Decomposition of Hardy functions into square-integrable wavelets of constant shape. SIAM J. Math. Anal. **15**, 723–736 (1984)
48. X. Shao, L. Shao, G. Zhao, Extraction of x-ray absorption fine structure information from the experimental data using the wavelet transform. Analyt. Commun. **35**, 135–137 (1998)
49. K. Yamaguchi, Y. Ito, T. Mukoyama, M. Takahashi, S.J. Emura, The regularization of the basic x-ray absorption spectrum fine structure equation via the wavelet-Galerkin method. Phys. B At. Mol. Opt. Phys. **32**, 1393–1408 (1999)
50. M. Muñoz, P. Argoul, F. Farges, Continuous Cauchy wavelet transform analyses of EXAFS spectra: a qualitative approach. Am. Mineralog. **88**, 694–700 (2003)
51. H. Funke, H.C. Scheinost, M. Chukalina, Wavelet analysis of extended x-ray absorption fine strucure data. Phys. Rev. B **71**, 094110 (2005)
52. C. Antoniak, Extended X-ray absorption fine structure of bimetallic nanoparticles. Beilstein J. Nanotechol. **2**, 237–251 (2011)
53. B.T. Thole, P. Carra, F. Sette, G. van der Laan, X-ray circular dichroism as a probe of orbital magnetization. Phys. Rev. Lett. **68**, 1943–1946 (1992)
54. P. Carra, B.T. Thole, M. Altarelli, X. Wang, X-ray circular dichroism and local magnetic fields. Phys. Rev. Lett. **70**, 694–697 (1993)
55. C.T. Chen, Y.U. Idzerda, H.-J. Lin, N.V. Smith, G. Meigs, E. Chaban, G.H. Ho, E. Pellegrin, F. Sette, Experimental confirmation of the X-ray magnetic circular dichroism sum rules for iron and cobalt. Phys. Rev. Lett. **75**, 152–155 (1995)
56. C. Ederer, M. Komelj, Magnetism in systems with various dimensionalities: a comparison between Fe and Co. Phys. Rev. B **68**, 052402-1–052402-4 (2003)
57. R. Nakajima, J. Stöhr, Y.U. Idzerda, Electron-yield saturation effects in L-edge x-ray magnetic circular dichroism spectra of Fe, Co. and Ni. Phys. Rev. B **59**, 6421–6429 (1999)
58. K. Fauth, How well does total electron yield measure x-ray absorption in nanoparticles? Appl. Phys. Lett. **85**, 3271–3273 (2004)
59. H. Ebert et al., The Munich SPR-KKR package, version 3.6; http://olymp.cup.unimuenchen.de/ak/ebert/SPRKKR; H. Ebert, Fully relativistic band structure calculations for magnetic solids formalism and application in electronic structure and physical properties of solids, ed. by H. Dreyssé, Lecture Notes in Physics, vol. 535, p. 191 (2000)
60. C. Antoniak, J. Lindner, K. Fauth, J.-U. Thiele, J. Minár, S. Mankovsky, H. Ebert, H. Wende, M. Farle, Composition dependence of exchange stiffness in Fe_xPt_{1-x} alloys. Phys. Rev. B **82**, 064403-1–064403-6 (2010)

11 Intrinsic Magnetism and Collective Magnetic Properties

61. H. Landolt, R. Börnstein, Numerical data and Functional Relationships in Science and Technology, New Series III/19a, (Springer, Berlin, 1986) and references therein
62. C. Antoniak, J. Lindner, M. Spasova, D. Sudfeld, M. Acet, M. Farle, K. Fauth, U. Wiedwald, H.-G. Boyen, P. Ziemann, F. Wilhelm, A. Rogalev, S. Sun, Enhanced orbital magnetism in $Fe_{50}Pt_{50}$ nanoparticles. Phys. Rev. Lett. **97**, 117201 (2006)
63. E. Goering, A. Fuss, W. Weber, J. Will, G. Schütz, Element specific X-ray magnetic circular dichroism magnetization curves using total electron yield. J. Appl. Phys. **88**, 5920 (2000)
64. M.P. Sharrock, Time dependence of switching fields in magnetic recording media. J. Appl. Phys. **76**, 6413 (1994)
65. O. Dmitrieva et al., Magnetic moment of Fe in oxide-free FePt nanoparticles. Phys. Rev. B **76**, 064414-1–064414-7 (2007)
66. C. Antoniak et al., A guideline for atomistic design and understanding of ultrahard nanomagnets. Nature Commun. **2**, 528 (2011)
67. M.E. Gruner, G. Rollmann, P. Entel, M. Farle, Multiply twinned morphologies of FePt and CoPt nanoparticles. Phys. Rev. Lett. **100**, 087203-1–087203-4 (2008)
68. J.B. Staunton, S. Ostanin, S.S.A. Razee., B.L. Gyorffy, L. Szunyogh, B. Ginatempo, E. Bruno, Temperature dependent magnetic anisotropy in metallic magnets from an Ab initio electronic structure theory: $L1_0$-ordered FePt. Phys. Rev. Lett. **93**, 257204-1–257204-4 (2004)
69. C. Antoniak, J. Lindner, M. Farle, Magnetic anisotropy and its temperature dependence in iron-rich Fe_xPt_{1-x} nanoparticles. Europhys. Lett. **70**, 250–256 (2005)
70. O. Mryasov, U. Nowak, K.Y. Guslienko, R.W. Chantrell, Temperature-dependent properties of FePt: effective spin Hamiltonian model. Europhys. Lett. **69**, 805–811 (2005)
71. E. Stavitski, F.M.F. de Groot, The CTM4XAS program for EELS and XAS spectral shape analysis of transition metal L edges. Micron **41**, 687 (2010)
72. R.D. Cowan, *The Theory of Atomic Structure and Spectra* (University of California Press, Berkeley, 1981), p. 307
73. F.M.F. de Groot, A. Kotani, *Core Level Spectroscopy of Solids* (Taylor & Francis, New York, 2008)
74. F.M.F. de Groot, X-ray absorption and dichroism of transition metals and their compounds. J. Electron Spectrosc. Relat. Phenom. **61**, 529 (1994)
75. P. Kuiper, B.G. Searle, L.-C. Duda, R.M. Wolf, P.J. van der Zaag, Fe $L_{2,3}$ linear and circular magnetic dichroism of Fe_3O_4. J. Electron Spectrosc. Relat. Phenom. **86**, 107 (1997)
76. T. Fujii, F.M.F. de Groot, G.A. Sawatzky, F.C. Voogt, T. Hibma, K. Okada, In situ XPS analysis of various iron oxide films grown by NO2-assisted molecular-beam epitaxy. Phys. Rev. B **59**, 3195–3202 (1999)
77. W.E. Henry, M.J. Boem, Intradomain magnetic saturation and magnetic structure of y-Fe2O3. Phys. Rev. **101**, 1253 (1956)
78. F. Kronast, N. Friedenberger, K. Ollefs, S. Gliga, L. Tati-Bismaths, R. Thies, A. Ney, R. Weber, C. Hassel, F.M. Römer, A.V. Trunova, C. Wirtz, R. Hertel, H.A. Dürr, M. Farle, Element-specific magnetic hysteresis of individual 18 nm Fe nanocubes. Nano Lett **11**, 1710–1715 (2011)
79. Object oriented micromagnetic framework, http://www.math.nist.gov/oommf
80. N. Friedenberger, C. Möller, C. Hassel, S. Stienen, F. Kronast, M. Farle, Single Nanoparticle Magnetism: Influence of morphology, to be published (2012)
81. N. Friedenberger, Ph.D. Thesis, Universität Duisburg-Essen, in preparation (2011)
82. W.H. Meiklejohn, C.P. Bean, New magnetic anisotropy. Phys. Rev. **102**, 1413–1414 (1956)
83. J. Nogués et al., Exchange bias in nanostructures. Phys. Rep. **422**, 65–117 (2005)
84. X. Batlle, Magnetic nanoparticles with bilklike properties (invited). J. Appl. Phys **109**, 07B524-1–07B524-6 (2011)
85. R. Fink, SMART: a planned ultrahigh-resolution spectromicroscope for Bessy II. J. Electr. Spectr. Rel. Phen. **84**, 231 (1997)
86. M. Farle, Ferromagnetic resonance of ultrathin metallic layers. Rep. Prog. Phys. **61**, 755–826 (1998)

87. A. Banholzer, R. Narkowicz, C. Hassel, R. Meckenstock, S. Stienen, O. Posth, D. Suter, M. Farle, J. Lindner, Visualization of spin dynamics in single nanosized magnetic elements. Nanotechnology **22**, 295713 (2011)
88. K. Baberschke, in Investigation of Ultrathin Ferromagnetic Films by Magnetic Resonance. Handbook of Magnetism and Advanced Magnetic Materials, vol. 3 (Wiley, New York, 2007), p. 1617 and references therein
89. J. Lindner, C. Hassel, A.V. Trunova, F.M. Römer, S. Stienen, I. Barsukov, Magnetism of single-crystalline Fe nanostructures. J. Nanosc. Nanotechn. **10**, 6161 (2010)
90. M. Mizuguchi, K. Takanashi, Ferromagnetic resonance of epitaxial Fe nanodots grown on MgO measured using coplanar waveguides. J. Phys. D Appl. Phys. **44**, 064007 (2011)
91. D, Bonneberg, H.A., Hempel, H.P.J. Wijn, Magnetic properties of 3d, 4d and 5d elements, alloay and compounds. in Landolt-Börnstein, New Series (Springer Berlin), vol. III/19a, ed. by K.-H. Hellwege, O. Madelung (1992), p. 178 and references therein
92. B.K. Kuanr, R.E. Camley, Z. Celinski, Relaxation in epitaxial Fe films measured by ferromagnetic resonance. J. Appl. Phys. **95**, 6610–6613 (2004)

Chapter 12
Optical Spectroscopy on Magnetically Doped Semiconductor Nanoparticles

Lars Schneider and Gerd Bacher

Abstract Semiconductor nanoparticles doped with magnetic ions represent an exciting class of materials with unique optical, electronic, and magnetic properties and potential applications in the field of spintronics. A key feature required is the exchange interaction between magnetic ions and charge carriers, which finally controls the magneto-optical response of these materials. In this contribution, some recent advances for two classes of magnetically doped nanoparticles, namely, ZnO doped with Cr and Co, respectively, and CdSe doped with Mn, are summarized. We found that chromium is incorporated as Cr^{3+} in ZnO. With increasing Cr concentration, the quantum efficiency is being reduced while the magnetic properties observed can be attributed to a phase separation between ZnO and $ZnCr_2O_4$. In contrast, cobalt apparently exists in the Co^{2+} configuration in the nanocrystals as demonstrated via optical spectroscopy. No enhanced magneto-optical properties have been obtained for both classes of magnetically doped ZnO nanoparticles. This is completely different in case of Mn-doped CdSe nanocrystals. A giant Zeeman effect is found as a consequence of a pronounced *sp–d* exchange interaction. The strong 3D carrier confinement finally results in a significantly enhanced exchange field leading to the observation of optically induced magnetism up to room temperature.

12.1 Introduction

Doping semiconductor nanostructures with magnetic ions has a long history. The idea behind is to combine unique properties of semiconductors—like tailoring the absorption/emission properties by band gap engineering or adjusting the electrical conductivity by introducing donor or acceptor atoms—with specific magnetic aspects

---

L. Schneider · G. Bacher (✉)
Electronic Materials and Nanodevices, Faculty of Engineering and CENIDE,
University of Duisburg-Essen, Bismarckstraße 81, 47057 Duisburg, Germany
e-mail: gerd.bacher@uni-due.de

A. Lorke et al. (eds.), *Nanoparticles from the Gas Phase*, NanoScience and Technology,
DOI: 10.1007/978-3-642-28546-2_12, © Springer-Verlag Berlin Heidelberg 2012

known from metallic magnets. While for tens of years, the magnetic properties of magnetically doped semiconductors have been restricted to cryogenic temperatures, experimental, and theoretical progress starting in the mid 1990s have triggered a large variety of research efforts to obtain magnetic semiconductors working up to room temperature.

One of the key developments on the experimental side was the first demonstration of III-Mn-As materials, like InMnAs or GaMnAs, with ferromagnetic properties at elevated temperatures [1, 2]. In these materials, manganese acts as an acceptor, and therefore III-Mn-As magnetic semiconductors are heavily p-doped. Theoretically, (i) ferromagnetism above room temperature for transition metal doped GaN and ZnO [3] and (ii) exchange fields on the order of 100–1,000 T in semiconductor nanoparticles with strong 3D carrier confinement [4] were predicted. The latter one is expected to strongly enhance the interaction between magnetic ions and charge carriers, and thus should extend the magneto-optical properties in such materials toward higher temperatures up to room temperature. All these developments strongly stimulated materials research in the field of bulk and nanoscale magnetically doped semiconductors during the last decade.

In this chapter, we will concentrate on two classes of magnetically doped nanoparticles. Section 12.2 is devoted to ZnO nanoparticles grown from the gas phase and doped with either chromium or cobalt. Combining optical spectroscopy with magnetic and structural characterization, a quite comprehensive picture of the optical and the magnetic properties of these nanomaterials is extracted. Section 12.3 summarizes some latest results on Mn-doped CdSe nanocrystals prepared by a liquid phase approach. The strong 3D carrier confinement is shown to result in a very large exchange field, which finally allows the observation of optically induced magnetization up to room temperature.

12.2 Magnetically Doped ZnO Nanoparticles

The research on magnetically doped ZnO was mainly initiated by theoretical predictions: a Zener model approach performed by Dietl et al. [3] showed that room temperature ferromagnetism can be expected in ZnO doped with Mn, provided a sufficiently large number of holes is available. Sato et al. [5] even predicted ferromagnetic ordering at room temperature for ZnO doped with different transition metals (TM) even without the need of additional doping. Since this time, numerous experimental studies on magnetically doped ZnO showing ferromagnetism above room temperature appeared in the literature [6–12] while other groups report on the absence of magnetic signatures [13, 14]. Even some indications of $sp-d$ exchange coupling between magnetic ions, and the conduction and valence band states have been observed [15–17], which is of strong importance for the usage of magnetically doped ZnO in spintronic applications. Frequently, isolated experiments like SQUID measurements are reported exhibiting pronounced features of ferromagnetic behavior, whereas a detailed and comprehensive structural, optical, and magnetic

12 Optical Spectroscopy on Magnetically Doped Semiconductor Nanoparticles 305

material analysis is missing. This, however, is quite important as the magnetism in TM-doped ZnO is expected to sensitively depend on synthesis route, defect structure, occurrence of secondary phases etc.

Here, we concentrate on ZnO nanoparticles prepared by chemical vapor synthesis in a hot wall reactor. For the preparation of ZnO:Cr, Zinc acetylacetonate ($Zn(acac)_2$) together with chromium acetylacetonate ($Cr(acac)_3$) have been used as precursors. The precursors are first thermally evaporated in a reactor at 423 and 493 K, respectively, and subsequently transferred into a second reactor ($T = 1,173$ K) under oxygen flow, where the nanoparticles are formed [18]. In case of ZnO:Co, the precursors (Zn and Co acetates) are evaporated using a CO_2 laser and then transferred into a hot wall reactor, where the nanoparticles are generated at a temperature of 1,373 K and a pressure of 20 mbar [19].

Time integrated and time-resolved photoluminescence (PL) experiments have been performed using a frequency doubled Ti-sapphire laser for excitation ($\lambda = $ 350 nm) and a synchroscan streak camera for detection. The spectral and temporal resolution of the system are 0.8 meV and $\sim$5 ps, respectively. A SQUID magnetometer was used for magnetization measurements and the crystal structure was determined by a X-ray diffractometer with Cu $K\alpha$ radiation in 2θ configuration. More experimental details can be found in Ref. [20].

12.2.1 ZnO Nanoparticles Doped with Chromium

In Fig. 12.1, a transmission electron microscope image of as-synthesized ZnO nanoparticles doped with 6 at.% of chromium is shown (left) and compared to a scanning electron microscopy image (right) of the same sample after annealing under oxygen at a temperature of 1,000 °C. It is obvious that the initial nanoparticle size before annealing is clearly below 20 nm. From X-ray diffraction (XRD) (see below), an average crystal diameter of 11 nm can be extracted. After annealing, the nanoparticle size significantly increases to about 100–200 nm due to coalescence and sintering.

The influence of doping ZnO with chromium on the optical properties of the nanoparticles can be seen in Fig. 12.2. Hereby, the time-integrated PL spectra (left) and the transient decay of the PL intensity (right) are shown for ZnO nanoparticles including a different concentration of chromium. All measurements have been performed at room temperature on the as-synthesized samples.

The time-integrated PL spectra are characterized by two prominent spectral features. The band at around 3.24 eV stems from near band gap emission (NBE) while the emission around 2.5 eV is related to deep defects, most likely related to oxygen defects [21, 22]. It is obvious that the incorporation of Cr significantly reduces the PL yield for the NBE, whereas no systematic dependence between the efficiency of the green defect emission and the Cr concentration is found. The loss of quantum yield for the NBE with increasing Cr concentration can also be seen in the time-resolved PL studies (see right part of Fig. 12.2). For the undoped ZnO reference sample, a

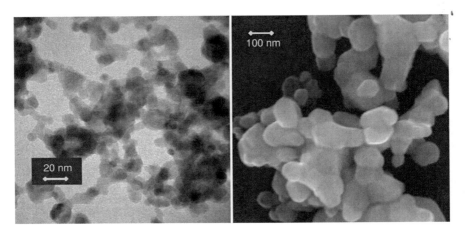

Fig. 12.1 Transmission electron micrograph of the as-synthesized ZnO nanoparticles doped with 6 at.% of Cr (*left*). Scanning electron micrograph of the same sample after annealing under oxygen for 30 min at 1,000 °C (*right*). Adapted from Ref. [20]

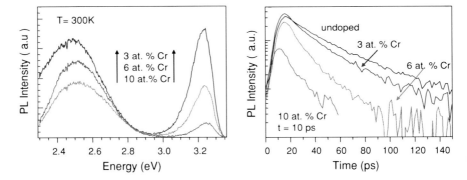

Fig. 12.2 *Left* time-integrated PL spectra of ZnO nanoparticles doped with different concentrations of chromium. The experiments have been performed at room temperature. *Right* PL intensity versus time recorded at room temperature for the same samples in comparison to an undoped ZnO reference sample [20]

recombination lifetime of the NBE of a few tens of ps can be extracted from the data. This time constant decreases with increasing Cr concentration and reaches about 10 ps in the case of ZnO nanoparticles doped with 10 at.% of Cr. This behavior—the reduction of the recombination lifetime with incorporation of transition metal ions into the crystal lattice—is related to non-radiative losses via defect states in the band gap caused by the TM. Note, however, that also the lifetime of the ZnO reference sample is quite short as compared to ZnO nanoparticles of similar size [23]. This is attributed to a certain amount of crystal defects, which limits the quantum yield in the as-synthesized nanoparticles.

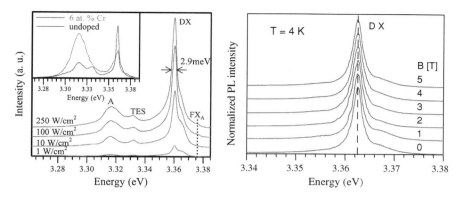

Fig. 12.3 *Left* low temperature ($T = 2.7$ K) PL spectrum of the annealed ZnO:Cr ($x_{at} = 6\%$) sample for various excitation power densities. In the *inset*, the PL spectrum at 250 W/cm^2 is compared to the undoped ZnO reference sample [20]. *Right* magneto-PL spectra for different magnetic fields applied in Faraday geometry. In the figure, the PL spectra are vertically shifted for clarity

In order to reduce the number of crystal defects and thus the number of non-radiative loss channels, the nanoparticles have been annealed at 1,000 °C under oxygen atmosphere. This results in an almost complete disappearance of the 2.5 eV defect emission and in a quantum yield enhancement of the NBE emission of more than one order of magnitude.

Figure 12.3 shows the low temperature NBE PL spectrum for the annealed sample for various excitation densities. Spectrally narrow emission peaks can be clearly resolved. With increasing excitation power, the donor bound exciton emission DX at $E = 3.36$ eV dominates the spectrum. The low spectral width of 2.9 meV (full width half maximum FWHM) indicates high sample quality. At the high energy tail of the DX transition at 3.376 eV, a significant contribution of the free A exciton emission develops, giving a donor binding energy of 16 meV. The emission at 3.332 eV exhibits the same power dependence as well as the same decay characteristics as the DX transition and is thus attributed to the two electron satellite of the DX [24]. The A-line occurring at 3.315 eV is under discussion in the literature and related to donor acceptor pair recombination, optical phonon replica of the free exciton transition, acceptor bound excitons, or excitons bound to surface states, respectively [25–29]. A more detailed analysis of our data shows a thermal broadening of the high energy tail of the A-line with increasing temperature, which indicates the contribution of free carriers to the recombination process. For that reason, we prefer an interpretation similar to what is reported by Schirra et al., who claim that the A-line is correlated to stacking faults and caused by a free electron to acceptor transition [30].

Surprisingly, the Cr-doped sample exhibits spectrally much narrower features than the undoped reference sample, indicating some improvement of the crystal quality when incorporating chromium, even though the quantum yield is reduced. By tracing the PL spectra up to room temperature it is found that the room temperature NBE, peaked at around 3.25 eV, is predominantly related to the A-line and the free exciton transition and its LO phonon replica [22].

Due to the spectrally narrow emission spectrum we are now able to analyze the magneto-optical properties of the ZnO:Cr nanoparticles. One of the signatures of a magnetic semiconductor is the occurrence of a pronounced $sp–d$ exchange interaction between charge carriers and magnetic ions [31], which can result in a giant Zeeman energy splitting if a magnetic field is applied in Faraday geometry. Indeed, huge effective g-factors far exceeding 100 can easily be achieved for a wide variety of II-Mn-VI chalcogenide semiconductors. As can be seen in the right plot of Fig. 12.3, no indication of a significant Zeeman energy shift is observed within our resolution. Comparing this to the literature it is remarkable that only a few reports on a 'giant' Zeeman effect have been published for TM-doped ZnO. Hereby, the Zeeman shift has been extracted by either magnetic circular dichroism [11, 16] or magneto-reflection [17] spectroscopy, respectively. No indication of an enhanced g-factor was found up to now in magneto-PL spectroscopy in these kinds of materials. Even more, no report of a giant Zeeman shift has been published for ZnO doped with chromium up to now.

In order to develop a further understanding of the magnetic and the magneto-optical properties of ZnO:Cr, SQUID measurements have been performed. In Fig. 12.4, the magnetization of the annealed ZnO nanocrystals doped with 6 at.% of chromium is depicted versus external magnetic field for both low temperature ($T = 5$ K) and room temperature. In each case, a pronounced hysteresis is observed, which can be regarded as a signature of at least partial ferromagnetic ordering. The signal is superimposed by a strong linear contribution related to a certain amount of Cr atoms showing paramagnetic behavior. A coercive field of about 80 Oe is obtained at room temperature, which is comparable to the literature data found for Cr-doped ZnO [10, 32], although also significantly higher values are reported for TM-doped ZnO [16, 33].

In the right part of Fig. 12.4, the magnetization curve is plotted versus temperature for two different experimental conditions. First, the sample was cooled down to $T = 5$ K at zero external magnetic field. After applying an external field of 1,000 Oe, the magnetization is then measured up to room temperature (zero-field cooling ZFC, lower curve in the figure). In a second experiment, the sample is cooled down in an external field of 1,000 Oe and subsequently measured up to room temperature under the applied field (field cooling, FC, upper curve in the figure). The experimental data (red) are compared to what is expected from a model paramagnet (black dashed line).

There are several distinct features which need to be discussed here. Below $T = 12$ K, the ZFC and the FC curves exhibit a pronounced difference, whereas for higher temperatures, this difference disappears and the magnetization decreases with increasing temperature. Above $T \sim 80$ K, the data can be well described by a paramagnetic susceptibility. Fitting the experimental data by a Curie–Weiss law, we extract a Curie–Weiss temperature of $T_C = -420$ K. This is in good agreement with the literature data for $ZnCr_2O_4$ [34, 35]. The behavior of the ZFC magnetization curve for $T < 12$ K is quite typical for an antiferromagnet. Again, a nice

12 Optical Spectroscopy on Magnetically Doped Semiconductor Nanoparticles

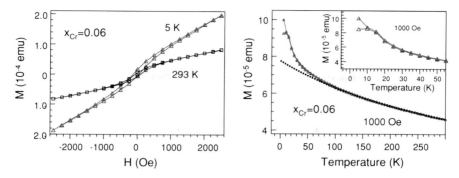

Fig. 12.4 *Left* magnetization measurements of the ZnO nanoparticle sample doped with 6 at.% of Cr for low temperature ($T = 5$ K) and room temperature ($T = 293$ K). At room temperature, a coercive field of 80 Oe is obtained. *Right* temperature dependent magnetization measured at an external field of 1,000 Oe (*red symbols*). The *black dotted line* indicates what is expected for a pure paramagnet. In the *inset*, an enhanced view of the low temperature regime is presented. The lower curve is obtained for zero-field cooling while the upper one is measured after field cooling. After Ref. [20] (colour figure online)

agreement with the typical Néel temperature of $T_N = 12$ K found in the literature is obtained [34, 35]. The fact that in case of FC the magnetization further increases below $T_N = 12$ K is a strong hint of a suppression of the antiferromagnetic coupling by the external field of 1,000 Oe.

The question on the origin of the hysteresis behavior found for low temperatures as well as for room temperature now arises. For an ideal antiferromagnet, such a hysteresis is not expected. It is discussed in the literature that excess spins at the surfaces of antiferromagnetic nanoparticles or domains can show ferromagnetic ordering [36, 37]. This would result in a measurable hysteresis in this kind of materials.

Detailed structural investigations using XRD and X-ray absorption near edge structure (XANES) measurements have been performed in order to proof the existence of a $ZnCr_2O_4$ phase, which can account for the magnetic behavior, in our samples. In the left part of Fig. 12.5, X-ray diffractograms of a typical Cr-doped ZnO nanoparticle sample are shown before (upper part of the figure) and after (lower part of the figure) annealing. The as-synthesized material consists of highly crystalline wurtzite ZnO. From the broadening of the individual peaks a typical crystal diameter of about 11 nm can be extracted. After annealing, the peaks become significantly narrower due to crystal growth to about 100–200 nm (see Fig. 12.1). The diffractogram indicates that the nanoparticles still mainly consist of ZnO in the wurtzite phase. However, a clear indication of the occurrence of a second phase with a fraction of about 2 at.% is present. This second phase consists of the cubic spinel $ZnCr_2O_4$, where the chromium atoms are incorporated on octahedral sites of the crystal. This is in very good agreement with the magnetic properties of the materials as discussed above. Apparently, the antiferromagnetic properties observed

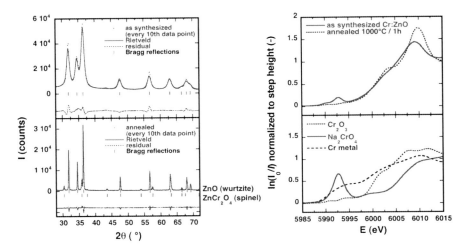

Fig. 12.5 *Left* X-ray diffractograms of as-synthesized (*top*) and annealed (*bottom*) ZnO nanoparticles doped with 3 at.% of chromium. The *vertical dashed lines* indicate the expected positions for the Bragg reflections from the wurtzite and the spinel phase, respectively. *Right* Cr K-XANES spectra of the as-synthesized and the annealed ZnO:Cr (3 at.% of Cr) nanoparticles (*top*) compared to XANES spectra from reference materials. Adapted from Ref. [20]

are related to the $ZnCr_2O_4$ spinel phase and the small hysteresis is stemming from ferromagnetically coupled excess spins at the surface of the $ZnCr_2O_4$ phase.

Further support of this interpretation is given by the XANES data presented on the right part of Fig. 12.5. The data measured at the Cr K-edge are compared to reference experiments done on Cr metal, Cr_2O_3 and Na_2CrO_4, respectively. Comparing the onset of the X-ray absorption spectra for the different materials at around 6,000 eV, it can be concluded that in our Cr-doped ZnO nanoparticles, the chromium is present in the Cr^{3+} configuration both, for the as-synthesized and the annealed sample. The pre-edge peak of the XANES spectra at 5,993 eV observed for the as-synthesized nanoparticles and for the Na_2CrO_4 reference corresponds to a partially allowed dipolar transition from the 1s to the hybridized $2p$–$3d$ states. It indicates the incorporation of the Cr^{3+} ions on crystal sites with tetrahedral symmetry, probably Zn places. After annealing, this pre-edge peak has vanished, which is in accordance to our findings from XRD, i.e., the formation of a second phase ($ZnCr_2O_4$ spinel), where chromium is sitting on an octahedrally coordinated crystal site.

This in summary leads to the conclusion that the Cr atoms are initially incorporated as Cr^{3+} ions in the ZnO nanoparticles. After annealing, they migrate from the original tetrahedral site to an energetically more favorable octahedral site forming a $ZnCr_2O_4$ spinel phase. The optical properties, thus, are most probably dominated by the ZnO wurtzite phase and the missing magneto-optical response is related to the spatial separation between the optically active ZnO and the magnetically active $ZnCr_2O_4$ phase.

12 Optical Spectroscopy on Magnetically Doped Semiconductor Nanoparticles 311

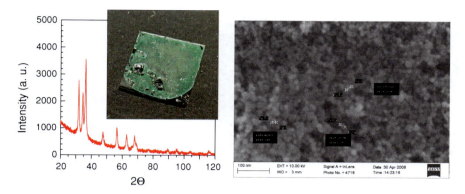

Fig. 12.6 *Left* X-ray diffractogram of ZnO nanoparticles doped with 10 at.% of Co. No second phase is observed (after [19]). The *inset* shows a photograph of a nanoparticle layer under ambient conditions. *Right* scanning electron micrograph of a ZnO:Co nanoparticle layer

12.2.2 ZnO Nanoparticles Doped with Cobalt

In contrast to chromium, which is incorporated into ZnO nanocrystals in the Cr^{3+} configuration, CoO is known as a stable chemical compound and thus cobalt might exist in the Co^{2+} configuration if included into ZnO [11]. Again, optical spectroscopy has been used in order to experimentally extract signatures of the incorporation of TM into ZnO nanocrystals.

In the inset of the left part of Fig. 12.6, a photograph of a dense layer of ZnO nanoparticles doped with 10 at.% of cobalt is depicted. The green color observed results from an efficient absorption in the UV, blue and red spectral range [38] and indicates an efficient incorporation of cobalt into the ZnO nanocrystals. From the X-ray diffractogram, no second phase can be recognized and a nanocrystal diameter of about 18 nm is estimated [19]. This is confirmed by the scanning electron micrograph of a nanoparticle layer as shown on the right part of Fig. 12.6.

As apparently the optical properties of ZnO nanocrystals are strongly influenced by cobalt atoms, we studied the room temperature PL spectra of ZnO nanocrystals doped with different concentrations of cobalt (see Fig. 12.7). The PL spectra consist of several emission bands. First, NBE emission dominates at around 3.3 eV with an intensity strongly decreasing with increasing Co concentration (see right plot of Fig. 12.7). Second, a broad peak is observed in the energy range around 2.2 eV, in particular, in case of the ZnO reference sample. This peak can be attributed to deep defects in the ZnO crystals and is most likely related to oxygen defects like interstitials (O_i) or vacancies (V_O) (see inset of Fig. 12.7, right). More interesting is the fact that in the Co-doped ZnO nanoparticles a pronounced emission peak at around 1.8 eV appears, which is absent in the undoped reference sample. The spectral position is in good agreement to the energy of the internal transition between the excited $^4T_1(P)$ energy state and the $^4A_2(F)$ ground state of Co^{2+} ions in the tetrahedral environment

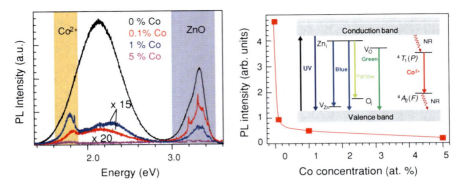

Fig. 12.7 Room temperature PL spectra of ZnO nanoparticles doped with different concentrations of cobalt (*left*). In the *right part* of the figure, the spectrally integrated PL intensity is plotted versus Co concentration. The *inset* schematically indicates the possible optical transitions according to Ref. [21]

Fig. 12.8 Low temperature PL spectrum of ZnO nanoparticles doped with 0.1 at.% of Co in the energy range of the internal $^4T_1$–$^4A_2$Co^{2+} transition. The peak around 1.85 eV stems from scattered laser light

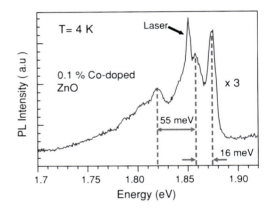

of the ZnO crystal lattice. This indicates a substitutional incorporation of the Co^{2+} ions on the tetrahedral Zn^{2+} sites of the ZnO wurtzite crystal lattice.

A more detailed picture of this internal Co^{2+} transition and thus of the incorporation of Co^{2+} into the crystalline structure of ZnO is obtained by low temperature PL spectroscopy. Figure 12.8 shows the low temperature ($T = 4$ K) PL spectrum of ZnO doped with 0.1 at.% cobalt in the spectral window between 1.7 and 1.94 eV. Several distinct emission peaks can be separated. In addition to the dominant emission line at 1.875 eV, low temperature replicas are observed, separated by 16 and 55 meV, respectively, from the main peak. In the literature, these satellites are often assigned to the E_2^{high} and E_2^{low} phonon replica of the internal Co^{2+} transition at 1.875 eV [39, 40]. As phonons are quite characteristic for the specific crystals, the occurrence of these phonon replicas is another strong hint for an efficient incorporation of Co^{2+} on Zn^{2+} sites.

12.3 Magnetically Doped CdSe Nanoparticles

In contrast to TM-doped ZnO nanoparticles, where the research efforts have been driven by the theoretical predictions of room temperature ferromagnetism even for bulk materials [3, 5], the motivation for the research on TM-doped CdSe nanoparticles is different: strong 3D quantum confinement is expected to drastically enhance the carrier exchange field [4] and thus might result in magneto-optical properties even up to room temperature.

Two strategies have been pursued in parallel. First, epitaxial approaches have been developed toward the fabrication of 'natural' [41] or self-organized [42–50] quantum dots with 3D carrier confinement. Second, chemically prepared chalcogenide nanocrystals are doped with Mn or other TM. While Mn could be incorporated into wide bandgap materials ZnSe or ZnS quite a long time ago [51–53], this remains extremely ambitious for CdSe until recently [53–56].

Concerning the optical properties of these materials, there is one important aspect to be considered: the incorporation of TM quite often results in discrete energy levels within the bandgap, and thus triggers an efficient non-radiative energy transfer from electrons and holes injected into conduction and valence band states into internal TM energy states [50, 51, 57]. While this process is dominant e.g., for ZnS:Mn or ZnSe:Mn, the choice of chalcogenides with a bandgap, which is lower than the energy of the internal $^4T_1-^6A_1$ transition of the Mn^{2+} ion, might suppress this energy transfer process. In case of epitaxially grown quantum dots, this was demonstrated for both, the CdTe and the CdSe material system [50, 57]. As a result, one of the most prominent consequences of the exchange interaction between charge carriers and TM ions, the exciton magnetic polaron (EMP), could be observed [41–43, 57, 58]. However, no signatures of a strongly enhanced exchange field due to 3D quantum confinement—as theoretically predicted—could be identified in these systems up to now. This might be related to limited quantum confinement as a consequence of (i) the larger lateral quantum dot diameter as compared to the vertical quantum dot extension parallel to the growth axis and (ii) the finite barrier height which causes wavefunction leakage into the barrier.

Here, we concentrate on chemically synthesized CdSe nanocrystals with strong quantum confinement in three dimensions. The nanoparticles have been prepared from a colloidal solution as outlined in detail in Ref. [59]. In short, a solution of $MnCl_2$ in hexadecylamine has been evaporated in vacuum at $130\,^\circ C$. After reducing the temperature to $<80\,^\circ C$, the precursor $((Me_4N)_2[Cd_4(SePh)_{10}])$ was added together with selenium powder under nitrogen atmosphere. After this, the temperature was fixed at $130\,^\circ C$ for 1.5 h and subsequently increased to $215\,^\circ C$ until the desired nanoparticle size was obtained. The successful incorporation of Mn^{2+} ions into the CdSe crystal lattice has been proved by electron paramagnetic resonance and magnetic circular dichroism [60]. By using atomic emission spectroscopy (AES), the effective Mn^{2+} concentration was measured.

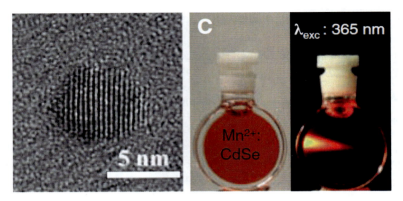

Fig. 12.9 Transmission electron micrograph of a CdSe:Mn nanoparticle (*left*). The *right part* of the figure shows a photograph of the nanoparticle solution under ambient light (*center*) and in the dark after unfocused laser excitation (*right*). Adapted from Ref. [55]

12.3.1 Characterization of Mn-Doped CdSe Nanoparticles

Figure 12.9 shows a transmission electron micrograph (TEM) of a single Mn-doped CdSe nanoparticle. The excellent crystallinity allows a resolution of the crystallographic planes of the wurtzite lattice. From TEM, the crystal diameter can be extracted and depending on the growth conditions, nanoparticle diameters between 2 and 5 nm have been achieved. In the right part of Fig. 12.9, photographs of a nanoparticle solution under ambient conditions (center) and in the dark with unfocused laser excitation at a wavelength of 365 nm (right) are shown. In that case, the particle diameter was 4.3 nm and the color is caused by bandgap emission of the nanoparticles. The intense red color indicates a remarkably high quantum yield at room temperature and is again an indication of the excellent crystalline quality.

For a detailed optical characterization, a dilute colloidal suspension of the nanocrystals has been spin coated onto a silicon wafer. The samples have been mounted onto the cold finger of a closed cycle cryostat for temperature variation. Optical excitation was performed using a picosecond laser pulse at $\lambda_{exc} = 400$ nm and a repetition rate of 76 MHz. A low excitation density of <0.1 W/cm^2 was chosen in order to avoid sample degradation. The transient PL signal was recorded using again a syncroscan streak camera providing an overall time resolution of ~ 5 ps. In all the experiments, a possible energy transfer from small to large nanocrystals was tested by studying samples with different quantum dot concentrations and found to be negligible in the highly diluted samples used here.

In Fig. 12.10, typical Streak camera images are depicted for CdSe nanoparticles with a diameter of 5 nm and a Mn^{2+} concentration of 4.2 % (left) in comparison to undoped CdSe nanoparticles with 4 nm in diameter (right). The experiments have been performed at low temperatures (3.2 and 5 K, respectively). These data give an

12 Optical Spectroscopy on Magnetically Doped Semiconductor Nanoparticles 315

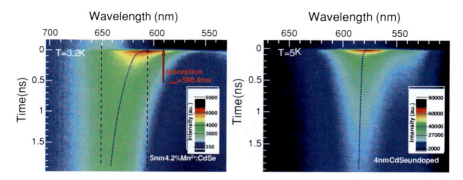

Fig. 12.10 Streak camera images of CdSe:Mn nanocrystals (diameter 5 nm, Mn concentration 4.2 %, *left*) and undoped CdSe reference nanocrystals (diameter 4 nm, *right*). The *horizontal axis* represents the emission wavelength, the *vertical one* the time after ps excitation and the PL intensity is *color-coded*

interesting first insight into the physics of the CdSe:Mn nanocrystals. Several features are remarkable, which will be discussed in the following.

The first important fact seen in the Streak camera images is the quite long recombination decay time. In colloidal nanocrystals, the electron–hole exchange interaction results in an energy splitting between bright ($m_{j,X} = \pm 1$, spins of the electron and the heavy hole are antiparallel) and dark ($m_{j,X} = \pm 2$, spins of the electron and the heavy hole are parallel) excitons up to several tens of meV. The optically forbidden dark exciton states is the energetically lower one, which results in a recombination lifetime on the order of a microsecond at low temperatures in CdSe nanocrystals [61, 62]. Note that, although this long lifetime cannot be accessed by our experimental setup because of the limited time window of the Streak camera and the highly repetitive laser system, it has been confirmed by other techniques in these samples [63]. It is important to mention that in spite of the blue shift of the CdSe bandgap due to quantum confinement the energy gap is still below the energy of the internal $^4T_1$–$^6A_1$ transition of the Mn^{2+} ion and thus, non-radiative carrier losses to internal Mn^{2+} states are completely suppressed. As will be discussed below, the resulting long recombination lifetime has a quite fundamental impact on the dynamics of optically generated magnetism in the CdSe:Mn nanocrystals.

The second remarkable feature is the fact that even directly after laser excitation a pronounced energy shift is obtained between absorption (see red arrow) and PL emission. This is found for both, the CdSe:Mn nanocrystals as well as the CdSe reference samples and is therefore apparently not related to the magnetic doping of the nanocrystals. Stokes shifts between absorption and emission are usually present in this kind of nanocrystals and different explanations are discussed [64, 65]. On the one hand, the fine structure splitting between (absorbing) bright exciton states and (emitting) dark exciton states must result in an energy splitting between absorption and emission maximum. Note that, the optically generated electron–hole pairs quite rapidly relax into the lowest state, which is the dark one. A mixture between

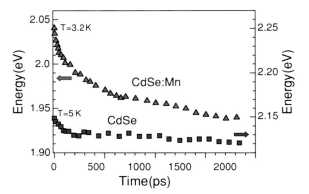

Fig. 12.11 PL energy versus time after ps excitation for CdSe:Mn (*left axis, red triangles*) and CdSe (*right axis, blue squares*) nanocrystals. The crystal diameter was 5 and 4 nm, respectively, and the Mn concentration was measured to be 4.2 % (colour figure online)

heavy and light hole states leads to a final oscillator strength even for dark excitons. On the other hand, a significant coupling of the excitons to acoustic phonons is expected, also resulting in an energy shift between absorption (accompanied by phonon generation) and emission (accompanied by phonon emission). Moreover, large nanocrystals within an ensemble are expected to absorb UV light stronger than smaller ones as the excitation probability is expected to be proportional to the volume of the nanocrystals. Thus, the PL emission of larger particles is enhanced leading to a Stokes shift between emission and absorption maximum.

Third and most important, there is a pronounced shift of the PL maximum to lower energies with time (indicated by the dotted line) for the CdSe:Mn nanocrystals, which seems to be strongly suppressed for the CdSe reference sample. A similar transient energy shift was found for epitaxially grown, self-organized CdSe/ZnMnSe quantum dots and attributed to the formation of an EMP, i.e., the ferromagnetic alignment of the Mn^{2+} ion spins in the exchange field of the optically generated exciton [57]. In the CdSe:Mn nanocrystals studied here, this energy shift amounts to ∼100 meV within the first 1.8 ns, which is almost one order of magnitude larger than observed for their epitaxially grown counterparts [55, 57].

Figure 12.11 compares the transient PL energy shift of the CdSe:Mn nanocrystals and the undoped reference samples. In case of the undoped reference, an initial energy shift within the first 100 ps of about 15–20 meV is obtained. This is attributed to energy relaxation of photo-generated charge carriers to lower states, probably within the energy levels of the excitonic fine structure. The PL energy of the Mn-doped CdSe nanocrystals shows a much stronger transient shift of ∼100 meV during the first 2 ns and its dynamics cannot be described by a single time constant. The initial energy shift of about 50–60 meV occurs on a time scale which is comparable to typical EMP formation times reported in the literature. Apparently, saturation is not yet achieved even after 2 ns, a time scale usually not accessible in epitaxially grown self-assembled quantum dots due to their much shorter recombination lifetime [57].

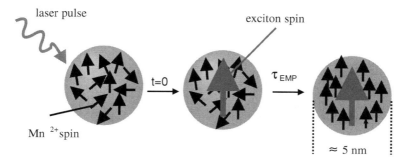

Fig. 12.12 Scenario of optically generated EMP formation dynamics in a nanocrystal

12.3.2 Exciton Magnetic Polaron Formation in Mn-Doped CdSe Nanoparticles

12.3.2.1 Theoretical Background

EMPs have a quite long history in the literature. Observed already in the 1960s in bulk Eu-chalcogenide semiconductors [66], EMPs were found in multiple magnetically doped chalcogenides starting from bulk semiconductors and quantum wells to even quantum dots. The localization of carriers plays a major role in EMP formation: an exciton bound to a donor or an acceptor is strongly localized and able to polarize the spins of the magnetic ions within their wavefunction via *sp–d* exchange interaction. Free excitons may also polarize the spins of the magnetic ions, leading to a self-localization. However, such free exciton magnetic polarons are much less stable in bulk semiconductors. The situation changes in quantum dots where the complete 3D carrier confinement results in a quite stable formation of free EMPs [67].

The main scenario occurring during EMP formation is schematically shown in Fig. 12.12. In case of low manganese concentrations (<1 %), the Mn^{2+} ion spins are disordered. The net magnetization is zero and the system can be described as a paramagnet.[1] At time $t = 0$, the laser pulse generates an electron–hole pair (exciton), which is able to polarize the magnetic environment due to *sp–d* exchange interaction within a typical time constant of τ_{EMP} This spin ordering reduces the energy of the whole system which is experimentally evident, e.g., by a transient red shift of the characteristic PL signal [57]. However, this only holds if the EMP formation time τ_{EMP} is shorter than the recombination lifetime τ_{rec} of the photo-generated carriers. After recombination of the excess carrier, the Mn^{2+} ion spins will relax into equilibrium, e.g., via spin–lattice relaxation.

Theoretically, the energy gain of the exciton during Mn^{2+} spin alignment can be described as follows. Considering the magnetization $M(B, T)$ of the paramagnetic

---

[1] For higher Mn^{2+} concentrations, antiferromagnetic interactions between neighboring Mn^{2+} spins have to be considered.

Mn^{2+} system given by

$$M(B, T) = g_{Mn}\mu_B x_{eff} N_0 S \cdot B_{5/2} \left(\frac{5 g_{Mn} \mu_B B}{2 k_B T_{eff}} \right) \qquad (12.1)$$

with the Lande g-factor of the Mn, $g_{Mn} = 2$, the Bohr magneton μ_B, the Boltzmann constant k_B, the number of cations per unit cell N_0 and the spin of the Mn^{2+} ions, $S = 5/2$. $T_{eff} = T_{bath} + T_0$ is an effective temperature, which is enhanced due to antiferromagnetic coupling between Mn^{2+} spins by T_0 as compared to the bath temperature T_{bath} [31, 68] and $x_{eff} < x$ represents an effective concentration of the Mn^{2+} ions contributing to the paramagnetic properties. $B_{5/2}$ is the Brillouin function for spin $= 5/2$ particles. The magnetic field $B = B_{ext} + B_{EMP}$ has two important contributions: first, an externally applied field B_{ext} and second, the exchange field B_{EMP} generated by the exciton. According to Kavokin et al. [67], the latter can be described for the bright exciton by

$$B_{EMP} = \frac{N_0(\alpha - \beta)}{2 g_{Mn} \mu_B} \cdot \frac{1}{N_0 V_{exc}} \qquad (12.2)$$

Here, $N_0\alpha > 0$ ($N_0\beta < 0$) represent the exchange constants of the conduction band (valence band) and V_{exc} is the exciton volume.

The resulting energy gain ΔE_{EMP} after EMP formation in case of $B_{ext} = 0$ and $\tau_{EMP} \ll \tau_{rec}$ is then simply given by

$$\Delta E_{EMP}(T) = M \cdot V_{exc} \cdot B_{EMP} = \frac{1}{2} N_0(\alpha - \beta) x_{eff} S \cdot B_{5/2} \left(\frac{5 g_{Mn} \mu_B B_{EMP}}{2 k_B T_{eff}} \right) \quad (12.3)$$

It is obvious that the temperature stability of the EMP mainly depends on the ratio $\mu_B B_{EMP} / k_B T_{eff}$, and therefore a large exchange field will enforce EMP stability at elevated temperatures. According to Eq. (12.2), a large exchange field can be achieved by reducing the volume of the exciton.

12.3.2.2 Experimental Results I: EMP Dynamics at Elevated Temperatures

According to theory, a strong influence of the EMP formation on temperature is expected. The left part of Fig. 12.13 compares transient PL spectra for various temperatures and two characteristic times: immediately after laser excitation ($t = 0$) and after $t \approx 13$ ns.

The data exhibit a quite remarkable behavior; whereas for $t = 0$, the peak energy systematically decreases with increasing temperature—as expected from the temperature dependent band gap shrinkage in a semiconductor—the opposite behavior is found for the peak energy at $t \approx 13$ ns. Here, an increase in temperature results in an increase of the PL energy. These findings are summarized in the right part of Fig. 12.13. At low temperature, the energy shift between the PL signal at

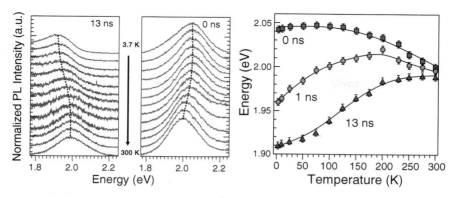

Fig. 12.13 *Left* temperature dependent PL spectra recorded at $t = 0$ and at $t \approx 13$ ns after pulsed laser excitation [77]. The *dotted lines* are guides to the eye and indicate the peak energy shift with temperature. *Right* peak energy versus temperature for $t = 0$, $t = 1$ and $t \approx 13$ ns

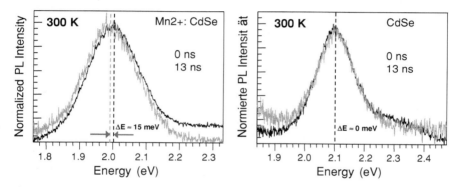

Fig. 12.14 Room temperature PL spectra of the Mn^{2+}-doped CdSe nanocrystals at $t = 0$ and $t \approx 13$ ns (*left*) compared to the transient PL spectra recorded for the undoped CdSe reference sample (*right*) [55]

$t = 0$ and $t \approx 13$ ns amounts to more than 130 meV, indicating a significant energy gain of the exciton during its lifetime. With increasing temperature, this energy difference decreases but even at room temperature a transient energy shift on the order of 15 meV is extracted experimentally [55].

For comparison, the room temperature PL spectra at $t = 0$ and at $t \approx 13$ ns are plotted for both, the CdSe:Mn nanocrystals and the undoped CdSe reference sample (see Fig. 12.14). While the transient energy shift is quite obvious for the magnetically doped nanocrystals, it is completely absent in the undoped CdSe reference sample. This again indicates the magnetic origin of the transient PL energy shift in CdSe:Mn.

To get a more fundamental insight into the mechanism of optically induced magnetization, the energy shift within the first nanosecond after laser excitation is depicted versus the inverse temperature in Fig. 12.15.

Fig. 12.15 PL energy shift of the Mn^{2+}-doped CdSe nanocrystals between $t = 0$ and $t = 1$ ns versus inverse temperature. The *dotted, dashed and solid lines* correspond to model calculations as outlined in the text

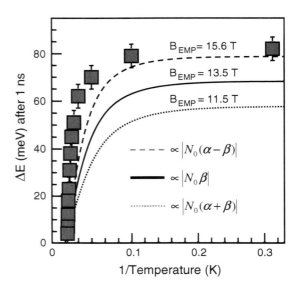

It is remarkable that the energy shift is virtually independent on temperature in the low temperature regime up to about 10 ... 20 K. This indicates magnetic saturation, i.e., the optically generated electron–hole pair completely aligns the magnetic ion spin system in the quantum dot. This is fundamentally different as compared to what was observed up to now in epitaxially grown diluted magnetic semiconductors where the carrier exchange field was not sufficiently large to completely align the Mn^{2+} spins [41, 43]. Note that, for the given manganese concentration of 4.2 % a strong antiferromagnetic coupling between neighboring Mn^{2+} pairs is expected. Large magnetic fields are required for breaking the antiferromagnetic coupled Mn^{2+} spins [31]. The observation of low temperature magnetic saturation due to the exchange field of the optically generated charge carrier spins thus indicates that the magnitude of the exchange field is sufficient for breaking antiferromagnetic coupled manganese spins [55].

We now calculated the expected exchange field and the resulting exciton magnetic polaron energy using Eqs. (12.1)–(12.3). The bulk literature data are used for $N_0\alpha = +0.23$ eV and for $N_0\beta = -1.31$ eV [69] and the exciton volume was calculated to be about 23.2 nm^3 according to the geometrical size of the nanoparticles. As magnetic saturation is achieved in the low temperature limit, we neglect antiferromagnetic coupling between neighboring Mn^{2+} spins, and thus assume $x_{\text{eff}} = x = 0.042$, $S = 5/2$ and $T_{\text{eff}} = T_{\text{bath}}$. The calculated value of B_{EMP} is included in Eq. (12.3) and the resulting excitonic magnetic polaron energy is compared to the experimental result.

In Fig. 12.15, three different scenarios are compared. The solid line represents the model calculations just considering the *p–d* exchange interaction, i.e., the interaction between the holes and the Mn^{2+} ion spins as it was proposed by Kavokin et al. [67]. The main experimental tendency is well reproduced although the theoretical curve is below the experimental data for the whole temperature range. Note that, no free parameters are used for the calculations. Taken into account the impact of the electron on the exciton magnetic polaron formation, one can distinguish two different cases: EMP related to the dark and to the bright exciton, respectively. In CdSe nanocrystals, the dark exciton state is considered to be the energetically lowest one. In that case, the spins of the electron and the hole are parallel and in Eq. (12.2), $N_0(\alpha - \beta)$ has to be replaced by $N_0(\alpha + \beta)$. The resulting EMP energy is included in Fig. 12.15 as dotted line and apparently, the deviation from the experimental data is increased. In case of the bright exciton magnetic polaron, Eq. (12.2) holds and the antiparallel orientation of the electron and the hole spin results in an enhanced EMP energy with respect to the dark exciton magnetic polaron. As can be seen in Fig. 12.15 by the dashed line, the agreement between experiment and theory is improved. This might be indicative for a reversal of bright and dark exciton energy states in CdSe:Mn quantum dots due to EMP formation.

Two important remarks have to be added here. First, independent on the detailed model, the exchange field significantly exceeds 10 T. It is therefore the highest value ever reported in the literature [41, 57, 67]. Thus, our data confirm at least qualitatively the theoretical prediction of an enhancement of the exchange field due to quantum confinement [55]. Second, the transient energy shift (see e.g. Fig. 12.11) cannot be described by a single time constant. For that reason, evaluating the energy shift after 1 ns can only be a rough estimate for the EMP energy according to Eq. (12.3). In fact, the transient energy shift proceeds in time even after 1 ns, suggesting a more complex dynamics of EMP formation in the CdSe:Mn nanocrystals.

12.3.2.3 Experimental Results II: Two-Step EMP Formation Process

In Fig. 12.16, the PL energy is plotted versus time for the CdSe:Mn nanocrystals at a temperature of 3.2 K. The data point at $t \approx 13$ ns is obtained by evaluating the transient PL spectra at $t < 0$, i.e., approximately 13 ns after the preceding laser pulse. The value at $t \to \infty$ corresponds to the stationary PL energy and is extracted from time-integrated PL experiments.

The transient energy shift has to be described by at least three different time constants. The origin of the fastest one with $\tau_1 = 10$ ps is not completely understood but probably related to an energy transfer between fine structure levels of the exciton states in the quantum dots. An initial fast transient energy shift is also observed for the non-magnetic reference sample (see Fig. 12.11). Most prominent is the transient energy shift described by τ_{MP}, which we relate to the initial step of the EMP formation, i.e., the alignment of the Mn^{2+} ion spins in the exchange field of the optically generated charge carriers.

Remarkably, the transient energy shift does not reach saturation even after 13 ns, indicating that equilibrium is not yet achieved. This is seen by comparing the time-resolved data to the PL energy measured in CW experiments (see Fig. 12.16). Such a behavior is not observed for the undoped CdSe reference and should thus also be of magnetic origin. The corresponding time constant τ_{slow} is much longer than the

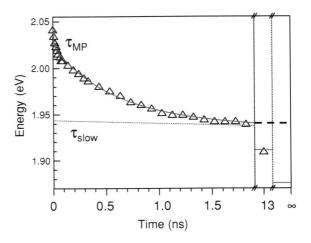

Fig. 12.16 PL energy shift of the Mn^{2+}-doped CdSe nanocrystals versus time. The value at $t \to \infty$ is extracted from the time-integrated PL experiments. The *red line* is a three-exponential fit with $\tau_1 = 10\,ps$, $\tau_{MP} = 570\,ps$ and $\tau_{slow} = 20\,ns$. After Ref. [55] (colour figure online)

typical exciton lifetime for direct bandgap diluted magnetic semiconductor quantum wells or quantum dots grown by epitaxy and can therefore only be accessed in our colloidal quantum dots because of their long recombination lifetime (~100 ns at cryogenic temperatures, ~30 ns at room temperature).

A possible interpretation can be given if one considers the anisotropy of the hole effective mass. After EMP formation, i.e., the initial alignment of the Mn^{2+} ion spins in the exchange field of the optically generated electron–hole pair, a directional reorientation of the complete spin complex may occur, thus gaining further energy. The anisotropy might be dominated by either the hexagonal crystal field asymmetry of the wurtzite lattice or by the crystal shape anisotropy. The time constant, which we attribute to the reorientation of the EMP complex, is comparable to time constants expected for directional EMP reorientation in CdMnTe epilayers in an applied external field [70], which supports our interpretation.

The scenario of the EMP formation dynamics in the colloidal Mn^{2+}:CdSe quantum dots is summarized schematically in Fig. 12.17. In the absence of any optical or electrical carrier injection, the time-averaged magnetization of the Mn^{2+} spin system is zero, having only small magnetization fluctuations around zero [71–74]. Because of the low excitation conditions, the laser pulse now generates at maximum one single electron–hole pair per quantum dot resulting in the formation of a magnetically ordered spin complex, the EMP, with a characteristic time τ_{MP}. The orientation of the EMP is expected to statistically vary from quantum dot to quantum dot within the ensemble because of both the random orientation of the magnetization of the paramagnetic Mn^{2+} system at the moment of optical excitation and the fact that non-resonant, linearly polarized laser excitation was used in the experiment. Thus, no net magnetization of the quantum dot ensemble can be expected. The hole anisotropy in the nanocrystal should result in an EMP anisotropy energy, similar to what was observed for digital DMS quantum wells [75]. Therefore, the whole spin complex reorients, minimizing its energy on a typical time scale of τ_{SLOW} ~ few tens

12 Optical Spectroscopy on Magnetically Doped Semiconductor Nanoparticles   323

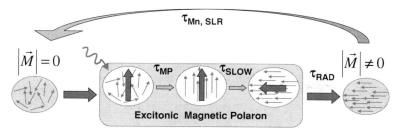

Fig. 12.17 Scenario of EMP formation in colloidal Mn^{2+}-doped CdSe nanocrystals after pulsed laser excitation [77]

of nanoseconds. Subsequently, the electron–hole pair recombines with τ_{RAD}, leaving a non-equilibrium, partially ordered Mn^{2+} spin system behind. Finally, thermalization to the completely disordered Mn^{2+} spin state occurs via spin–lattice relaxation on a time scale of $\tau_{Mn,SLR} \sim 0.1\,\mu s$ [76]. As the repetition rate of our laser system is 76 MHz, i.e., the time between two subsequent pulses is about 13.15 ns, the nanocrystals can in principle be excited again before the Mn^{2+} spin system returned to thermal equilibrium. We avoid this kind of spin accumulation by low excitation conditions with an estimated time of about 100 μs between subsequent excitation events of each quantum dot.

12.4 Conclusion

In conclusion, the strength of optical spectroscopy for analyzing magnetically doped semiconductor nanoparticles is demonstrated. Our results manifest the challenges in achieving a magnetic semiconductor based on ZnO nanoparticles. Although doping the nanoparticles with chromium is shown to result in an incorporation of Cr^{3+} ions in a tetrahedral site, annealing triggers the formation of a ZnCr$_2$O$_4$ spinel phase. Thus, the optical properties is dominated by the ZnO wurtzite phase while the ferromagnetism observed even up to room temperature is most likely caused by aligned excess spins close to the surface of the antiferromagnetic spinel phase. No 'giant' magneto-optical response is therefore found. ZnO doped with Co^{2+} appeared to be more promising as photoluminescence data indicate the incorporation of Co^{2+} in the tetrahedral Zn^{2+} sites of the ZnO nanocrystal. As a very successful route toward room temperature magnetism in semiconductors, CdSe:Mn^{2+} nanocrystals in the strong quantum confinement regime are investigated. Optically induced magnetization is found up to saturation at cryogenic temperatures as a consequence of the enhanced *sp–d* exchange interaction. Most importantly, photo-magnetic signatures are observed up to room temperature. The time-resolved experiments indicate a two-step exciton magnetic polaron formation process—an initial spin alignment in the carrier exchange field is followed by a directional reorientation of the whole spin complex on a time scale of tens of nanoseconds. These results are quite promising with respect to spintronic device applications operating under ambient conditions.

Acknowledgments The authors are indebted to Wei Jin, Ruzica Djenadic and Markus Winterer for preparing the Cr^{3+}- and Co^{2+}-doped ZnO nanoparticles and performing XRD and XANES measurements and Mehmet Acet for SQUID investigations of $ZnO:Cr^{3+}$ nanoparticles. The preparation of excellent $CdSe:Mn^{2+}$ nanocrystals by Remi Beaulac, Paul I Archer, and Daniel R. Gamelin, and the very fruitful collaboration in that field is gratefully acknowledged. We thank Rachel Fainblat-Padua for a critical reading of the manuscript and the German Research Foundation for financial support within the SFB 445—Nanoparticles from the gas phase.

References

1. T. Hayashi, M. Tanaka, K. Seto, T. Nishinaga, K. Ando, III-V based magnetic (GaMnAs)/nonmagnetic(AlAs) semiconductor superlattices. Appl. Phys. Lett. **71**, 1825 (1997)
2. S. Koshihara, A. Oiwa, M. Hirasawa, S. Katsumoto, Y. Iye, C. Urano, H. Takagi, H. Munekata, Ferromagnetic order induced by photogenerated carriers in magnetic III-V semiconductor heterostructures of (In, Mn)As/GaSb. Phys. Rev. Lett. **78**, 4617 (1997)
3. T. Dietl, H. Ohno, F. Matsukura, J. Cibert, D. Ferrand, Zener model description of ferromagnetism in zinc-blende magnetic semiconductors. Science **287**, 1019 (2000)
4. D.M. Hoffman, B.K. Meyer, A.I. Ekimov, I.A. Merkulov, A.L. Efros, M. Rosen, G. Couino, T. Gacoin, J.P. Boilot, Giant internal magnetic fields in Mn doped nanocrystal quantum dots. Solid State Commun. **114**, 547 (2000)
5. H. Sato, H. Katayama-Yoshida, Material design for transparent ferromagnets with ZnO-based magnetic semiconductors. Jpn. J. Appl. Phys. **39**, L555 (2000)
6. J.M.D. Coey, M. Venkatesan, C.B. Fitzgerald, Donor impurity band exchange in dilute ferromagnetic oxides. Nat. Mat. **4**, 173 (2005)
7. K. Ueda, H. Tabata, T. Kawai, Magnetic and electric properties of transition metal doped ZnO films. Appl. Phys. Lett. **79**, 988 (2001)
8. P. Sharma, A. Gupta, K.V. Rao, F.J. Owens, R. Sharma, R. Ahuja, J.M.O. Guillen, B. Johansson, G.A. Gehring, Ferromagnetism above room temperature in bulk and transparent thin films of Mn-doped ZnO. Nat. Mat. **2**, 673 (2003)
9. H.J. Lee, S.Y. Jeong, C.R. Cho, C.H. Park, Study of diluted magnetic semiconductor: Co-doped ZnO. Appl. Phys. Lett. **81**, 4020 (2002)
10. H. Liu, X. Zhang, L. Li, Y.X. Wang, K.H. Gao, Z.Q. Li, R.K. Zheng, S.P. Ringer, B. Zhang, X.X. Zhang, Role of point defects in room-temperature ferromagnetism of Cr-doped ZnO. Appl. Phys. Lett. **91**, 072511 (2007)
11. D.A. Schwartz, N.S. Norberg, Q.P. Nguyen, J.M. Parker, D.R. Gamelin, Magnetic quantum dots: synthesis, spectroscopy, and magnetism of Co^{2+} and Ni^{2+} doped ZnO nanocrystals. J. Am. Chem. Soc. **125**, 13205 (2003)
12. S.T. Ochsenbein, Y. Feng, K.M. Whitaker, E. Badaeva, W.K. Liu, X. Li, D.R. Gamelin, Charge controlled magnetism in colloidal doped semiconductor nanocrystals. Nat. Nanotechnol. **4**, 681 (2009)
13. G. Lawes, A.S. Risbaud, A.P. Ramirez, R. Seshadri, Absence of ferromagnetism in Co and Mn substituted polycrystalline ZnO. Phys. Rev. B **71**, 045201 (2005)
14. C.N.R. Rao, F.L. Deepak, Absence of ferromagnetism in Mn- and Co-doped ZnO. J. Mat. Chem. **15**, 573 (2005)
15. Z. Jin, T. Fukumura, M. Kawasaki, K. Ando, H. Saito, T. Sekiguchi, High throughput fabrication of transition metal doped epitaxial ZnO thin films: a series of oxide-diluted magnetic semiconductors and their properties. Appl. Phys. Lett. **78**, 3824 (2001)

12 Optical Spectroscopy on Magnetically Doped Semiconductor Nanoparticles

16. K.R. Kittilstved, J. Zhao, W.K. Liu, J.D. Bryan, D.A. Schwartz, D.R. Gamelin, Magnetic circular dichroism of ferromagnetic Co2+-doped ZnO. Appl. Phys. Lett. **89**, 062510 (2006)
17. W. Pacuski, D. Ferrand, J. Cibert, C. Deparis, J.A. Gaj, P. Kossacki, C. Morhain, Effect of the s, p-d exchange interaction on the excitons in ZnCoO epilayers. Phys. Rev. B **73**, 035214 (2006)
18. W. Jin, I.K. Lee, A. Kompch, U. Dörfler, M. Winterer, Chemical vapor synthesis and characterization of chromium doped zinc oxide nanoparticles. J. Eur. Ceram. Soc. **27**, 4333 (2007)
19. R. Djenadic, G. Akgül, K. Attenkofer, M. Winterer, Chemical vapor synthesis and structural characterization of nanocrystalline $Zn_{1-x}Co_xO$ ($x = 0$–0.5) particles by X-ray diffraction and X-ray absorption spectroscopy. J. Phys. Chem. C **114**, 9207 (2010)
20. L. Schneider, S.V. Zaitsev, W. Jin, A. Kompch, M. Winterer, M. Acet, G. Bacher, Fabrication and analysis of Cr-doped nanoparticles from the gas phase. Nanotechnology **20**, 135604 (2009)
21. Q. Ou, T. Matsuda, M. Mesko, A. Ogino, M. Nagatsu, Cathodoluminescence property of ZnO nanophosphors prepared by laser ablation. Jpn. J. Appl. Phys. **47**, 389 (2008)
22. L. Schneider, S.V. Zaitsev, G. Bacher, W. Jin, M. Winterer, Recombination dynamics in ZnO nanoparticles produced by chemical vapor synthesis. J. Appl. Phys. **102**, 023524 (2007)
23. S. Polarz, A. Roy, M. Merz, S. Halm, D. Schröder, L. Schneider, G. Bacher, F.E. Kruiss, M. Driess, Chemical vapour synthesis of size-selected zinc oxide nanoparticles. Small **1**, 540 (2005)
24. K. Thonke, T. Gruber, N. Teofilov, R. Schönfelder, A. Waag, R. Sauer, Donor acceptor pair transitions in ZnO substrate material. Phys. E **308**, 945 (2001)
25. T. Voss, C. Bekeny, L. Wischmeier, H. Gafsi, S. Börner, W. Schade, A.C. Mofor, A. Bakin, A. Waag, Influence of exciton-phonon coupling on the energy position of the near-band-edge photoluminescence in ZnO nanowires. Appl. Phys. Lett. **89**, 182107 (2006)
26. Y. Zhang, B. Lin, X. Sun, Z. Fu, Temperature dependent photoluminescence of nanocrystalline ZnO thin films grown on Si(100) substrates by the sol-gel process. Appl. Phys. Lett. **86**, 131910 (2005)
27. B.D. Zhang, N.T. Binh, Y. Segawa, K. Wakatsuki, N. Usami, Optical properties of ZnO rods formed by metalorganic chemical vapour deposition. Appl. Phys. Lett. **83**, 1635 (2003)
28. D.C. Look, D.C. Reynolds, C.W. Litton, R.L. Jones, D.B. Eason, G. Cantwell, Characterization of homoepitaxial p-type ZnO grown by molecular beam epitaxy. Appl. Phys. Lett. **81**, 1830 (2002)
29. J. Fallert, R. Hauschild, F. Stelzl, A. Urban, M. Wissinger, H. Zhou, C. Klingshirn, H. Kalt, Surface-related luminescence in ZnO nanocrystals. J. Appl. Phys. **101**, 073506 (2007)
30. M. Schirra, R. Schneider, A. Reiser, G.M. Prinz, M. Feneberg, J. Biskupek, U. Kaiser, C.E. Krill, R. Sauer, K. Thonke, Acceptor-related luminescence at 3.314 eV in zinc oxide confined to crystallographic line defects. Physica B **401**, 362 (2007)
31. J.K. Furdyna, Diluted magnetic semiconductors. J. Appl. Phys. **64**, R29 (1998)
32. H.-J. Lee, S.-Y. Jeong, J.-Y. Hwang, C.R. Cho, Ferromagnetism in Li co-doped ZnO:Cr. Europhys. Lett. **64**, 797 (2003)
33. A.C. Tuan, J.D. Bryan, A.B. Pakhomov, V. Shutthanandan, S. Thevuthasan, D.E. McCready, D. Gaspar, M.H. Engelhard, J.W. Rogers Jr., K. Krishnan, D.R. Gamelin, S.A. Chambers, Epitaxial growth and properties of cobalt-doped ZnO on α-Al_2O_3 single crystal substrates. Phys. Rev. B **70**, 054424 (2004)
34. H.X. Chen, H.T. Zhang, C.H. Wang, X.G. Luo, P.H. Li, Effect of particle size on magnetic properties of zinc chromite synthesized by sol-gel method. Appl. Phys. Lett. **81**, 4419 (2002)
35. H. Martinho, N.O. Moreno, J.A. Sanjurjo, C. Rettori, A.J. García-Adeva, D.L. Huber, S.B. Oseroff, W. Ratcliff II, S.W. Cheong, P.G. Pagliuso, J.L. Sarrao, G.B. Martins, Magnetic properties of the frustrated AFM spinel $ZnCr_2O_4$ and the spin-glass $ZnCdCr_2O_4$. Phys. Rev B **64**, 024408 (2001)
36. Y. Wang, C.-M. Yang, W. Schmidt, B. Spliethoff, E. Bill, F. Schüth, Weakly ferromagnetic ordered mesoporous Co3O4 synthesized by nanocasting from vinyl-functionalized cubic Ia3d mesoporous silica. Adv. Mat. **17**, 53 (2005)

37. T. Dietl, T. Andrearczyk, A. Lipinska, M. Kiecana, M. Tay, Y. Wu, Origin of ferromagnetism in ZnCoO from magnetization and spin-dependent magnetoresistance measurements. Phys. Rev. B **76**, 155312 (2007)
38. S. Colis, H. Bieber, S. Begin-Colin, G. Schmerber, C. Leuvrey, A. Dinia, Magnetic properties of Co-doped ZnO diluted magnetic semiconductors prepared by low-temperature mechanosynthesis. Chem. Phys. Lett. **422**, 529 (2006)
39. Z. Xiao, H. Matsui, N. Hasuike, H. Harima, H. Tabata, Systematic investigation on structures and excitonic-related transitions: an evidence fro ZnCoO alloy film as a wide gap semiconductor. J. Appl. Phys. **103**, 043504 (2008)
40. P. Koidl, Optical absorption of Co^{2+} in ZnO. Phys. Rev. B **15**, 2493 (1977)
41. A.A. Maksimov, G. Bacher, A. Mc Donald, V.D. Kulakovskii, A. Forchel, C.R. Becker, G. Landwehr, L.W. Molenkamp, Magnetic polarons in a single diluted magnetic semiconductor quantum dot. Phys Rev. B **62**, R7767 (2000)
42. Y. Oka, J. Shen, K. Takabayashi, N. Takahashi, H. Misu, I. Souma, R. Pittini, Dynamics of excitonic magnetic polarons in nanostructure diluted magnetic semiconductors. J. Lumin. **83**, 83 (1999)
43. G. Bacher, H. Schömig, M.K. Welsch, S. Zaitsev, V.D. Kulakovskii, A. Forchel, S. Lee, M. Dobrowolska, J.K. Furdyna, B. König, W. Ossau, Optical spectroscopy on individual CdSe/ZnMnSe quantum dots. Appl. Phys. Lett. **79**, 524 (2001)
44. P.R. Kratzert, J. Puls, M. Rabe, F. Henneberger, Growth and magneto-optical properties of sub 10 nm (Cd, Mn)Se quantum dots. Appl. Phys. Lett. **79**, 2814 (2001)
45. M. Scheibner, T.A. Kennedy, L. Worschech, A. Forchel, G. Bacher, T. Slobodskyy, G. Schmidt, L.W. Molenkamp, Coherent dynamics of locally interacting spins in self-assembled CdMnSe/ZnSe quantum dots. Phys. Rev. B **73**, 081308 (2006)
46. S. Mackowski, J. Wrobel, K. Fronc, J. Kossut, F. Pulizzi, P.C.M. Christianen, J.C. Maan, G. Karczewski, Exciton spectroscopy of single CdTe and CdMnTe quantum dots. Phys. Stat. Sol. (b) **229**, 493 (2002)
47. M. Goryca, T. Kazimierczuk, M. Nawrocki, A. Golnik, J.A. Gaj, P. Kossacki, P. Wojnar, G. Karczewski, Optical manipulation of a single Mn spin in a CdTe-based quantum dot. Phys. Rev. Lett. **103**, 087401 (2009)
48. Y. Terai, S. Kuroda, K. Takita, Self-assembled formation and photoluminescence of CdMnTe quantum dots grown on ZnTe by atomic layer epitaxy. Appl. Phys. Lett. **76**, 2400 (2000)
49. L. Maingault, L. Besombes, Y. Leger, C. Bougerol, H. Mariette, Inserting one single Mn ion into a quantum dot. Appl. Phys. Lett. **89**, 193109 (2006)
50. S. Mackowski, S. Lee, J.K. Furdyna, M. Dobrowolska, G. Prechtl, W. Heiss, J. Kossut, G. Karczewski, Growth and optical properties of Mn-containing II-VI quantum dots. Phys. Stat. Sol. (b) **229**, 469 (2002)
51. R.N. Bhargava, D. Gallagher, X. Hong, A. Nurmikko, Optical properties of manganese-doped nanocrystals of ZnS. Phys. Rev. Lett. **72**, 416 (1994)
52. D.J. Norris, N. Yao, F.T. Charnock, T.A. Kennedy, High-quality manganese-doped ZnSe nanocrystals. Nano Lett. **1**, 3 (2001)
53. S.C. Erwin, L.J. Zu, M.I. Haftel, A.L. Efros, T.A. Kennedy, D.J. Norris, Doping semiconductor nanocrystals. Nature **436**, 91 (2005)
54. R. Beaulac, P.I. Archer, X.Y. Liu, S. Lee, G.M. Salley, M. Dobrowolska, J.K. Furdyna, D.R. Gamelin, Spin-polarizable excitonic luminescence in colloidal Mn^{2+}-doped CdSe quantum dots. Nano Lett. **8**, 1197 (2008)
55. R. Beaulac, L. Schneider, P.I. Archer, G. Bacher, D.R. Gamelin, Light-induced spontaneous magnetization in doped colloidal quantum dots. Science **325**, 973 (2009)
56. J.H. Yu, X.Y. Liu, K.E. Kweon, J. Joo, J. Park, K.T. Ko, D. Lee, S.P. Shen, K. Tivakornsasithorn, J.S. Son, J.H. Park, Y.W. Kim, G.S. Hwang, M. Dobrowolska, J.K. Furdyna, T. Hyeon, Giant Zeeman splitting in nucleation-controlled doped $CdSe:Mn^{2+}$ quantum nanoribbons. Nat. Mat. **9**, 47 (2010)
57. J. Seufert, G. Bacher, M. Scheibner, A. Forchel, S. Lee, M. Dobrowolska, J.K. Furdyna, Dynamical spin resonse in a semimagnetic quantum dots. Phys. Rev. Lett. **88**, 027402 (2002)

12 Optical Spectroscopy on Magnetically Doped Semiconductor Nanoparticles

58. P. Wojnar, J. Suffczynski, K. Kowalik, A. Golnik, M. Aleszkiewicz, G. Karczeswski, J. Kossut, Size-dependent magneto-optical effects in CdMnTe diluted magnetic quantum dots. Nanotechnology **19**, 235403 (2008)
59. P.I. Archer, S.A. Santangelo, D.R. Gamelin, Direct observation of sp-d exchange interactions in colloidal Mn^{2+}- and Co^{2+}-doped CdSe quantum dots. Nano Lett. **7**, 1037 (2007)
60. R. Beaulac, P.I. Archer, S.T. Ochsenbein, D.R. Gamelin, Doped CdSe quantum dots: new inorganic materials for spin-electronics and spin-photonics. Adv. Func. Mat. **18**, 3873 (2008)
61. O. Labeau, P. Tamarat, B. Lounis, Temperature dependence of the luminescence lifetime of single CdSe/ZnS quantum dots. Phys. Rev. Lett. **90**, 257404 (2003)
62. S.A. Crooker, T. Barrick, J.A. Hollingsworth, V.I. Klimov, Multiple temperature regimes of radiative decay in CdSe nanocrystal quantum dots: intrinsic limits to the dark-exciton lifetime. Appl. Phys. Lett. **82**, 2793 (2003)
63. R. Beaulac, P.I. Archer, J. van Rijssel, A. Meijerink, D.R. Gamelin, Exciton storage by Mn^{2+} in colloidal Mn^{2+}-doped CdSe quantum dots. Nano Lett. **8**, 2949 (2008)
64. A.L. Efros, M. Rosen, The electronic structure of semiconductor nanocrystals. Ann. Rev. Mat. Sci. **30**, 475 (2000)
65. T.J. Liptay, L.F. Marshall, P.S. Rao, R.J. Ram, M.G. Bawendi, Anomalous Stokes shift in CdSe nanocrystals. Phys. Rev. B **76**, 155314 (2007)
66. T. Kasuya, A. Yanase, Anomalous transport phenomena in Eu-chalcogenide alloys. Rev. Mod. Phys. **40**, 684 (1968)
67. A.V. Kavokin, I.A. Merkulov, D.R. Yakovlev, W. Ossau, G. Landwehr, Exciton localization in semimagnetic semiconductors probed by magnetic polarons. Phys. Rev. B **60**, 16499 (1999)
68. J.A. Gaj, R. Planel, G. Fishman, Relation of magneto-optical properties of free excitons to spin alignment of Mn^{2+} ions in CdMnTe. Sol. State. Commun. **29**, 435 (1979)
69. O. Goede, W. Heimbrodt, Optical properties of (Zn, Mn) and (Cd, Mn) chalcogenide mixed crystals and superlattices. Phys. Stat. Sol. (b) **146**, 11 (1988)
70. I.A. Merkulov, D.R. Yakovlev, K.V. Kavokin, G. Mackh, W. Ossau, A. Waag, G. Landwehr, Hierarchy of relaxation times in the formation of an excitonic magnetic polaron in (CdMn)Te. JETP Lett. **62**, 335 (1995)
71. T. Dietl, J. Spalek, Effect of thermodynamic fluctuations of magnetization on the bound magnetic polaron in dilute magnetic semiconductors. Phys. Rev. B **28**, 1548 (1983)
72. I.A. Merkulov, G.R. Pozina, D. Coquillat, N. Paganotto, J. Siviniant, J.P. Lascaray, J. Cibert, Parameters of the magnetic polaron state in diluted magnetic semiconductors Cd-Mn-Te with low manganese concentration. Phys. Rev. B **54**, 5727 (1996)
73. G. Bacher, A.A. Maksimov, H. Schömig, M.K. Welsch, V.D. Kulakovskii, P.S. Dorozhkin, A. Forchel, S. Lee, M. Dobrowolska, J.K. Furdyna, Monitoring statistical magnetic fluctuations on the nanometer scale. Phys. Rev. Lett. **89**, 127201 (2002)
74. P.S. Dorozhkin, A.V. Chernenko, V.D. Kulakovskii, A.S. Brichkin, A.A. Maksimov, H. Schoemig, G. Bacher, A. Forchel, S. Lee, M. Dobrowolska, J.K. Furdyna, Longitudinal and transverse fluctuations of magnetization of the excitonic magnetic polaron in a semimagnetic single quantum dot. Phys. Rev. B **68**, 195313 (2003)
75. R. Fiederling, D.R. Yakovlev, W. Ossau, G. Landwehr, I.A. Merkulov, K.V. Kavokin, T. Wojtowicz, M. Kutrowski, K. Grasza, G. Karczewski, J. Kossut, Exciton magnetic polarons in (100)- and (120) oriented semimagnetic digital alloys (Cd, Mn)Te. Phys. Rev. B **58**, 4785 (1998)
76. T. Dietl, P. Peyla, W. Grieshaber, Y. Merle D'Aubigné, Dynamics of spin organization in diluted magnetic semiconductors. Phys. Rev. Lett. **74**, 474 (1995)
77. G. Bacher, L. Schneider, R. Beaulac, P.I. Archer, D.R. Gamelin, Magnetic polaron formation dynamics in Mn^{2+}-doped colloidal nanocrystals up to room temperature. J. Korean Phys. Soc. **58**, 1261 (2011)

Chapter 13
Gas Sensors Based on Well-Defined Nanostructured Thin Films

A. Nedic and F. E. Kruis

Abstract The ability to prepare nanoparticles having well-defined size and narrow size distribution is an important advantage for optimising and understanding nanoparticulate gas sensors. It allows to monitor the size effect of SnO_2 particles as well as that of the addition of the noble metal particles on sensing behaviour. The synthesis of monodisperse SnOx, Pd and Ag nanoparticles and the development the thin films deposition technology as well as suitable microchip platforms are described. Sensing results of SnO_x:M mixed nanoparticle layers are presented, especially the effects of operating temperature, particle size, type of noble metal additive and electrode distance are investigated. Sensor to sensor reproducibility as well as long-term stability is investigated. Finally, pure Pd nanoparticle layers are demonstrated to show concentration-specific H_2 sensing at room temperature.

13.1 Introduction

The increasing demand of clean and safe environments in industrial and domestic ambients necessitates the development of fast, sensitive and selective gas sensors. The first semiconductor gas phenomenon was observed by Seiyama in 1962 [1], since then there is a widespread interest in the area of solid state gas sensors. Inorganic and especially semiconducting oxide nanostructures are the most promising materials for solid state gas sensor applications in terms of sensitivity and selectivity. Semiconductor oxide sensors are based on the change in conductivity when exposed to the gas at elevated temperatures. This change in conductivity is due to interaction between chemically bonded surface oxygen species and reducing or oxidising gases [2–4]. Among all the different semiconducting oxide materials, SnO_2 was one of the first, and still is the most extensively used material in gas

---

A. Nedic · F. E. Kruis (✉)
Faculty of Engineering and CENIDE, University of Duisburg-Essen,
Bismarckstraße 81, 47057 Duisburg, Germany
e-mail: einar.kruis@uni-due.de

A. Lorke et al. (eds.), *Nanoparticles from the Gas Phase*, NanoScience and Technology, 329
DOI: 10.1007/978-3-642-28546-2_13, © Springer-Verlag Berlin Heidelberg 2012

sensing research fields. Well-known advantages of gas sensors consisting of this material are their high sensitivity, rapid response and recovery times and especially low cost in comparison to others [5]. Till date, different synthesis methods e.g. (sputtering, laser ablation, sol–gel) have been introduced to produce SnO_2 in the form of thick or thin films and nanostructures [6–8]. Particularly, in nanostructure forms SnO_2 is drawing special attention due to a variety of effects due to quantum confinement of charge carriers and the larger surface to volume ratio. In case of nanoparticles, the percentage of atoms on the surface increases with decreasing particle size due to its inverse proportionality to the grain radius. The large surface area allows to detect changes caused by chemisorption more easily, which is a predominant gas sensing mechanism [9]. Marked drawbacks of SnO_2 sensors are drift in sensor response and low selectivity due to high cross-sensitivities to varieties of reducing or oxidising gases. Drift in sensor response is due to ageing effects of the SnO_2 sensing layers [10]. These mentioned drawbacks can be overcome by modifying structural and chemical properties of the sensing layers or by doping and mixing additives [11–13]. In the present study, aerosol technology is utilised to prepare SnO_2 nanoparticle layers for sensing application. The SnO_2 nanoparticles prepared by this method are reported to have high stability against grain growth [14], especially due to an in-flight preannealing step. Therefore, it is possible to overcome the problem related to drift in the response signal. Aerosol technology allows to homogeneously add catalytic nanoparticles during synthesis of SnO_2. Noble metals e.g. Au, Pd and Ag are well-known catalytic materials, which are often used for enhancing gas sensing properties [15–18]. The ability to prepare nanoparticles having well-defined size and narrow size distribution is an additional important advantage of our synthesis technique. Therefore, it provides us with the possibility to monitor the size effect of SnO_2 particles as well as that of the addition of the noble metal particles on sensing behaviour.

13.2 Experimental Details

A gas phase synthesis technique by means of the aerosol route was used in the present work to synthesise SnO_x and SnO_x:M (with M=metallic: Pd, Ag) mixed nanoparticle layers. A schematic diagram of the used synthesis setup is shown in Fig. 13.1. In this method SnO powder as host material is evaporated inside a tube furnace in presence of a carrier gas flow (N_2). The obtained polydisperse aerosol formed by nucleation, condensation and subsequent coagulation (agglomeration) pass through a neutralizer (Kr^{85}—radioactive β source) for achieving the Boltzmann charge distribution. These most singly charged agglomerates are lead into a differential mobility analyzer (DMA) to perform the desired size selection in order to obtain monodisperse particles. The DMA separates the size of the agglomerates based on their electrical mobility diameter corresponding to a particular voltage. These particles pass through a subsequent sintering furnace which is kept at temperature $T_s = 873\,K$. During sintering, 10 vol% oxygen is also introduced inside the hottest zone of sintering furnace, simultaneously. This so-called in-flight annealing leads to the formation of monocrystalline and spherical nanoparticles, whereas the additional oxidation step

13 Gas Sensors Based on Well-Defined Nanostructured Thin Films

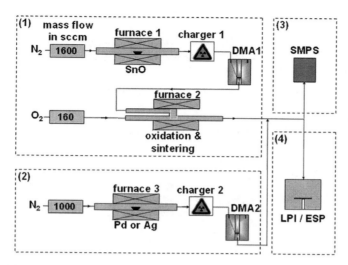

Fig. 13.1 Schematic of the synthesis setup for the fabrication of monodisperse mixed SnO$_x$:Pd (SnO$_x$:Ag) nanoparticle based gas sensors. It can be divided in four main sections: *1* synthesis of monodispersed SnO$_x$ nanoparticles, *2* synthesis of monodisperse metallic nanoparticles, *3* SMPS for measuring size distributions and concentrations of SnO$_x$ and Pd (Ag) particles before *4* deposition by means of LPI / ESP

controls the chemical composition and stoichiometry of the obtained oxide particles. Auger electron spectroscopy (AES) measurements on SnO$_x$ layers prepared by this method reveal a stoichiometry of $x = 1.8$ [19]. The metallic nanoparticles are synthesised in a similar way, by using tube furnace and DMA but without applying the in-flight annealing step.

For online monitoring of size distribution and concentration of both, SnO$_x$ and Pd (Ag), a scanning mobility particle sizer (SMPS) was utilised. The SMPS measures the mobility-based particle size and consists of a bipolar charger (Kr^{85}), a Nano-DMA (model 3085, TSI) and a condensation particle counter with lower detection limit of 4 nm (CPC, model 3775, TSI). To produce sensors made of SnO$_x$:M, two aerosols containing monosized SnO$_x$ and M nanoparticles are homogeneously mixed in desired concentration and directly deposited onto Si substrates equipped with gold electrodes by use of low pressure impaction (LPI) technique [20]. The electrode structures are bonded to a DIL-24 chip carrier through gold bond wires. The thickness of the deposited films on Si substrate was between 500 nm and 1 μm as confirmed by stylus profilometer (Ambios XP200) measurements. For transmission electron microscopy (TEM) (Philips CM12 twin microscope) and high resolution transmission electron microscopy (HRTEM) (Philips Tecnai F20 supertwin microscope) studies, the particles have been deposited on carbon coated Cu grids using an electrostatic precipitation route (ESP) [21, 22]. The mobility diameter (D_m) for SnO$_x$ nanoparticles was chosen to be 10–20 nm and for Pd or Ag in a range of 5–15 nm with concentrations from 0–5 % based on the number concentration.

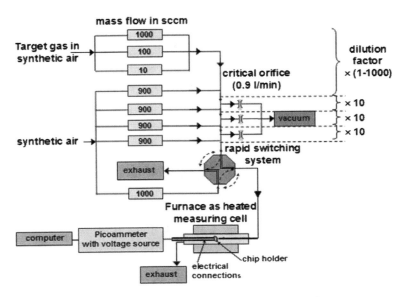

Fig. 13.2 Schematic diagram of the dilution and sensor response measuring system

The setup used for the dilution of the target gases and the corresponding sensing measurements on the manufactured gas sensors is shown in Fig. 13.2. It is a fully automated assembly consisting of a system of mass flow controllers, magnetic valves and a Keithley 487 picoammeter with internal voltage source attached via IEEE bus to a computer. The software was compiled in visual basic which allows a complete adjustment for different gas environments in one step. Seven different gas concentrations in a range of 10 ppb to 10^4 ppm can be set, with specified times (cleaning time duration with synthetic air, process time duration with target gas in a specific concentration) and number of repetitions.

Gases used in this work are as follows: Ethanol, CO, CO_2, propane, butane, SO_2 and H_2. Before the sensing measurements, all samples are annealed for 2 h at 673 K in dry synthetic air and afterwards cooled down to room temperature in order to obtain stable and well-reproducible resistance values independent of previous investigation states. Basically, this annealing step must be performed to cause an initial activation of the SnO_x particles. A further investigation concerning the influence of the electrode distance to the sensitivity of the sensor was also done. The sensors with seven "nano-gap" electrodes of all different gap sizes (S: 10, 3, 1μm, 300, 100, 60 and 30 nm) were fabricated on Si substrate by means of electron beam-lithography (EBL) and photolithography techniques. At first 200 nm thermal oxide is grown on Si substrate, and then the fine gap lines with thickness of 50 nm (5 nm Cr/45 nm Au) and width of 1μm were deposited by EBL. Afterwards the Au bond pads and the Au wire

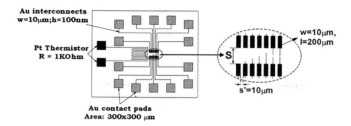

Fig. 13.3 Schematic drawing of nano-gap sensor processed at the NanoFabrication Center from the University of Minnesota

interconnects as well as the Pt thermistor are manufactured by a photolithography process. The dimensions of the thermistor were set to a width of 10 μm, a thickness of 45 nm and a total length from pad to pad of $L = 0.424$ cm. A magnified view of the microsensor can be found in Fig. 13.3. An augmented inset on the right side gives a detailed description of the electrode assembly with seven face to face electrode pairs and gap sizes varying from 10 μm (left) down to 30 nm (right). Films of about 1 mm^2 and 500 nm—1 μm thickness are deposited on top of the gap electrodes by means of LPI.

For reliable data extraction from these measurements it is necessary to have a deposition method in which we are able to deposit the films every time nearly on the same point and to have a control over the generated film thickness. It is known from the literature that the sensing mechanism depends on the film thickness [23]. In order to produce reliable and first of all comparable data it is indispensable to ensure the ability of reproducible sensors in thickness and spot profile. Because of the used synthesis method it is very easy to control the film thickness by permanent online monitoring of the particle concentration entering the deposition chamber. The deposition time for one monolayer can be calculated, as the particles are monodisperse (geometric standard deviation, $\sigma_g \leq 1.10$) and their diameter is known. Thereby, the deposition spot thickness can be determined. The challenge to overcome the exact positioning problem has been solved through the design of a new LPI in which a vacuum compatible miniature X–Y–Z positioning table with piezoelectric inertial drive (model MX35, Mechonics) was integrated. The travel range is up to 10 mm in any three directions with a resolution ∼100 nm and a maximum speed of 0.7 mm/s. The table can be controlled from outside via handheld controller (model CN30, Mechonics) or by the serial interface by computer. To ensure deposition on sensor at exact spot, a laser diode ($\lambda = 635$ nm, $P_{opt,max} = 3.6$ mW) is placed centred above the orifice inlet ($d_{or} = 0.5$ mm). The housing of the LPI consists of acrylic glass that provides visibility of deposition spot from outside.

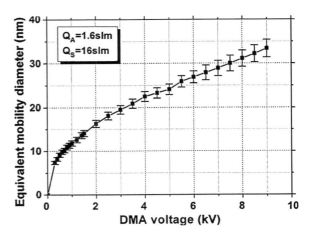

Fig. 13.4 Calibration characteristic of the used DMA for an aerosol to sheath flow rate ratio of 1:10. The DMA provides narrow size distributions with a geometric standard deviation σ_g of ≤ 1.06 as measured with a SMPS system (model 3080, TSI)

13.3 Results and Discussion

13.3.1 Differential Mobility Analyser (DMA) Measurements

Before starting the particle synthesis, a homemade DMA was calibrated with different flow conditions and voltages. Calibration of a DMA was carried out with the aerosol and sheath gas flows of 1.6 and 16 slm, respectively. Size distributions of nanoparticles passing through on applying different voltages in the range of 100 V–9 kV are measured by means of SMPS. The selected equivalent mobility diameters by DMA on applying different voltages are shown in Fig. 13.4.

13.3.2 Sintering Behaviour of Generated SnO_x Nanoparticles

To optimise synthesis parameters for the generation of compacted nanoparticle sizes of 10, 15 and 20 nm, sintering was performed of SnO_x agglomerates produced by tube furnace. Figure 13.5 shows the sintering behaviour of SnO_x agglomerates of initially selected sizes of 30, 20 and 10 nm. On sintering, equivalent mobility diameter decreases for all the three sizes as shown in Fig. 13.5a. Sintering above 610 °C, reevaporation of particles was observed, leading to formation of much smaller particles, as shown in Fig. 13.5b.

Therefore, the optimum sintering temperature was found around 600 °C, which is a good compromise between obtaining crystalline particles and avoiding reevaporation of particles. Optimized synthesis parameters for getting 10, 15 and 20 nm size particles are mentioned in Table 13.1.

13 Gas Sensors Based on Well-Defined Nanostructured Thin Films

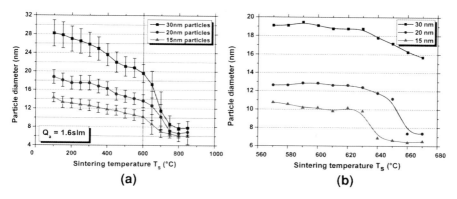

Fig. 13.5 Sintering characteristics of in-flight annealed SnO_x nanoparticles with primary diameters of 30, 20 and 15 nm by concurrent oxygen addition inside the tube furnace, over a range of 100–850 °C (**a**) and a small section of temperature region in which the particles start to reevaporate (**b**)

Table 13.1 Summary of optimised synthesis parameters for three different aerosol flow rates

Desired particle size (nm)	Preset flow rate					
	1 slm		1.6 slm		2 slm	
	U_{DMA} [V]	T_S [°C]	U_{DMA} [V]	T_S [°C]	U_{DMA} [V]	T_S [°C]
10	1250	570	1350	580	1500	580
15	3000	580	4000	595	4200	600
20	5000	590	7300	605	7500	610

13.3.3 Synthesis of Monodispersed SnO_x, Pd and Ag Nanoparticles

By introducing sintering at optimised temperatures, well-defined monosized nanoparticles are prepared. Figure 13.6a, b demonstrates the size distributions of size-selected particles before and after sintering at sintering temperature (T_S) 600 °C. The size distributions become narrower after the sintering step, as clear from diagrams shown in Fig. 13.6a, b. This can be explained by means of doubly charged particles which are more compacted than the singly charged ones. The doubly charged particles are larger than the singly charged ones (second peak appearing at 32 nm on the left and at 44 nm at the right) but they exhibit the same electrical mobility so they were also selected by the DMA. The larger the selected particles are, the higher the fraction of doubly charged particles. During the sintering step, the total concentration decreases by at least 50 % due to thermophoretic losses at the tube walls. The TEM images in Fig. 13.6c, d show the as-deposited SnO_x particles of diameter 15 (Fig. 13.6c) and 20 nm (Fig. 13.6d), respectively. It is important to mention, that the in-flight sintering at high temperatures 650 °C stabilises the SnO_x against grain growth during post-deposition

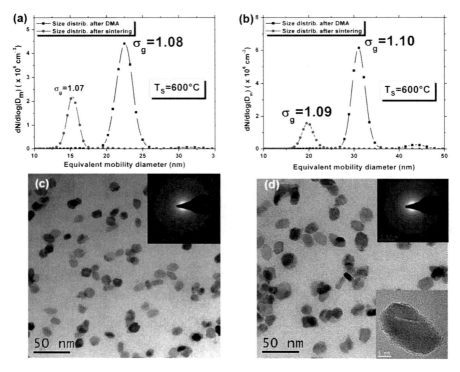

Fig. 13.6 The size distributions of SnO_x nanoparticles with mean diameter of 15 nm **a** and 20 nm **b** synthesised by in-flight sintering and oxidation, as measured online by using SMPS prior to deposition. The geometric standard deviation is ≤ 1.10 for both the cases. This was also confirmed by TEM images of 15 nm **c** and 20 nm **d** size nanoparticle samples. The corresponding diffraction patterns and a HRTEM image of a 20 nm SnO_x particle are shown as insets

annealing, resulting in thermal long-term stability [14]. The concentration generated via this process remains nearly stable over a period of minimum 24 h.

As mentioned before, for the synthesis of Pd and Ag nanoparticles sintering and oxidation steps are not performed. Pd nanoparticles of sizes of about 5 nm are spherical and highly monosized ($\sigma_g \leq 1.05$) as clear from TEM image in Fig. 13.7. However, to prepare larger (>5 nm), quasispherical metallic particles, a further sintering step at temperatures of around 700 °C would be required.

13.3.4 Low Pressure Impaction (LPI)

Before the film deposition, the working behaviour of the designed LPI is examined. The optimum distance from nozzle outlet to substrate must be found in order to generate Gaussian-shaped profiles of desired thickness. Figure 13.8a shows the simulated

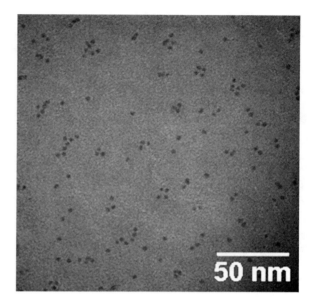

Fig. 13.7 TEM image of 5 nm Pd particles as synthesised by the evaporation method at 1450 °C

contours of the velocity magnitude profile for an orifice with diameter (d_{or}) of 0.5 mm and two different nozzle to substrate distances.

Set table distances shown in Fig. 13.8a correspond to the maximum (left: D_S = 11.3 mm) and to the minimum (right: D_S = 1.3 mm) distance range that can be managed by the piezo-driven positioning table. The simulations are done with help of the computational fluid dynamics software Fluent 6.2 based on a 2D axisymmetric model. The input data implies information about the exact geometry dimensions, the properties of used inert gas (N_2), pressure conditions inside (10 mbar) and outside (1013 mbar) of the LPI and for calculations of particle stream trajectories (shown in Fig. 13.8c) the appropriate characteristics of the material used (here: SnO_2). Thereby, the local particle drag force is modelled, considering a Stokes–Cunningham model with variable local slip correction factor and Brownian diffusion. The particles are assumed as highly diluted, having no influence on the gas stream and do not interact with each other. Pressure inside the LPI was measured with a pressure gauge while a rotary pump is connected to the LPI. Fluent calculates the corresponding flow through the nozzle based on the existent pressure conditions. Dependent on d_{or} and thereby on the flow rate entering the LPI, a mach disc appears at certain distances (here at 3.2 mm) which slowdown the particle stream, leading to a widening of the deposited spot due to turbulences occurred at the mach disc. The appearing mach disc can be clearly seen below the red region (Fig. 13.8a). At that place, the velocity magnitudes are dragged to small velocities (∼50 m/s) indicating a drag in particle velocity. Figure 13.8b shows simulation results for the expected deposition spot at certain orifice to substrate distances. 98 % of the particles impact on the substrate even at maximum distance, but the spot (>7 mm^2) appears covering nearly the complete

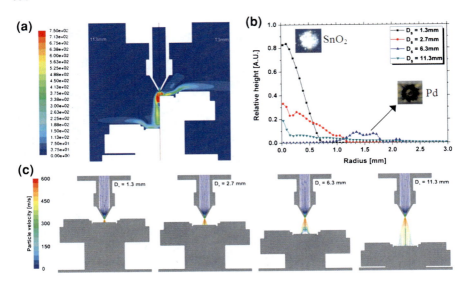

Fig. 13.8 a Contours of simulated gas velocity magnitudes (m/s) arising inside the LPI for the case of maximum or minimum nozzle to substrate distances. **b** Simulated, standardised deposition spot profiles based on the calculations of sampled particle trace trajectories for a constant stream of N = 3000 particles of 20 nm at different distances with accordingly expected spot-size radius. **c** Shape of N = 10 particle trajectory stream lines (assuming of only N = 10 particles for better illustration purposes) with belonging velocity scale on the left, for the four distances depicted in (**b**). Simulations are done by D. Kiesler

sensor substrate. Interestingly, at distance 6.3 mm the mach disc appears strongly, so that the midpoint of the substrate is not penetrated by particles, which results in a ring-shaped deposition spot (demonstrated in the inset of Fig. 13.8b). For our case, the best working distance (D_S) is 1.3 mm in which mach disc formation is not observed, and the particles reach velocities of more than 500 m/s shortly above the substrate. The impaction takes place at very high velocities and leads to spot sizes smaller than 1.5 mm. The nozzle with d_{or} of 0.5 mm permits a maximum flow rate of 1.88 slm which is matching the synthesis setup.

13.3.5 Gas Sensor Preparation

The sensors in the present work have been prepared by directly depositing the as-prepared monosized SnO_x, SnO_x:Pd, SnO_x:Ag or pure Pd nanoparticles in the gas phase using the LPI and the nano-gap sensor arrangement mentioned in Sect. 13.2. A typical DIL-24 chip carrier with deposited SnO_x (20 nm): Pd (5 nm, 3 %) mixed nanoparticle layer is shown in Fig. 13.9a. Low magnification SEM image showing the deposited layer and the pipe wiring of the seven distinct electrode

Fig. 13.9 a DIL-24 chip carrier with mixed SnO$_x$:Pd nanoparticle layer, b low magnification SEM image of the deposited spot, c high magnification SEM image revealing the extreme layer porosity, and d corresponding STEM image showing the SnO$_x$ and Pd nanoparticles (Pd particles are pointed to by the arrows). Reprinted with permission from [39]

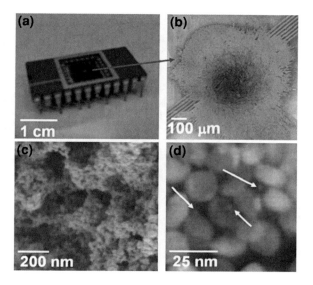

distances (Fig. 13.9b). At a higher magnification, the porous structure of the deposited layer becomes evident (Fig. 13.9c). The corresponding STEM image clearly shows the SnO$_x$ particles a small amount of Pd particles (3 %) dispersed over the layer (Fig. 13.9d). The high layer porosity leads to effective diffusion of gaseous species, resulting in enhanced interaction assisting a complete adsorption/desorption [4, 24, 25] process, as observed in the interaction of the samples for nearly all examined gases. Further advantage of porous layers, is the lesser influence of the electrode contact properties on the sensor resistance [26].

One goal in the sensor preparation is the control of film thickness. By permanent online monitoring of the particle concentration (UCPC, model 3025, TSI), entering into the deposition chamber the reached thickness of the deposited layer could be estimated and the deposition process could be stopped when the desired height was reached. To prove the accuracy of the applied method, multiple samples with 500 nm estimated thickness are prepared on Si substrates under identical conditions. The effective thickness deposited was measured subsequently by means of profilometer. In profilometer, a stylus moves over the deposited area during the thickness measurements, which may cause scratch in our samples. In order to avoid this damage, a 100 nm thick Al film was evaporated on top of complete Si substrate and afterwards the measurements are done. To be sure that the evaporated thickness (100 nm) of Al is correct, an interdigital finger structure was also evaporated on a clean, blank Si substrate and afterwards also measured by the profilometer. The thickness of the fingers was found to be ∼97 nm and thereby diverges only by 3 %. A corresponding 3D scan of a sample prepared in such a way is shown in Fig. 13.10.

The peaks appearing on the left are coming from splinter inclusions underneath the evaporated Al layer, because the Si substrates were cut afterwards to fit into the

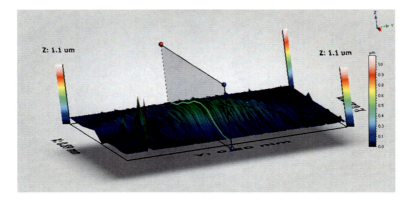

Fig. 13.10 3D profile obtained from sample with estimated thickness of 500 nm and post evaporated 100 nm aluminium layer, scanned by a profilometer

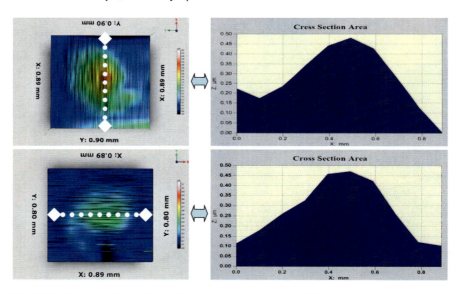

Fig. 13.11 Top view of two 3D profile scans performed on two different samples, prepared under the same conditions. The estimated layer thickness was 500 nm, whereas the measured was found to be ∼5 % smaller (∼480 nm). The dashed lines on the left represent the location belonging to the cross-sectional areas shown on the right

substrate holder in the evaporation chamber. The top views for two different sensors and their cross-sectional areas are given in Fig. 13.11. It was indeed observed that the maximum peak of the Gaussian-shaped deposition spots lies in the considered region. The heights were measured to be around 470–480 nm, which is a deviation of about 5 %. A very good sample to sample reproducibility was found.

13 Gas Sensors Based on Well-Defined Nanostructured Thin Films 341

Pure Pd sensors have been prepared in a different manner as explained before. Whereas the SnO_x sensors were deposited on preassembled Si substrates (furnished with gold electrodes and bonded onto a chip carrier), the palladium sensors were deposited onto blank glass substrates without any electrode pattern. Pd particles with a geometric mean mobility diameter of 15 nm and a geometric standard deviation of $\sigma_g = 1.05$, synthesised in the gas phase were deposited onto the glass substrate by use of LPI. Two Al metallic finger contacts, acting as electrodes are post-precipitated inside an evaporation chamber. H_2 sensing measurements at room temperature and additionally at moderate elevated temperatures were also carried out [27].

13.3.6 Sensing Results on SnO_x:M Mixed Nanoparticle Layers

The common working principle of a semiconductor resistive gas sensor is very simple. In presence of oxidising gas, electrons from the semiconductor conduction band are removed through the reduction of molecular oxygen, which leads to a formation of O^- species at the surface at sufficiently high temperatures. The negative charge trapped in these oxygen species causes an upward band bending which is equivalent in forming an electron-depleted space charge region near the surface, and hence leading to higher resistive films compared to the flat band situation. In the presence of a reducing gas, it underlies a chemical reaction (oxidation) because of the O^- found at the surface. The conduction band will be refilled by the released electrons whereby the resistance decreases. A broad variety of metal oxides show sensitivity towards different oxidising and reducing gases by means of their changes in electrical properties. SnO_2 was known as n-type semiconductor. In presence of a reducing gas, such as propane (C_3H_8), butane (C_4H_{10}) or CO, the conductance of SnO_2 sensors increases [28]. The sensitivity of a gas sensor is defined traditionally as the ratio of measured resistance in air (R_a) to that in target gas (R_g) [29]. Thus, the n-type response is commonly defined as R_a/R_g. For the case of decreasing conductance in presence of reducing species, the sensor signal is defined as R_g/R_a (p-type response).

Figure 13.12a shows a typical behaviour of the fabricated SnO_x:Pd nanoparticle layers at 400 °C in 10^3 ppm propane and dry synthetic air for three consecutive cycles. A reproducible base line stability and resistance saturation in the presence of the sensing gas are evident, equivalent to complete adsorption/desorption ability of the highly porous mixed layers.

That this is not applicable for every gas species can be seen in Fig. 13.12b. Neither the resistance in air nor that in SO_2 was reproducible. It seems to be equivalent for both loops at 100 ppm, but second time measurement reveals different results. This can be understood in terms of a poisoning of the layer in presence of SO_2. After the contamination with SO_2, the films were also inoperative towards different reducible hydrocarbons.

From the literature it is known that the sensing characteristics of SnO_2 sensors can be improved with different modifications of the sensing layer, such as decreasing

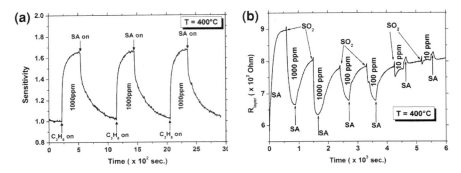

Fig. 13.12 a Sensor signal (sensitivity) from a SnO$_x$:Pd mixed layer in presence of a reducing gas (C$_3$H$_8$) and measured over three consecutive cycles at 400 °C. The sensor signal remains unchanged in each repeated cycle meaning no sensor drift is evident. The time point at which the gas and respectively the dry synthetic air were inserted, are indicated by the arrows. **b** Measured layer resistance at three different SO$_2$ concentrations

the particle size [30–33], sensitisation with various types of noble metal additives acting as catalyst and accelerating the oxidation rate of the gas on the semiconductor surface [17, 30, 34]. It is also known that the sensing of SnO$_x$ films can be improved when working with higher operating temperatures [35]. In the present study, the films fabricated were prepared of different SnO$_x$ particle sizes (10–20 nm), miscellaneous noble metal nanoparticles as mixing additives (Pd, Ag / 5 nm) and operated at elevated temperatures (300–400 °C) to study their effect on the sensor signal.

13.3.6.1 Effect of Operating Temperature

The sensor characteristics were measured on SnO$_x$:Pd mixed samples [SnO$_x$ (20 nm): Pd (5 nm, 5 %)] placed inside a tube furnace and exposed to various gas species in different concentrations at varying temperatures. It is to be noted that all the mixed sensors investigated in the present study were produced on the basis of experimental results found in [17, 30]. Hence, the concentration of noble metal additives was fixed to 3–5 %. If not explicit mentioned, the measurements were carried out on layers deposited on substrates as shown in Fig. 13.3 and the chosen sensor element was that with an electrode distance of $D = 300$ nm.

The measured sensitivities of such samples towards butane and propane show clearly the increasing sensing behaviour with temperature rising (Fig. 13.13a, b). It is observed that the temperature of maximum sensing increases with increase in gas concentration. For temperatures ≤ 300 °C, the sensor was practically not reacting at concentrations of 10 ppm and beneath. Heating up to temperatures ≥ 350 °C leads to a strong enhancement in its responding behaviour. Clearly, the response to butane was much higher compared to that in propane. Still at 350 °C, a reaction to 10 ppm butane takes place, whereas no changes were measured towards propane even at higher temperatures. First changes in sensor response towards propane were noticeable with

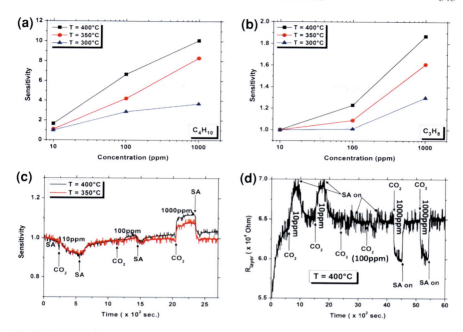

Fig. 13.13 Variation in sensor signal as a function of concentration for three different temperatures with target gas butane **a** and propane **b**. Sensitivity measured towards CO_2 with concentrations ranging from $10–10^3$ ppm at 350 °C and 400 °C **c**, and the corresponding resistance response at 400 °C obtained for two consecutive cycles **d**

concentrations ≥ 100 ppm. Another interesting observation is the response towards CO_2 shown in Fig. 13.13c, d. CO_2 is known to be a difficult gas to detect with sensors made of semiconducting materials because of their ineffective response to the highly oxidised CO_2. In contrast, the sensor in this study shows obvious markedly and quite interesting response behaviour. Indeed, the response was very small and the signal measured noisy but a clear tendency is evident. For 10 ppm, the sensor exhibit p-type behaviour whereas at 10^3 ppm n-type response appears. At a concentration of 100 ppm nearly no reaction was evident. This dual response behaviour provides a capability to discriminate among high and low traces of CO_2. The dual response behaviour will be discussed in more detail in Sect. 13.3.6.6.

13.3.6.2 Effect of Particle Size and Type of Noble Metal Additive

Reducing the SnO_x particle size from 20 to 10 nm leads to a strong increase in the sensor response as shown in Fig. 13.14a, b. The sensors existing of [SnO_x (20 nm): Pd (5 nm, 5%)] and [SnO_x (10 nm): Pd (5 nm, 5%)] were investigated towards butane and propane at operating temperatures about 350–400 °C. In general, it can be seen that the sensitivities are enhanced through the size reduction even at reduced

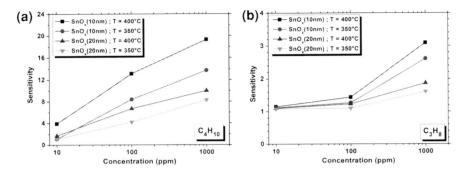

Fig. 13.14 Sensing characteristics obtained from SnO$_x$:Pd mixed sensors in butane **a** and propane **b** ambient at 350–400 °C. The SnO$_x$ particle size varies from 10–20 nm, whereas the Pd particle size was fixed (5 nm, 5 %). The layer thickness is estimated to be 1 μm

temperatures. The sensor with SnO$_x$ (10 nm) for instance yields higher sensitivities at 350 °C compared to that measured for SnO$_x$ (20 nm) at 400 °C in case of butane as target gas, increasing the concentration from 10–10^3 ppm. Same trend could be observed in propane ambient although an appreciable increase takes place at higher gas concentrations. In case of 10 ppm of propane and an operating temperature of 400 °C, the sensitivity change of the sensor made of 10 nm particles compared to that consisting of 20 nm sized particles is about 5 % (butane: 29 %). The sensors in this study show a higher response towards butane among all gas concentrations, compared to that measured for propane. Increasing the temperature from 350 up to 400 °C and a decrease in particle size from 20–10 nm leads to a maximum sensing enhancement for butane and propane varying the concentration from 10–10^3 ppm.

The decrease in particle size offers a larger effective surface area, and hence a higher generation rate of O$^-$ species at the surface. Consequently, the change in resistance increases, leading to enhanced sensitivities. The sensors prepared show a higher sensitivity towards butane with respect to that of propane.

The response and recovery times were also determined. The response time gives the required period for a sensor to respond to the target gas. It is defined as the time needed to reach 90 % (τ_{90}) of the saturated signal for a given concentration. On the other hand, the recovery time represents the time for the signal to fall below 10 % (τ_{10}) of the value measured without gas. One of the main requirements of a "good" sensor is to respond quickly to any change in an environment and to recover immediately. The found response and recovery times for the sensors above are given in Fig. 13.15a–d. It can be seen that with the rise in temperature the response times in both the sizes decreases. In case of 10 ppm butane, on reducing the size of nanoparticles from 20 to 10 nm, the response time reduces from 146 to 51 s. Similarly, for both sizes response time is observed to reduce on increasing temperature. The response time for 10 ppm butane decreases by 44 and 28 s in case of nanoparticles of sizes 10 and 20 nm, respectively. With identical operating temperature, on increasing the gas concentration from 10–10^3 ppm, the difference between response time decreases for

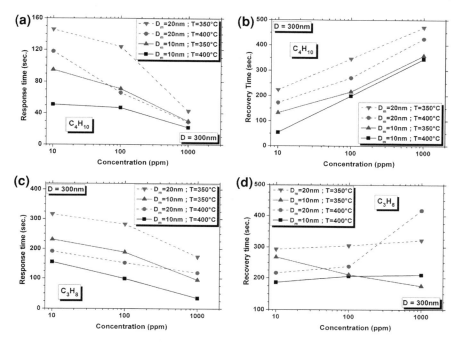

Fig. 13.15 Response and recovery times measured for sensors consisting of SnO$_x$ (20 nm) (*dashed lines*) and SnO$_x$ (10 nm) (*solid lines*) particles at different operating temperatures in butane **a**, **b** and propane **c**, **d** diluted ambient

20 and 10 nm size particles. An almost similar trend of reduction in the difference between response times has been observed on increasing the operating temperature for both the sizes with increasing gas concentration.

The recovery time of sensors decreases with the reduction in size of the nanoparticles and with the increase in the operating temperature. The sensors made of 10 and 20 nm sized particles, recovery times reduce by 78 and 51 s, respectively, with increase in temperature from 350–400 °C for 10 ppm butane. As the decrease in the recovery time is more than the decrease in response time, therefore, the sensors show better improvement in recovery time in comparison to response time. On utilising propane as a target gas, similar behaviour in response time for both 20 and 10 nm size particles has been observed as becomes clear from Fig. 13.15c. The only observed difference between butane and propane is the time scale of the response time, which is higher in case of propane. This pronounced difference is due higher sensitivity of butane towards the sensor as mentioned in Fig. 13.14. For propane, sensing the recovery time does not show any systematic trend. Only the difference in recovery times for sensors of 10 and 20 nm sized particles increases with the increase in propane concentration for both the operating temperatures 350 and 400 °C. A small increase in recovery time is observed for 20 nm sized sensor, which is operating at temperature 350 °C, on increasing propane concentration from 10–10^3 ppm.

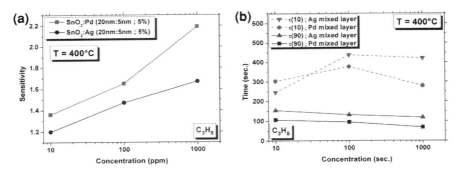

Fig. 13.16 Sensing response signal as a function of propane concentration for SnO$_x$ samples mixed with Pd and Ag nanoparticles, respectively, measured at 400 °C **a**. Belonging response (*solid line*) and recovery (*dashed line*) times **b**

Similar sensor operating at 400 °C first shows a small increase in recovery time up to 100 ppm propane concentration, after that relatively large increase in recovery time is observed. For sensor made of 10 nm-sized particles recovery time follows opposite trend when operated at 350 and 400 °C. Recovery time increases in case of operating temperature of 350 °C, whereas decreases for 400 °C with the increase in propane concentration. For the present study, sensor made of 10 nm size particles operating at 400 °C is the optimum condition for sensing behaviour of 10 ppm target gas. Therefore, the sensing properties of the mentioned sensor towards butane and propane are compared. The sensor shows faster response and recovery towards butane in comparison to propane. The response and recovery times for butane are 51 and 54 s, respectively, whereas for propane these times are 156 and 186 s, respectively. Interestingly, small difference is observed between response and recovery times for both the gases, especially in case of butane which is about 3 s.

Different noble metal additives (Pd, Ag) in nanoparticle form were added to the SnO$_x$ layers by homogeneous mixing in the synthesis route. The dependency of the sensing performance affected by the type of noble metal additive was investigated. Samples consisting of 20 nm SnO$_x$ particles with homogeneously dispersed 5 nm Pd or Ag particles and a layer thickness of 500 nm. The concentration of Pd or Ag nanoparticles is about 5 % by number concentration. Figure 13.16a compares the two different sensor signals obtained at an operating temperature of 400 °C for propane concentrations in a range from 10–10^3 ppm. The obtained response and recovery times are shown in Fig. 13.16b. It is important to note that the difference in sensitivities measured on sample SnO$_x$ (20 nm): Pd (5 nm/5 %) depicted in Fig. 13.14 to the Pd dispersed sensor here can be explained due to the different layer thicknesses.

In case of SnO$_x$:Pd mixed nanoparticle sensor, the sensing response towards propane at 400 °C is improved for all the concentrations compared to that of SnO$_x$:Ag mixed nanoparticle sensor. The improvement is increased with increasing propane concentration. Clearly, the response times in both the cases are decreased

Fig. 13.17 Sensor response at 400 °C and a butane concentration of 10^3 ppm in dry synthetic air measured over a period of 1 year. The sensors investigated are prepared of SnO_x nanoparticle layers homogeneously mixed with Pd nanoparticles (5 nm, 5%). The sizes of the SnO_x nanoparticles are 10 and 20 nm

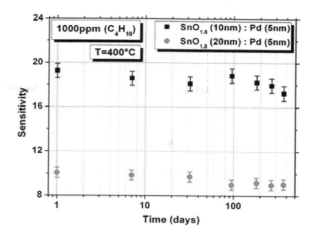

with increase in propane concentration, whereby the response time for sensor dispersed with Pd nanoparticles is decreased compared to that of the sensor dispersed with Ag nanoparticles. The recovery time for the sensor dispersed with Pd nanoparticles is enhanced towards propane concentrations higher of about 30 ppm compared to that of Ag dispersed particles. At concentrations <30 ppm, the recovery time of the sensor dispersed with Ag nanoparticles decreases more than in case of Pd dispersed sensor. In both the cases, the recovery time reaches a maximum for a propane concentration of 100 ppm. The maximum recovery times are found to be 437 s in case of Ag and 376 s in case of Pd. In conclusion, for propane concentrations <30 ppm better recovery times are achieved mixing the sensing layer with Ag nanoparticles. With rise in propane concentration, Pd nanoparticles become more effective.

13.3.6.3 Long-Term Stability

Repeatable and reproducible response is one of the most important requirements for a gas sensor. The fabricated sensors in this study are stable over a long period towards butane as shown in Fig. 13.17. The SnO_x (10 nm): Pd (5 nm/ 5 %) sensor and the SnO_x (10 nm): Pd (5 nm/ 5 %) sensor shows a sensitivity drift of 3.5 % and 4.9 %, respectively, over a period of 1 year. Every point represents a single measure made under same environmental conditions.

First, the measuring cell was heated up to 400 °C and flooded with synthetic air for 1 hour to obtain an initial resistance state of the film. After that, butane in a concentration of 10^3 ppm was inserted till the resistance value no further shows markedly changes. The ratio of both values provides the sensitivity outlined above. In effect, the drift of the resistance is much higher because the measured resistance in synthetic air was each time different and varies over a few tens to kΩ but the change

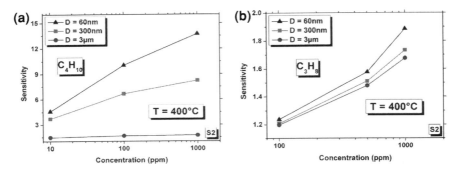

Fig. 13.18 Sensor response of sample S2 to dilute butane **a** and propane **b**, respectively, measured at 400 °C and different electrode gap sizes

in resistance differed apparently in the same manner so that at least the sensitivity drift becomes marginal.

13.3.6.4 Influence of Electrode Distance on the Layer Sensitivity

As in semiconductor gas sensors the electrical resistance of the sensing layer between the metallic electrodes is measured, it is interesting to observe what happens with the layer resistance and thereby with the sensitivity of the sensor, when the gap size between the electrodes is decreased down to a nanometer level. The effect of gap size on sensing properties to dilute ($10–10^3$ ppm) butane and propane was investigated at 400 °C. The sensor (denoted as S2) is made of a SnO_x:Pd mixed nanoparticle layer of 500 nm thickness [SnO_x (20 nm): Pd (5 nm/ 5 %)]. A "nano-gap" effect has been found as clearly shown in the measured sensor sensitivities towards butane and propane as depicted in Fig. 13.18a, b.

Propane shows less sensitivity at concentrations below 100 ppm, therefore, concentrations above 100 ppm are selected for investigation. In both the cases sensitivity tends to increase with decreasing gap sizes. In case of 10^3 ppm of butane, the sensitivity is increased from 1.7 to 13.7 in decreasing the electrode gap size from 3 μm to 60 nm. The sensitivity increases by almost 700 %. Even at a concentration of 10 ppm of butane, the sensitivity is enhanced by almost 200 %. This shows that increase in gas concentration leads to stronger enhancement in sensitivity for both the cases. In case of propane, the maximum sensitivity is measured for the electrode distance of 60 nm at 10^3 ppm. This sensitivity (1.87) is almost 12 % higher compared to that measured in 1000 ppm of propane at the distance of 3 μm. As propane is less reactive with the film, the enhancement in sensitivity is smaller compared to that in butane, as expected. A reason for the increase in sensitivity could be the decreased number of particles deposited inside the gap between the electrodes, which leads to a decrease in the number of grain boundaries resulting in a decrease in measured layer resistance. The sensor response can be divided into two different parts, one at

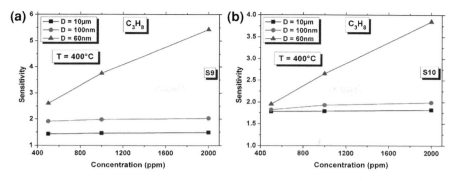

Fig. 13.19 Sensitivities upon exposure to dilute propane of sensors S9 **a** and S10 **b** at 400 °C and different electrode distances as a function of propane concentration

oxide–electrode interface (S_i) and the other at oxide–grain boundary (S_{gb}). As the response at the grain boundary is more pronounced compared to the oxide–electrode interface response, the S_{gb}/S_i ratio was found to be responsible for the gap effect [3].

13.3.6.5 Sensor to Sensor Reproducibility

The ability of producing almost equivalent sensing layers with sufficient accuracy was shown before in Sect. 13.3.5, but what about the sensor response obtained from sensors, manufactured under completely same conditions and parameters. To clarify this question, two different sensors consisting of SnO_x:Pd mixed layers [SnO_x (20 nm): Pd (5 nm/ 5 %)] and a thickness of 500 nm are prepared under same conditions (denoted as S9 and S10). Sensing response measurements to dilute propane at different electrode distances and 400 °C operating temperature are carried out. The observed sensor signals are given in Figs. 13.19a, b. Both the sensors show that an increase in propane concentration and decrease in electrode distance increases the sensitivity. For the distances of 10 μm and 100 nm, the enhancement in sensitivity remains almost constant, independent of the gas concentration.

However, a difference in sensitivity with increase of the propane concentration is observed at the distance of 60 nm for both the sensors. A great difference is observed in the measured sensitivity values on comparing the results for S9 and S10. For instance, the sensor signal in case of S9 at a concentration of 2000 ppm of propane and a distance of 60 nm is about 5.4, whereas in case of S10 sensitivity of 3.85 was observed. Switching from distance 10μm to that of 60 nm at 500 ppm propane, an increase in sensor response to about 82 % was observed in case of S9 while S10 shows an increase of only 10 %. From this one can conclude that the control in depositing the films is not enough to achieve sensors with identical sensitivity values under equal conditions, but is sufficient to be able to recognise trends in sensor behaviour when changing processing conditions, as shown in the preceding sections. A more reliable deposition process seems to be required when industrial grade sensors which do not require individual calibration have to be produced.

13.3.6.6 Dual Conductance Response of SnO_x:Pd Mixed Layers Towards CO and Ethanol

The sensing behaviour of SnO_x:Pd mixed nanoparticle layers towards different CO and ethanol ambient ($10\,ppb-10^3$ ppm) in dry synthetic air and temperatures ranging from $250-400\,°C$ has been examined in detail. In the mixed layers, the Pd nanoparticle size ($D_m = 5-15nm$) and number concentration ($0-3\,\%$) have been varied keeping the SnO_x nanoparticle size constant at $D_m = 20nm$. The variation in sensor signal, as a function of operating temperature dilute in different CO concentrations ($10\,ppb-10^3$ ppm) for three different composed samples is shown in Fig. 13.20a–c. Sample SP0 consists only of monodispersed SnO_x nanoparticles, whereas in samples SP5 and SP15, Pd particles in a concentration of $3\,\%$ and sizes of 5 nm and 15 nm, respectively, were homogeneously mixed with the SnO_x (20 nm) nanoparticles before deposition. Clearly, SP5 exhibits enhanced sensitivity as compared to SP15 due to its reduced Pd nanoparticle size. It is observed that the temperature of maximum sensing decreases, increasing the CO concentration for both the samples with added Pd nanoparticles. For 10 ppb CO, maximum sensitivity is obtained at 623K, whereas for 10^3 ppm a maximum is obtained at 523K. Nevertheless, the temperature of maximum sensitivity is always smaller in the mixed nanoparticle samples as compared to the sample without Pd (SP0), at all concentrations. Further, a decrease in the sensitivity of the Pd loaded samples can be observed, with increasing CO concentrations. In case of SP5, the sensor response decreases from 6.5 to 3 with increase in CO concentration from 10 ppb up to 10^3 ppm. At low CO concentrations, the reaction rate is controlled by CO adsorption, whereas at higher concentrations it is controlled by dissociative oxygen adsorption, which is significantly detained by the CO adsorption, leading to lower reactions at high CO concentrations.

Figure 13.20d shows a comparison of the lowest detected CO concentration reported in the literature and the belonging sensitivities, with those found in this study. A lowering of the detectable concentration towards CO as well as a substantial increase in sensitivity is observed in sample SP5. Figure 13.20e shows the measured response of sample SP5 at 623K in 1 ppm ethanol/CO and synthetic air for four consecutive cycles. Whereas in CO the normalised conductance (R_a/R_g) decreases (p-type response), in ethanol it increases (n-type response). It was expected that in both the cases n-type response should appear, both gas species being reducing ones. Interestingly, a dual inductance response could be observed. At temperatures $\geq 573K$ Pd oxidation is possible in oxygen [37]. Pd particles $\leq 10\,nm$ show incomplete oxidation with interface oxide, required for the ion transport from the particle boundary along the metal–oxide interface. As the samples in this study were initially annealed in synthetic air, the presence of surface or interface PdO could be possible. Therefore, detailed XPS studies have been carried out on samples SP0, SP15 directly after pretreatment in synthetic air, followed by CO or ethanol (100 ppm)

13 Gas Sensors Based on Well-Defined Nanostructured Thin Films 351

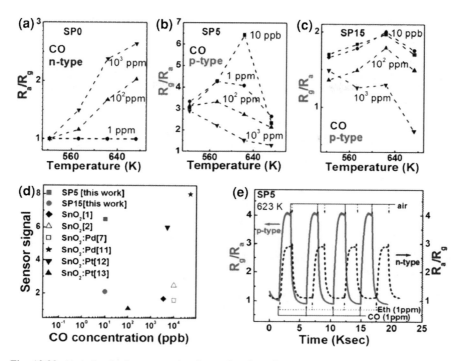

Fig. 13.20 Variation in the sensor signal as a function of temperature and CO concentration in dry air for SnO$_x$ nanoparticle sample **a** SP0 and SnO$_x$:Pd mixed nanoparticle samples **b** SP5 and **c** SP15. **d** Comparison of the lowest detected CO concentrations (in dry synthetic air) reported in the literature and the corresponding sensor signal. **e** Variation of sensor signal for sample SP5 as a function of exposure time in four cycles in 1 ppm of ethanol (*dotted curve*)/CO (*solid curve*) and synthetic air (air) at 623K. Reprinted with permission from [39]

exposure at 623K for 30 min. In Fig. 13.21, the normalised Pd $3d$ core level spectra for SP15, pretreated in synthetic air (Fig. 13.21a) and after exposure to CO (Fig. 13.21b) are shown. The sample pretreated in synthetic air reveals mainly PdO (336.6eV, dashed line) and in a small contribution of Pd (335.4eV, solid line). Upon exposure to CO, only a Pd contribution is observed. The interfacial/surface PdO formed upon exposure to synthetic air is continuously consumed and reformed upon CO exposure (Mars van Krevelen mechanism). In situ time resolved infrared reflection absorption spectroscopy measurements on SP15 reveal a shift in the Fermi level towards the valence band edge upon CO exposure (i), whereas an opposite directed shift is observed upon exposure to ethanol (ii). These shifts correspond to an increase (i) and decrease (ii) in work function [38] resulting in an increase or rather decrease in surface resistance accordingly to the p-type response in CO and n-type response in ethanol. More details can be found elsewhere [39, 40].

Fig. 13.21 Normalised Pd 3d core level spectra of sample SP15, pretreated in **a** synthetic air and after exposure to **b** CO (100 ppm in synthetic air). The vertical solid lines correspond to the binding energy positions for metallic Pd and the dotted lines to that of PdO. Reprinted with permission from [39]

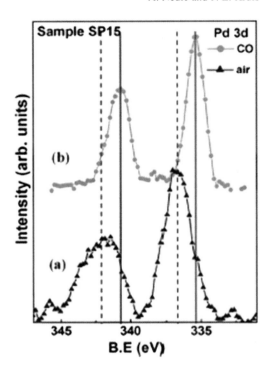

13.3.7 Pure Pd-Nanoparticle Layers for Concentration Specific H_2 Sensing at Room Temperature

Monosized, quasispherical and monocrystalline Pd nanoparticles with geometric mean diameter of 15 nm and a geometric standard deviation of $\sigma_g = 1.05$ (obtained from TEM results) were synthesises and deposited on glass substrates with a thickness of about 100 nm by means of LPI. The sensing response of such a Pd layer towards hydrogen has been studied as a function of H_2 concentration and operating temperature. Two types of sensing response are observed as shown in Fig. 13.22a, b. In Fig. 13.22a, insertion of hydrogen in a concentration of 5 % leads to a sharply increase in layer resistance followed by an abruptly decrease and subsequently saturation, whereas in Fig. 13.22b the normally in gas sensing observed saturated response with increase in resistance appears for a hydrogen concentration of 0.5 %. Both the sensing behaviours are stable and reproducible over several hydrogen incorporation cycles at room temperature. Figure 13.22c, d shows the different steps in both types of behaviour. The resistance increase between the points a and b is due to an electronic effect (EE) and the resistance decrease in between points b and c is due to a geometric effect (GE) caused by lattice expansion when the hydride is formed. The absence of GE for H_2 concentrations of 0.5 % leads to the saturated response behaviour. Adsorption of hydrogen on the Pd surface leads to a dissociation

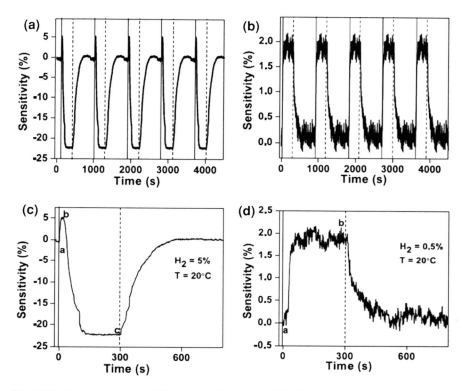

Fig. 13.22 Sensing response of Pd nanoparticle layers at 5 % of hydrogen concentration (**a** and **c**) and 0.5 % (**b** and **d**), respectively, both measured at room temperature. The solid lines represent the H$_2$ "on", whereas the dashed lines represent the H$_2$ "off" state. Sensitivity is defined as the percentage change in resistance [$(R_f - R_i) \times 100/R_i$] in which $i = a$ and $f = b$ for EE and $i = b$ and $f = c$ for GE, as depicted in **c** and **d**, respectively. Reprinted with permission from M. Khanuja, S. Kala, B.R. Mehta and F.E. Kruis, Concentration-specific hydrogen sensing behaviour in monosized Pd nanoparticle layers, *Nanotechnology* **20** (2009), p. 015502 (7pp.), Copyright [2009], Institute of Physics Publishing Ltd

of molecular to atomic hydrogen followed by incorporation of the atomic hydrogen at the Pd surface [41]. This very fast effect requires smallest traces of H at the Pd surface. A decrease in electron mobility leads to a resistance increase, which is equivalent to the pronounced EE. Incorporating of higher H concentrations leads to the formation of PdH, leading to an increase in lattice constant of Pd by about 6 % [41]. As the thermally activated hopping across the interparticle barrier is the dominant electron transport mechanism [42], the PdH formation decreases this barrier due to the interparticular gap reduction and therefore leads to a decrease in resistance. This effect is understood as GE. The EE will occur first and GE will follow if enough H was offered. If GE is present it will always succeed the EE.

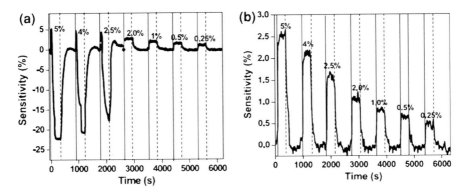

Fig. 13.23 Sensing response at **a** room temperature and **b** 80 °C for H_2 concentrations ranging from 5–0.25 %. Solid and dotted lines show the H_2 "on" and "off" states, respectively. Reprinted with permission from M. Khanuja, S. Kala, B.R. Mehta and F.E. Kruis, Concentration-specific hydrogen sensing behaviour in monosized Pd nanoparticle layers, *Nanotechnology* **20** (2009), p. 015502 (7pp.), Copyright [2009], Institute of Physics Publishing Ltd

The sensing behaviour measured at different H_2 concentrations operated at temperatures of 20 [(room temperature) (Fig. 13.23a)] and 80 °C (Fig. 13.23b) are shown in Fig. 13.23.

At 20 °C and H_2 concentration ≥ 2.5 %, pulsed response is observed, whereas for concentration ≤ 2 % saturated response can be found. At 80 °C the pulsed response vanishes completely. At all H_2 concentrations (0.25–5 %) saturated response is observed. By this, it is evident that H_2 concentrations ≤ 2 % are insufficient to cause dimensional nanoparticle changes required for GE. The sensitivity and response time due to GE is found to be larger in comparison to EE at all H_2 concentrations. A more detailed report can be found in [27].

References

1. T. Seiyama, A. Kato, K. Fujiishi, M. Nagatani, A new detector for gaseous components using semiconductive thin films. Anal. Chem. **34**, 1502–1503 (1962)
2. P.T. Moseley, B.C. Tofield (eds.), *Solid State Gas Sensors* (Adam Hilger, Bristol, 1987)
3. K. Ihokura, J. Watson. The Stannic Oxide Gas Sensor—Principles and Applications (CRC Press, Boca Raton, 1994)
4. M.J. Madou, S.R. Morrison, *Chemical Sensing with Sold State Devices* (Academic Press, New York, 1989)
5. K. Ihokura, Application of sintered tin (IV) oxide for gas detector, NTG—Fachberichte. **79**, 312–317 (1982)
6. L.I. Poopova, M.G. Michailov, V.K. Gueorguiev, Structure and morphology of thin SnO_2 films. Thin Solid Films **186**, 107–112 (1990)
7. S. Nicoletti, L. Dori, G. Cardinali and A. Parisini, Gas sensors for air quality monitoring: realisation and characterisation of undoped and noble metal-doped SnO_2. Thin sensing films deposited by the pulsed laser ablation. Sens. Actuators B **60**, 90–96 (1999)

13 Gas Sensors Based on Well-Defined Nanostructured Thin Films

8. J. Zhang, L. Gao, J. Solid State Chem. **177**, 1425 (2004)
9. S.R. Morrison, Mechanism of semiconductor gas sensor operation. Sens. Actuators **11**, 283–7 (1987)
10. P. Nelli, G. Faglia, G. Sberveglieri, E. Cereda, G. Gabetta, A. Dieguez, A. Romano-Rodriguez, J.R. Morante, The aging effect on SnO_2-Au thin film sensors: electrical and structural characterization. Thin Solid Films **371**, 249–253 (2000)
11. N. Yamazoe, Y. Kurokawa, T. Seiyama, Effects of additives on semiconductor gas sensors. Sens. Actuators **4**, 283–289 (1983)
12. N. Yamazoe, New approaches for improving semiconductor gas sensors. Sens. Actuators B **5**, 7–19 (1991)
13. C. Xu, J. Tamaki, N. Miura, N. Yamazoe, Stabilization od SnO_2 ultrafine particles by additives. J. Mater. Sci. **27**, 963–971 (1992)
14. R. Ramamoorthy, M.K. Kennedy, H. Nienhaus, A. Lorke, F.E. Kruis, H. Fissan, Surface oxidation of monodisperse SnO_x nanoparticles. Sens. Actuators B **88**, 281–285 (2003)
15. M. I. Ivanovskaya, P.A. Bogdanov, D.R. Orlik, A.Ch. Gurlo, V.V. Romanovskaya, Structure and properties of sol–gel obtained SnO_2 and SnO_2-Pd films. Thin Solid Films. **296**, 41–43 (1997)
16. S. Harbeck, A. Szatvanyi, N. Barsan, U. Weimar, V. Foffmann, DRIFT studies of thick film un-doped and Pd-doped SnO_2 sensors: temperature changes effect and CO detection mechanism in the presence of water vapour. Thin Solid Films **436**, 76–83 (2003)
17. R. K. Joshi, F. E. Kruis, O. Dmitrieva, Gas sensing behavior of $SnO_{1.8}$: Ag films composed of size-selected nanoparticles, J. Nanoparticle Res. **8**, 797–808 (2006)
18. R. K. Joshi, F. E. Kruis, Size-Selected $SnO_{1.8}$: Ag Mixed nanoparticle films for ethanol, CO and CH_4 detection, J. Nanomaterials ID67072 (2007)
19. H. Nienhaus, V. Kravets, S. Koutouzov, C. Meier, A. Lorke, H. Wiggers, M. K. Kennedy, F. E. Kruis, Quantum size effect of valence band plasmon energies in Si and SnO_x nanoparticles. J Vac. Sci. Technol. B **24**, 1156 (2006)
20. P. Biswas, R.C. Flagan, High-velocity inertial impactors. Environ. Sci. Technol. **18**, 611–616 (1984)
21. M.K. Kennedy, F.E. Kruis, H. Fissan, B.R. Mehta, S. Stappert, G. Dumpich, Tailored Nanoparticle Films from Monosized Tin Oxide Nanocrystals: Particle Synthesis, Film Formation, and Size-dependent Gas-sensing Properties, J. Applied Physics, 93, pp. 551–560, (2003)
22. M. K. Kennedy, F. E. Kruis, H.Fissan, and B. R. Mehta, Fully Automated, Gas Sensing, and Electronic Parameter Measurement Setup for Miniaturized Nanoparticle Gas Sensors. Rev. Sci. Instrum. **74**(11), 4908–15 (2003)
23. W.S. Hu, Z.G. Liu, J.G. Zheng, X.B. Hu, X.L. Guo, Preparation of nanocrystalline SnO_2 thin films used in chemisorption sensors by pulsed laser reactive ablation, J. of Materials Science: Materials in. Electronics **8**, 155–158 (1997)
24. R. Dolbec, M.A. El Khakani, Sub-ppm sensitivity towards carbon monoxide by means of pulsed laser deposited SnO_2: Pt based sensors. Appl. Phys. Lett. **90**(17), 173114 (2007)
25. R. Dolbec, M.A. El Khakani, Pulsed laser deposited platinum and gold nanoparticles as catalysts for enhancing the CO sensitivity of nanostructured SnO_2 sensors. Sens. Lett. **3**, 216–221 (2005)
26. N. Barsan, U. Weimar, Conduction Model of Metal Oxide Gas Sensors. J. Electroceramics **7**, 143–167 (2001)
27. M. Khanuja, S. Kala, B. R. Mehta, F. E. Kruis, Concentration-specific hydrogen sensing behavior in monosized Pd nanoparticle layers, Nanotechnology, **20**, 015502 (7 pp) (2009)
28. G. Heiland, Homogeneous semiconducting gas sensors. Sens. Actuators **2**, 434–361 (1982)
29. J. Watson, K. Ihokura, Coles G.S.V, The tin oxide gas sensor, Measurement Sci. Technol. **4**(7), 711–719 (1993)
30. R.K. Joshi, F.E. Kruis, Influence of Ag particle size on ethanol sensing of $SnO_{1.8}$: Ag nanoparticle films: a method to develop parts per billion level gas sensors. Appl. Phys. Lett. **89**, 153116 (2006)

31. C. Xu, J. Tamaki, N. Miura, N. Yamazoe, Grain size effects on gas sensitivity of porous SnO_2-based elements. Sens. Actuators B **3**, 147–155 (1991)
32. C. Xu, J. Tamaki, N. Miura, N. Yamazoe, Correlation between Gas Sensitivity and Crystallite Size in Porous SnO_2-Based Sensors, Chem. Lett. 441–444 (1990)
33. N. Barsan, Conduction models in gas-sensing SnO_2 layers: grain-size effects and ambient atmosphere influence. Sens. Actuators B **17**, 241–246 (1994)
34. B. Gautheron, M. Labeau, G. Delabouglise, U. Schmatz, Undoped and Pd-doped SnO_2 thin films for gas sensors. Sens. Actuators B **15–16**, 357–362 (1993)
35. G. Sakai, N. Matsunaga, K. Shimanoe, N. Yamazoe, Theory of gas-diffusion controlled sensitivity for thin film semiconductor gas sensor. Sens. Actuators B **80**, 125–131 (2001)
36. J. Tamaki, Y. Nakataya, S. Konishi, Micro gap effect on dilute H_2S sensing properties on SnO_2 thin film microsensors. Sens. Actuators B **130**, 400–404 (2008)
37. W. Prost, F.E. Kruis, F. Otten, K. Nielsch, B. Rellinghaus, U. Auer, A. Peled, E.F. Wassermann, H. Fissan, F.J. Tegude, Microelectron. Eng. **41–42**, 535 (1998)
38. Y.-J. Lin, C.-L. Tsai, J. Appl. Phys. **100**, 113721 (2006)
39. I. Aruna, F.E. Kruis, S. Kundu, M. Muhler, R. Theissmann, M. Spasova, CO ppb sensors based on monodispersed SnO_x:Pd mixed nanoparticle layers: Insight into dual conductance response. J. Appl. Phys. **105**, 064312 (2009)
40. I. Aruna, F. E. Kruis, Temperature dependent sensitivity inversion in SnO1.8: Pd mixed nanoparticle layer based CO sensors, Mater. Res. Soc. Symp. Proc. **1056**, HH04-11
41. F. A. Lewis, *The Palladium-Hydrogen System* (Academic Press, London, 1967)
42. J.B. Pelka, M. Brust, P. Glertowski, W. Paszkowicz, N. Schell, Appl. Phys. Lettt. **89**, 063110 (2006)

Chapter 14
III/V Nanowires for Electronic and Optoelectronic Applications

Christoph Gutsche, Ingo Regolin, Andrey Lysov, Kai Blekker, Quoc-Thai Do, Werner Prost and Franz-Josef Tegude

Abstract III/V semiconductor nanowires are grown by the vapour–liquid solid growth mode from Au seed particles in an industrial type metal–organic vapour phase epitaxial apparatus. For electronic applications InAs nanowires with very high electron were developed on InAs (111), InAs (100), and GaAs (111) substrates. The wires were deposited on insulating host substrate for metal–insulator–semiconductor FET fabrication. Their excellent DC and RF performance are presented. For optoelectronic applications the focus is on selective n- and p-type doping. GaAs nanowires with an axial p–n junction are presented. Pronounced electroluminescence at room temperature reveals the quality of the fabricated device. Moreover, spatially resolved photocurrent microscopy shows that optical generation of carriers took place only in the vicinity of the p–n junction. A solar conversion efficiency of 9 % was obtained. In summary, III/V semiconductor nanowires are emerged to high performance and versatile nanoscaled building blocks for both electronic and optoelectronic applications.

14.1 Introduction

Nanoparticles give access to the high volume exploitation of the nanoscale. They offer a wide variety of materials with a precise size control at very low cost. One powerful application of nanoparticles is their use for the fabrication of nanoscaled structures and devices. A very promising example is growth of crystalline

---

C. Gutsche · I. Regolin · A. Lysov · K. Blekker · Q.-T. Do · W. Prost (✉) · F.-J. Tegude
Faculty of Engineering, Solid-State Electronics Department,
University of Duisburg-Essen, Lotharstraße 55,
47057 Duisburg, Germany
e-mail: werner.prost@uni-due.de

Q.-T. Do
Wacker Siltronic, Burghausen, Germany

A. Lorke et al. (eds.), *Nanoparticles from the Gas Phase*, NanoScience and Technology, DOI: 10.1007/978-3-642-28546-2_14, © Springer-Verlag Berlin Heidelberg 2012

semiconductor needles based on nanoparticle seeds defining the diameter of the needle in the nm range. These structures are widely named nanowires.

Nanowires are extremely fast epitaxially growing nanostructures with an almost perfect crystal structure (cf. Sect. 14.2). Their diameter range is starting at a few nanometers and may go up to some micrometer; their length can easily extend from tens up to hundreds of microns. The growing surface of nanowires, the diameter of the nanowire is that small that even a very large lattice mismatch to the growth substrate can be accommodated within the first atomic layers of the growing nanowire [1]. Therefore, new material combinations with huge lattice mismatch becomes feasible and open up new applications. This aspect is very important for the use of III/V semiconductors which are uniquely suited for:

(1) high carrier mobility electronic devices
(2) efficient transformation, both from electrical current-to-light and vice versa.

Based on the nanowire approach, the high carrier mobility semiconductor InAs can be grown or deposited on any substrate including Silicon. A field-effect transistor using an InAs nanowire as a channel hold the promise to overcome some of the most stringent limitations due to the more than ten-fold higher carrier mobility compared to Silicon CMOS (cf. Sect. 14.3).

The light/current conversion efficiency and the high material consumption of expensive materials are the limiting factors for room lighting and solar cell applications. Nanowires with an extreme surface-to-volume ratio are very promising candidates to provide high electrical/optical output power with reasonable conversion efficiency but at record low material consumption and weight. In Chap. 3, the doping of nanowires is presented as a difficult but absolutely necessary prerequisite for light-to-current conversion. In addition spatially resolved photoluminescence and electroluminescence data give detailed insights in high power, low volume conversion of current, and light.

14.2 Growth of III/V Nanowires

The growth of nanowires started back in the 1950s based on metallic particle seeded mercury [2] and moved forward to epitaxial semiconductor silicon needles in the 1960s. Thereby, Wagner and Ellis [3] initiated the vapour–liquid–solid (VLS) growth mode and the corresponding model was substantially pushed forward by Givargisov [4] in 1975. The device oriented growth of III/V semiconductor nanowires in a commercial metal–organic vapour-phase epitaxy apparatus (MOVPE) was started by Hiruma and coworkers in the 1990s [5]. The current nanowire research was stimulated by the results from Liebers group around 2000 [6]. Since then basically, all types of modern III/V growth apparatus like molecular beam epitaxy [7], chemical beam epitaxy [8], and MOVPE [5] have been adopted for nanowire growth. Besides the VLS mechanism used here, there are other methods like the vapour–solid–solid, the oxide-assisted, as well as the selective area growth.

14.2.1 Vapour–Liquid–Solid Growth

The initial element for the VLS mechanism for nanowire growth is a metallic seed nanoparticle. This work is restricted to the widely used gold (Au) seeds. The Au nanoparticle can be fabricated by various means such as:

(1) aerosol deposition of monodisperse nanoparticles,
(2) deposition of a thin Au-film with subsequent annealing,
(3) deposition of colloidal Au nanoparticles,
(4) lithography such as electron beam lithography, stamp lithography, and various kinds of templates.

The choice of the method shall be made according to the application in mind. The more control and flexibility about size, density, and position are needed, the higher are the efforts; i.e., electron beam lithography gives the ultimate control and flexibility at the highest effort level while the annealing of a thin Au-film results in dense, but polydisperse Au nanoparticles basically without any size and position control but lowest efforts and costs. Within the range of detection limits no impact of the various methods was found on the crystalline quality or the purity of the nanowire material.

Prior to growth the Au seed nanoparticle has been deposited on a crystalline substrate that defines the crystal of the nanowire. In Fig. 14.1, the principles of GaAs nanowire growth are depicted. A (111)B oriented GaAs substrate is selected for orthogonal grown free standing GaAs wires because the nanowires favor to grow in (111)B direction. Prior to growth the Au seed particle on the GaAs substrate is annealed $T_A \cong 600\text{--}650\,^\circ\text{C}$ at in order to achieve an Au/Ga alloy due to solid diffusion from Ga atoms from the substrate into the seed particle. Next the temperature is ramped down to the growth temperature of about $T = T_G \approx 400\,^\circ\text{C}$ down to $450\,^\circ\text{C}$. At temperatures below $491\,^\circ\text{C}$, the solvable amount of Ga in the seed particle is limited as depicted in the simplified phase diagram of Fig. 14.1b. Ga-adatoms reaching the Au-seed will result in a supersaturation such that Ga will fall out into the solid phase. This effect results in the growth of the nanowire underneath the Au-seed.

Ga adatoms hitting the substrate surface in the vincinity of the Au-seed within the surface diffusion length may reach the Au-seed, too. This way a much higher growth rate can be obtained in comparison to standard layer growth conditions. The model in Fig. 14.1 describes the route of the Ga-adatoms. The path of the arsenic is subject of current discussion. An excess availability of As at the growing interface is assumed that, however, does not diffuse through the Au-seed particle but gets incorporated via diffusion along the liquid–solid interface underneath the seed droplet. Givargizov [4] described the growth process by the following four main steps (cf Fig. 14.1a): mass transport in the gas phase (1), chemical reaction at the interface (2), diffusion in the liquid phase (3), and incorporation in the solid crystal (4). Today's models [9, 10] consider also the absorption on the whole sample surface (5), as well diffusion processes along the facets towards the nanowire top (6). The possibility of additional layer growth on the side facets (7) as well as on the substrate surface (8) is also

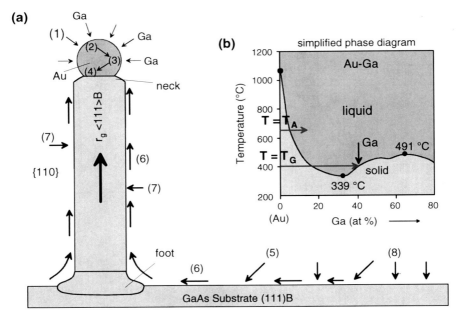

Fig. 14.1 Principal description of Vapour–Liquid–Solid growth of a GaAs nanowire based on a Au seed nanoparticle. **a** surface adhesion and mobility of Ga-adatoms for GaAs nanowire growth, and **b** simplified Au–Ga phase diagram

included. The tapered nanowire foot and the typical neck regions are formed during temperature reduction. During cool down, the solubility in the Au particle decreases, resulting in shrinking of the droplet size and therefore a smaller foot and/or nanowire diameter.

The creation of axial heterojunctions in III/V nanowires can be realized generally in two different ways, changing the group III or group V element. The change of the group III element, which is definitely stored in the Au droplet, prevents the creation of atomically abrupt heterojunctions. A group V change in general leads to sharp junctions, demonstrated by published results on GaAs/GaP [11], as well as on InAs/InP [12] heterojunctions. Besides, axial material changes it is also possible to create radial (core–shell) nanowire heterojunctions, with the defined combination of both, VLS and conventional growth [13]. After successful nanowire growth in the VLS mode, the conditions have to be switched to conventional growth, enabling the formation of defined shells around the pre-grown nanowire structures. It should be kept into account that the growth on different facets with their different orientations can lead to inhomogeneous growth results. In addition, the lattice mismatch can be neglected no more due to the enlarged interface area compared to axial heterojunctions.

The controlled doping of nanowires is a substantial challenge. Up to now it is not clear how dopants are incorporated into the nanowire structure during VLS growth. Depending on the dopant species the atoms could be incorporated through the seed particle, via diffusion through the nanowire sidewalls or along the liquid–solid interface underneath the seed particle. In addition, the determination of the doping concentration is a difficult task. Conventional methods like van der Pauw Hall cannot be transferred to nanowire structures. Up to now there are only few publications describing controlled doping of different III/V nanowires [14–16]. Moreover, an apparent n- or p-conduction may also be generated by unintended core–shell structures, whereas the conductivity is only provided by grown shell.

Growth experiments for this work were carried out on (111)B GaAs or (001) InAs substrates at a total pressure of 50 mbar in a MOVPE apparatus (AIX200 RF) with a fully non-gaseous source configuration [17]. The total gas flow of 3.4 l/min was provided by N_2 as the main carrier gas, while H_2 was used for the precursors. Trimethyl gallium (TMGa), Trimethyl indium (TMIn), and tertiary butyl arsine (TBAs) are used as precursors. The investigated doping sources for p-type doping are diethyl zinc (DEZn), or carbon tetra bromide (CBr$_4$), and for n-type tetraethyl tin (TESn) or ditertiarybutyl silane (DitBuSi).

14.2.2 InAs

InAs is a low band gap material (0.35 eV) with a high mobility for electrons. Because of the low band gap, the intrinsic carrier concentration of InAs is nearly nine orders of magnitude higher compared to GaAs for example, leading to a much lower electrical resistance [18]. Furthermore, the Fermi level at the InAs surface is pinned in the conductance band leading to an accumulation of electrons at the surface, enabling a high conductivity in the smallest wires. Therefore, InAs is a very promising candidate for low voltage, low power, and high speed electronic devices (cf. Sect. 14.3).

Prior to growth, commercial colloidal Au nanoparticles are deposited on the InAs (001) substrate as seed particles. The Au nanoparticles are monodisperse with a small size distribution. In our experiments, nanoparticles of 30–100 nm diameter were used. The InAs surface is hydrophobic which substantially reduces their adhesion. Therefore, no wafer spinning but a hot plate was used in order to dry the surface. Unfortunately, the hot plate process results in a more inhomogeneous distribution of nanoparticles compared to the spinning. After loading into the low pressure MOVPE system, the samples were annealed at 620 °C in order to form Au–In composite nanoparticles. The annealing was carried out under TBAs flow to prevent InAs substrate surface decomposition. After annealing, the temperature was ramped down to a growth temperature of 480 °C $\leq T_g \leq 400$ °C. TMIn was used as group III precursor and TBAs for group V at a constant V/III ratio of six. The nanowire diameter is equal to the monodisperse nanoparticle size which may be selected in a wide range. Sometimes, seed particles may partially merge with each other during the annealing step giving rise for higher diameter seeds and nanowires. In the following, the VLS growth mode in a low pressure MOVPE is described in more detail as a common example of InAs nanowires growth.

At a growth temperature of 480 °C, the decomposition of TMIn is still quite effective and results in an additional conventional epitaxial layer growth on the substrate and on the side walls resulting in tapered nanowires. If the growth temperature is reduced to 400 °C the tapering effect is suppressed. VLS growth is generally preferred in (111)B directions such that a (111)B oriented substrate is needed for perpendicular growth. As an exception on InAs (001) substrate both perpendicular and a tilted growth is observed. The maximum growth rate of 1.3 μm/min is achieved at $T_G = 420$ °C. At the typical growth temperature for devices ($T_G = 400$ °C) the rate is reduced to 0.8 μm/min which is still about 40 times higher than in conventional layer growth without any degradation of the crystal. The carrier mobility in unintentionally doped n-type InAs nanowires exceeds 10,000 cm^2/Vs [19]. Under the presented conditions the growth under DiTBuSi supply as a possible n-type dopant has shown no impact on transport data.

The crystalline quality of the InAs nanowires is investigated by transmission scanning electron microscopy [20]. A bright-field image of an individual InAs nanowire and its corresponding electron diffraction pattern and HRTEM micrograph are shown in Fig. 14.2a–c, representing tens of these observed (001) oriented InAs nanowires. The identified (1-10) zone axis and these sharp diffraction spots indicate the cubic zinc blende structure of the NWs and their high crystalline nature, respectively. By checking the image contrast, diffraction patterns, and HRTEM micrographs of different parts of the entire nanowire, it is confirmed that the nanowire is stacking-faults-free and (111)-twinning-free. The highly defect-free state of the InAs NWs is of particular importance for exploiting fabrication of nanosized electronic devices. More detailed investigations on InAs nanowire growth can be found elsewhere [21, 22].

14.2.3 GaAs

14.2.3.1 p-Type Doping

A sufficiently high doping of nanowires at the relatively low growth temperatures ($\sim$400 °C) turns out to be a difficult task. In this chapter, we will report on controlled p-type doping of GaAs nanowires using DEZn as precursor. In all experiments, the initial growth starts with a nominally undoped nanowire stump at 450 °C for better nucleation process. The DEZn dopant to TMGa (II/III ratio) was varied in the range $0.0008 \leq$ II/III ≤ 0.008 while the V/III ratio was kept constant at 2.5. Since the nanowires were grown untapered, we can exclude the formation of an additive shell around the VLS grown nanowire. Using a transport model which includes a space charge depletion region at the wire surface, the carrier concentration of doped GaAs:Zn wires could be determined in a wide conductivity range. As growth seeds both, monodisperse and polydisperse Au nanoparticles were deposited prior to growth. Monodisperse nanoparticles with a diameter of 150 nm were taken from a colloidal solution. Polydisperse nanoparticles were produced by annealing of a thin

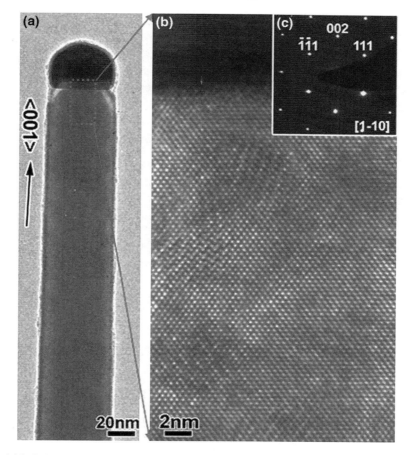

Fig. 14.2 InAs nanowire grown on an InAs (001) substrate: **a** A low magnification bright-field TEM micrograph of a perpendicular grown nanowire, **b** high resolution image at the InAs/Au interface, **c** electron diffraction pattern [20]

evaporated Au layer of nominally 2.5 nm thickness. The anneal step was carried out at 600 °C for 10 min under group V overpressure and results in nanoparticles with diameters from 30 nm to some 100 nm. First, an undoped nanowire stump is grown for 3 min at a growth temperature of 450 °C. Next, the growth temperature is reduced to 400 °C which suppresses the conventional layer growth on the side facets [23] leading to a very high aspect ratio up to $g_{r,VLS}/g_{r,VS} > 1,000$ for the upper DEZn doped part of the wire.

Figure 14.3a–d shows scanning electron microscopy micrographs of GaAs nanowires. The wire given in (a) is grown from a colloidal Au seed nanoparticle with 150 nm diameter and is non-intentionally doped. This sample is used as a reference. Nanowires grown from polydisperse seed particles, formed by annealing of

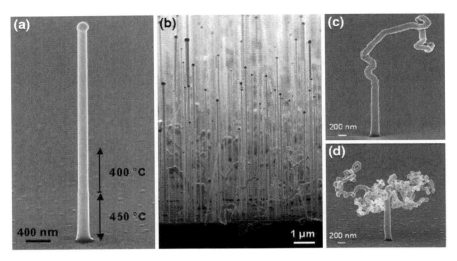

Fig. 14.3 Scanning electron micrographs of GaAs nanowires grown on GaAs (111)B substrate at 400 °C with an initial stump grown at 450 °C **a** from colloidal Au nanoparticles with 75 nm radius and **b** grown with a II/III ratio of 0.002 from polydisperse seed particles formed by annealing of a 2.5 nm Au layer [24]. **c, d** The impact of increased DEZn supply on nanowire morphology [25]

a 2.5 nm thick Au layer, are depicted in Fig. 14.3b. This approach provides a wide range of wire diameters. In the first series of experiments we studied the effect of DEZn flow on the properties of the nanowire. We observed good structural properties up to a II/III ratio of 0.004 while at higher II/III ratio wire kinking (Fig. 14.3c) and at even higher (II/III = 0.067) a seed splitting was observed (Fig. 14.3d) [24, 25]. The number of structural defects increased with wire radius and II/III ratio, respectively. However, below a II/III ratio of 0.004, nanowires up to 350 nm show no kinking effects, and are therefore best suited for further investigations. Most of the wires grown at II/III = 0.008 show kinking effects, enabling an only limited analysis.

For electrical measurements, the nanowires were scratched off the growth substrate and transferred to a carrier. The carrier consists of a semi-insulating GaAs substrate which was covered with 300 nm silicon nitride (SiN_x) for improved isolation. The ohmic contacts were formed by evaporation of Pt (5 nm)/Ti (10 nm)/Pt (10 nm)/Au (400 nm). All patterning was done with electron beam lithography, evaporation, and liftoff. If no DEZn is supplied during growth, the current is in the pA range. The *I–V* characteristic of a GaAs nanowire grown with DEZn at II/III = 0.004 is perfectly ohmic, exhibiting a high current of 400 μA at 1 V applied bias ($d = 150$ nm). A possible side wall diffusion effect attributed to DEZn in the gas phase was studied using nominally undoped wires which were exposed to DEZn under TBAs overpressure at $T_{Diff} = 400$ or 500 °C for 20 min. At these temperatures, no increase of the wire conductivity could be observed. Therefore, we conclude that the incorporation of the doping atoms happens via the Au seed particle or via the seed/wire interface.

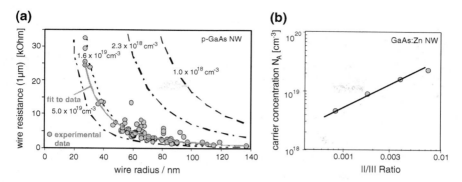

Fig. 14.4 DEZn doped p-GaAs nanowires: **a** measured wire resistances normalized to one micron length versus the wire radius for II/III = 0.004, the fitted trend line and additionally modeled data for different acceptor concentrations (*dashed lines*). **b** carrier concentration as function of the II/III ratio (based on electrical measurements of about 80 single nanowires per sample) [24]

The electrical conductivity of p-GaAs:Zn nanowires and thus the carrier concentration of a large number of nanowires with various radii ($20\,\text{nm} < r_0 < 150\,\text{nm}$) were analyzed. We used transmission line structures with three, four, or five contacts and different contact spacing's along the wires, leading into a statistical evaluation. Exemplary the corresponding experimental wire resistances for a II/III ratio of 0.004, normalized to a contact spacing of $L = 1\,\mu\text{m}$ are depicted in Fig. 14.4a. By fitting the data to our model to about 80 single nanowires for each samples grown under different Zn supplies, we were empirically able to demonstrate a linear dependence between carrier concentration and II/III ratio in p-GaAs:Zn nanowires (Fig. 14.4b) [24].

The use of DEZn implies a p-type conductivity of the wire with a higher source efficiency compared to standard layer growth. To verify the type of doping and the transport model accuracy, metal–insulator field-effect transistor (MISFET) devices were fabricated [26] (see also Sect. 14.3). Figure 14.5 shows the transfer characteristic of a GaAs nanowire MISFET grown with a II/III ratio of 0.004. The transfer characteristic exhibits typical p-channel behaviour as the channel conductance increases with negative gate bias. This experiment proves the p-type doping effect of DEZn. Based on the analysis of the drain current I_D and the transconductance g_m of the MISFET, the transport data of a DEZn doped nanowire are estimated as follows: $\mu = 90\,\text{cm}^2/\text{Vs}$ at a hole concentration of $1.7 \times 10^{19}\,\text{cm}^{-3}$. This is in excellent agreement with the values evaluated from the transport model [24].

Although DEZn was injected into the reactor chamber simultaneously with TMGa, we noticed a delayed Zn incorporation. Compared to the TMGa the DEZn partial pressure is very low. Therefore, we suggest that the saturation of the Au particle with Zn proceeds slowly. The resistance and carrier concentration along the wire are depicted in Fig. 14.6 for two single nanowires grown at II/III ratios of 0.0008 and 0.002, respectively. It is evident that it takes a few microns wire length till constant

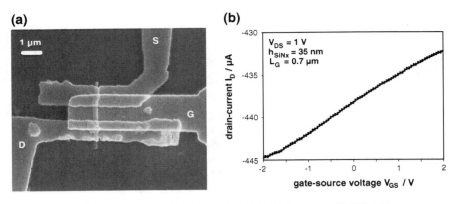

Fig. 14.5 SEM image (**a**) and transfer characteristic (**b**) of fabricated MISFET. [24]

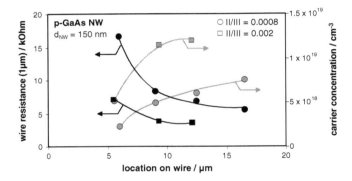

Fig. 14.6 Carrier concentration and wire resistance along the wire for two single nanowires grown at II/III ratios of 0.0008 and 0.002, respectively [24]

values are reached. For the lowest used DEZn supply these variations extend along the whole wire.

A possible explanation for the doping gradient is the absence of super saturation of Zn in an Au/Zn alloy. Hence, there is no maximum amount of Zn within the Au seed particle. With growth time the amount of Zn will increase and proportional to the amount of Zn there will be fixed ratio transferred into the growing nanowire. Therefore, the amount of p-type doping is assumed to follow the amount of Zn dissolved in the seed particle. The upper limit is reached if the same amount of Zn collected in the seed particle from the gas phase is transferred into the growing nanowire.

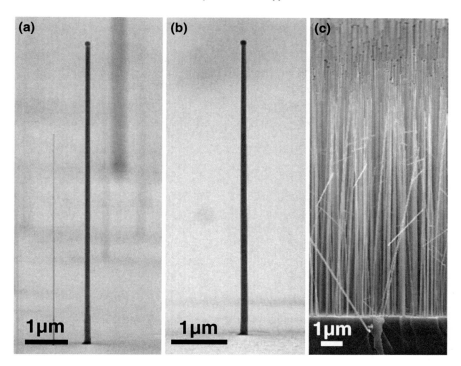

Fig. 14.7 SEM micrographs of 150 nm diameter GaAs nanowires grown on GaAs (111)B substrates: **a** from colloidal nanoparticles under TESn supply (IV/III = 0.08), **b** from colloidal nanoparticles with under DitBuSi supply (IV/III = 0.52), and **c** like **a** but polydisperse seed particles formed by annealing of a 2.5 nm Au layer. [27]

14.2.3.2 n-Type Doping

Next, we report on the analogous experiments for n-type doping using two different precursors. n-type GaAs nanowires were synthesized using the same setup and methods as already described in Sect. 14.2.3.1. This time TESn ($0.02 \leq$ IV/III ≤ 0.16) or DitBuSi (IV/III ≤ 0.52) was supplied during growth. The SEM micrographs of the corresponding nanowires are given in Fig. 14.7. Au seed particles with 150 nm diameter were used for (a) and (b), while the sample in (c) was grown from polydisperse seed particles stemming from a metallic Au layer. Wires in (a, c) are grown under TESn supply of (IV/III = 0.08) while the wire in (b) under supply of DitBuSi (IV/III = 0.52), respectively.

All of the nanowires adopted the crystal orientation of the growth substrate and are upstanding in (111)B direction. Furthermore, in contrast to p-type doping with DEZn (see Sect. 14.2.3.1), no wire kinking or other structural defects, even at higher TESn supply up to IV/III = 0.16, were observable. We assume from the lower doping density that there is much less Sn or Si in the Au seed particle compared to Zn. This might be caused by a lower source efficiency of DitBuSi and DESn compared to DEZn. On the other hand, the solubility of Sn and Si in the Au particle is much lower than for Zn at the selected growth parameters. The phase diagrams [28] of Au–Sn, Au–Si, and Au–Zn confirm that there is no eutectic point for Au–Zn alloy at 400 °C such that the amount of Zn in the Au seed can follow the II/III ratio. In case of Au–Sn and Au–Si there is a given maximum of dopant in the seed possibly reducing both, the doping effect and the kinking probability, respectively.

Representative I–V characteristics for nanowires grown without dopant supply, with supply of DitBuSi (IV/III = 0.52), and with supply of TESn (IV/III = 0.08) were carried out. The ohmic contacts were formed by evaporation of Ge (5 nm)/Ni (10 nm) /Ge (25 nm) /Au (400 nm). To improve the contact properties a rapid thermal annealing was carried out for 30 or 300 s at 320 °C. The non-intentionally doped (nid) GaAs nanowires show a current of a few pA at 1 V applied bias, corresponding to a resistance in the GΩ range. Adding DitBuSi to the gas phase during growth has no effect on the conductivity of nanowires, even at relatively high IV/III ratios. The growth temperature of 400 °C might be too low for a sufficient cracking of the DitBuSi precursor [29]. In addition, in (111) oriented GaAs growth Si is an amphoteric impurity [30]. First principle calculations claim that this also holds for nanowires [31]. With TESn at IV/III = 0.08 the current is about six orders of magnitude higher than for the nid sample, giving evidence of the doping effect. The corresponding I–V characteristic is not perfectly ohmic, which indicates a remaining contact barrier. However, annealing at temperatures higher than 320 °C leads to an increased out diffusion of Ga into the Au contact layer. This effect is also reported for bulk material [32], but gets crucial in the nanoscale.

In order to determine the carrier concentration of the Sn-doped GaAs nanowires, we adopted the model used for p-GaAs. We conclude that the doping density N_D varies in the range of 7×10^{17} cm$^{-3} \leq N_D \leq 2 \times 10^{18}$ cm^{-3} [27]. The spreading is attributed to a doping inhomogeneity similar to GaAs:Zn [24]. TESn enables heavily n-type doped GaAs nanowires in a relatively small process window while no doping effect could be found for ditertiarybutylsilane. TESn implies a n-type conductivity of the GaAs nanowires. We again fabricated multichannel MISFET devices with the field assisted self-assembly (FASA) approach [33, 34], to verify the type of doping. By plotting the drain current I_D versus gate-source voltage V_{GS} the n-channel behaviour was proven [27]. To our knowledge this experiment is the first successfully n-doped GaAs nanowire grown by VLS in an MOVPE apparatus. An additive proof will be provided by measuring the electroluminescence of axial p–n junctions in single GaAs nanowires (Sect. 14.4.2).

14.2.3.3 Axial p–n Junctions Formed by MOVPE Using DEZn and TESn in GaAs Nanowires

This chapter describes the combination of the previously discussed p-type doping with DEZn and n-type doping with TESn to realize axial GaAs nanowire p–n junctions. The impact of dopant supply was investigated both on structural properties and

14 III/V Nanowires for Electronic and Optoelectronic Applications

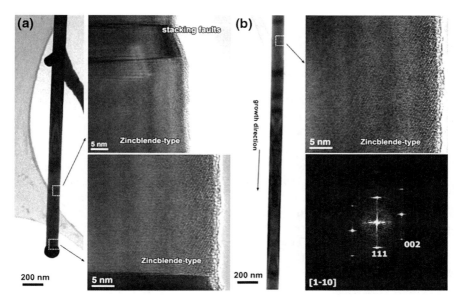

Fig. 14.8 TEM micrographs of pn-doped GaAs nanowire: **a** low magnification morphology and HRTEM lattice images near the nanowire tip (n-doped), two selected regions as outlined by dotted frames, showing the zinc blende type cubic structure and stacking fault segments, and **b** low magnification morphology and HRTEM lattice image near the bottom of the nanowire (p-doped) and its corresponding Fourier transform images, revealing the growth direction of ⟨111⟩ [25]

carrier density (see Sect. 14.4.1). Growth experiments were carried out as described before using polydisperse nanoparticles formed from a nominally 2.5 nm thin Au-layer during the annealing step prior to all growth runs. The doped nanowire growth was performed at 400 °C with a V/III ratio of 2.5. A total growth time of about 50 min was chosen to realize structures up to 20 μm in length. After 25 min, the dopant precursor supply was switched from DEZn (II/III = 0.004) to TESn (IV/III = 0.08).

For transmission electron microscopy the nanowires were peeled off the substrate and dispersed in isopropanol. A drop of suspension containing nanowires was cast on a Cu-grid coated with a thin film of carbon. The nanowire suspension was also dropped onto special prepared carrier substrates and some nanowires contacted via E-beam lithography to enable I–V measurements.

TEM studies were performed to investigate the influence of DEZn and TESn precursors on the crystal structure of pn-doped GaAs nanowires with a (111)B growth direction. Micrographs of the morphology and HRTEM lattices images are presented in Fig. 14.8. Careful examinations of the structure along the whole length of the wires indicate that no structural transition, which could depend either on the nanowire length or on the dopant precursors, is observed. We suppose that the formation enthalpy in GaAs nanowires is dominated by the group III precursor and not by the doping species for the investigated II/III and IV/III ratios. The presence of the doping species in the growth seed was below the detection limit of the energy dispersive x-ray spectroscopy (1.5 atomic percentage) and thus could not be quantified.

14.3 InAs Nanowire MISFET

The transistor is probably the most important invention of the twentieth century. This three-terminal device provides high-speed switching and very high frequency signal amplification at very low power consumption. Especially, the field-effect transistor is today's ultimate answer to the needs of both digital and analog data/signal processing. Current research is focusing on (i) reducing the features size (Moore's rule), and (ii) to address specific needs such as speed, gain, power, or efficiency, often using III/V semiconductor heterostructures. Semiconductor nanowires [35, 36, 40] may be used as the channel of a field-effect transistor device. In contrast to carbon nanotubes [37] the charge polarity can be selected and a metallic phase that may inhibit the channel depletion is avoided. The nanowire approach provides large material diversity and enables a fully surrounding gate contact. The high degree of freedom from lattice mismatch constrains offers a wide range of substrate/nanowire combinations such as III/V on Si [38]. The surrounding gate arrangement gives optimum charge control and results routinely in ultra-high transconductance devices [26]. In this chapter the electronic properties of InAs nanowire Metal–Insulator–Semiconductor Field-Effect Transistor (MISFETs) are discussed.

14.3.1 Nanowire FET Design

Nanowire transistors are formed as core–shell structures consisting of the nanowire channel core and the surrounding gate metal isolated by a dielectric shell. This surrounding gate geometry allows for an optimum field-effect control of the channel carriers and provides the most important advantage of the nanowire FET design which is also of key importance of for other "beyond CMOS devices" like tunnel FETs. In this sense, the vertical all-around-gate version is the ultimate approach (Fig. 14.9b), but it requires ultra-high resolution lithography and a highly sophisticated three-dimensional contact technology. The obtained performance [26, 39] is impressive and underlines the improved charge control. The transistors in Fig. 14.9a may be produced on any substrate without any ultra-high resolution lithography using self-assembly technologies and will be discussed thoroughly in this chapter.

Important to note is that these nanowire core–shell structures are not that seriously restricted by the lattice match requirements, allowing for materials combination best suited for FET function, like high conduction band offsets and Schottky barriers. In this context, the surface and interface properties of the channel material have to be considered. Besides properties like density of states, mobility, carrier type and concentration, the surface properties, mainly the Fermi level position, are important:

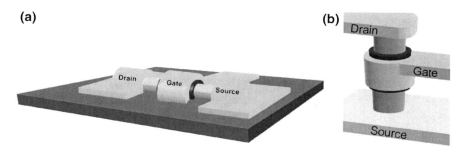

Fig. 14.9 Concepts of nanowire transistors [40]. **a** Omega shaped gate MISFET **b** vertical all-around-gate MISFET

it determines the type, accumulation or depletion, of the device. Further, it determines the quality of the source and drain ohmic contacts, which are especially important for a downscaled device with small contact areas. For low power and simultaneously high speed FETs, materials with low effective mass like InAs and InSb, are especially well suited.

The drain current of any type of field-effect transistor is controlled by a potential applied to the gate-channel capacitance. The strategy towards ultimate device performance is threefold:

(1) reducing the gate length and scaling other device parameters properly,
(2) restricting the capacitive gate control to mobile carriers in the channel, and
(3) selecting a channel material offering excellent mobile carrier transport properties.

The ultra-short gate length (strategy 1) is a general need for high integration and high speed device development. The nanowire approach gives easy access to a nanometer scale channel depth, basically the nanowire diameter. Moreover, the drain-induced barrier lowering is reduced by a surrounding gate that relaxes scaling needs in terms of oxide and nanowire thickness as compared to the gate length [8].

To restrict the capacitive charge control to mobile carriers in the channel (strategy 2) is the major motivation for the nanowire approach. The nanowire channel is coaxially surrounded by the gate and any escape into a substrate as in the case of epitaxial grown layered devices is inhibited. On the other hand, strategy (2) also gives rise to the main challenges of III/V nanowire design. The gate metallization should be confined at the intrinsic gate-channel capacitance in order to avoid a parasitic capacitance having in mind that the intrinsic gate capacitance is about 1 fF or less.

III/V semiconductors offer a huge number of materials with excellent transport properties (strategy 3). Moreover, according to the specific application the optimum material may be selected; i.e., the low band gap InAs with very high mobility is very well suited for high speed but low voltage applications while other materials may address other applications. However, this material has to fulfil the requirements of the other strategies too, which reduces the number of promising candidates substantially; i.e., the GaAs surface and its interface to a gate dielectric exhibits a high density of parasitic surface and interface states. The high surface potential results in a deep space charge region which may fully deplete the nanowire channel. Due to the limited n-type doping capability of GaAs in general and GaAs nanowires in particular only thick GaAs nanowires exhibit conductivity in the non-gated area. Therefore, despite the huge number of reports on GaAs nanowires ([41] and references therein) there are very few reports on FET [42]. In contrast to GaAs the Fermi level at the InAs surface is pinned within the conduction band. This enhances the nanowire conductivity down to very low diameters and simplifies ohmic contact formation substantially. In addition, InAs offers a shallow n-type background doping which is high enough for FET operation and does not require intentional doping. Therefore, InAs is today the preferential III/V semiconductor nanowire channel material. A few reports are provided for alternatives like GaN [43], InGaAs [44] or InSb [35]. Beyond III/V materials Si [36] is the major nanowire channel material and among the II/VI semiconductors many reports are available for ZnO nanowire FET [45].

14.3.2 Device Performance

High performance nanowire FET devices are designed as a MISFET. InAs has a low band gap with a carrier accumulation at the surface such that an insulting thin film as a gate dielectric is indispensable. The deposition of high k-materials on sophisticated nanowires like InAs or in the presence of Au-contacts requires deposition techniques offering relatively low substrate temperatures such as room temperature electron–cyclotron resonance plasma enhanced chemical vapour deposition (ECR-CVD) [46], atomic layer deposition (ALD) [38], or even molecular beam epitaxy [47].

Here, the performance of omega shaped InAs nanowire MISFET (Fig. 14.10a) is presented. The InAs nanowires were deposited on an insulating substrate. Ti/Au ohmic contacts were formed by liftoff and the contacts were annealed at 300 °C for 30 s. The contacted nanowires were covered by SiN_x gate dielectric with various thicknesses investigated (20–90 nm). Finally, a Ti/Au gate metal was evaporated forming an omega-shaped gate partly wrapped around the InAs wire (Fig. 14.10a). Figure 14.10b shows a typical room temperature DC output characteristics $I_D - V_{DS}$ with a SiN_x dielectric layer thicknesses $h_{SiNx} = 30$ nm. The devices show good pinch-off behaviour. The threshold voltage is $V_T = -0.25$ V and shifts to positive values with decreasing silicon nitride thickness. At high gate bias ($V_{GS} = 1.9$ V) a maximum drain current of 130 μA is measured corresponding to a current density in the wire of $6.6 \times 10^6 A/cm^2$. The measured maximum transconductance of the FET is $g_m = 97.5$ μS (cf. Fig. 14.11a) [26].

For FET devices the normalized drain current $I_D^* = I_D/W_g$ and transconductance $g_m^* = g_m/W_g$ are important figures of merit, where w_g is the channel width. In the case of a cylindrical channel the normalization may be obtained by replacing the channel width by the channel diameter which is fixed here to $d = 50$ nm: $I_D^* = I_D/d$

14 III/V Nanowires for Electronic and Optoelectronic Applications

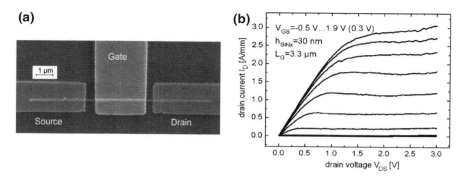

Fig. 14.10 Omega-shaped single n-InAs nanowire field-effect transistor, $d_{NW} = 50$ nm, $h_{SiN_x} = 30$ nm. **a** SEM micrograph, **b** I–V characteristics [26]

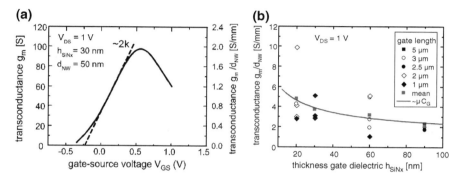

Fig. 14.11 Transconductance of single n-InAs nanowire field-effect transistor with SiN_x gate dielectrics. **a** transconductance versus gate bias, **b** max transconductance normalized to the nanowire diameter versus the gate dielectric thickness [26]

and $g_m^* = g_m/d$. The single n-InAs nanowire channel FET provides a high output current $I_D^* = 3$ mA/mm and a very high transconductance of $g_m^* \geq 2$ S/mm if the gate dielectric thickness is less than $h_{SiN_x} = 30$ nm and the gate length is $L_G = 2$ μm or less given (cf. Fig. 14.11b). The measured maximum transconductance of about 2 S/mm [26] is among the highest ever reported for any semiconductor nanowire FET.

The high performance obtained is attributed to the excellent transport properties of the InAs NW channel. Long-channel MOSFET equations may be used to model the device characteristics and to extract the carrier mobility of the channel which is not easily measured by other means [19]. The transconductance g_m of a long channel MOSFET in saturation is given as $g_m = 2k(V_{GS} - V_T)$ with the scaling factor $2k$, the applied gate bias V_{GS}, and the threshold voltage V_T. Experimentally, the scaling factor $2k$ may be taken from the slope of the transfer characteristic $g_m - V_{GS}$ according to Fig. 14.11b. For this purpose, a batch of nanowire MISFET were fabricated with a gate length varied from 1 to 5 μm and with various thickness of the gate dielectric

(20–90 nm). Following the long-channel MOSFET equations, the scaling factor is given as $2k = \mu \cdot C_G \cdot L_G^{-2}$ where C_G is the gate capacitance, μ the low field mobility, and L_G is the gate length, respectively. The capacitance of a cylindrical shaped gate-channel configuration is determined as

$$C_G = \frac{2\pi \cdot \varepsilon_0 \varepsilon_r L_G}{ln\left(\alpha + \frac{2h}{d}\right)}, \tag{14.1}$$

where ε_0 is the vacuum permittivity and $\varepsilon_r = 7.5$ is the dielectric constant of the SiN_x gate isolation. In case of a coaxial gate all-around structure the correction factor is $\alpha = 1$. Here, the SiN_x dielectric and the gate metal is only partly surrounding the nanowire and the effective gate capacitance is reduced. An evaluation using an electrostatic field simulation results to $\alpha = 1.55$. So the transconductance of the InAs nanowire channel FET can be modelled in dependence of the geometrical device parameters and the results can be fitted to experimental data using the low-field mobility as the only fitting parameter. Figure. 14.11 shows both experimental and modelled data of the transconductance versus the dielectric thickness for different gate lengths. The best agreement to the average measured data is obtained if a low field mobility of $\mu = 13,000\,cm^2/Vs$ is assumed. This result confirms the ultra-high mobility available in InAs nanowires.

Figure 14.11b also shows that there is a substantial scattering of experimental data attributed to variations in the fabrication process. In addition, the InAs nanowires grown both in (100) and (111)B direction are randomly used for device fabrication and there might be a contribution from crystal orientation dependent transport properties.

The major challenges toward a reliable RF characterization of nanowire transistors, regardless of the material system, are the dominant parasitic capacitances and the small signal power. In addition, on-wafer RF scattering parameter measurements with a coplanar waveguide pattern with characteristic impedance of 50 Ω provide currently the best measurement conditions but result in severe incompatibilities. The nanowire FET device impedance is much too high and its output power is still too low to feed the 50 Ω waveguide resulting in huge reflections and ultra-low signal levels. Therefore, high frequency measurements of nanowire FETs are barely reported [43, 48, 49].

With use of the generic open and short structures the parasitics of InAs nanowire transistors could be deembedded from the measurements resulting in corrected S-parameters exhibiting typical field-effect transistor behavior. The maximum stable gain (MSG) of the investigated NW-FET is shown in Fig. 14.12a. A MSG higher than 30 dB at low frequency and a maximum oscillation frequency of 15 GHz are obtained. However, these data are still preliminary due to the high parasitic capacitances of the contact pattern. The InAs nanowire MISFET characterized does not reflect the performance limits of InAs nanowire MISFETs due to the large gate length of 1.4 μm. The cutoff frequency is expected to increase with decreasing gate length as a result of lower gate capacitance and higher transconductance, but in this case parasitics are increasingly dominating [49].

14 III/V Nanowires for Electronic and Optoelectronic Applications

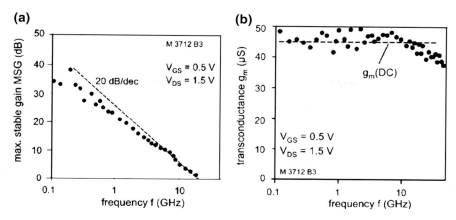

Fig. 14.12 Extracted RF parameters of a single self-aligned InAs nanowire MISFET with 1.4 μm gate length: **a** maximum stable gain MSG, **b** transconductance g_m versus frequency with a good agreement with the DC in a wide frequency range. [49]

It is well-known that the III/V interface to dielectric materials suffers from a high density of low-mobile surface states which result in a degraded sub threshold swing and in a hysteresis of the I–V characteristics. The SiN_x gate dielectric enables the highest gain [26] but it suffers from severe hysteresis effect in the I–V characteristics. Therefore, a study on its impact on the high frequency performance is important. The measured DC transconductance of a single InAs nanowire FET with SiN_x dielectric is compared to the small signal parameter transconductance deduced from a de-embedded from measured RF parameters in the frequency range from 0.1 to 50 GHz. Figure 14.12b shows that the small signal parameter g_m is constant up to at least 50 GHz and that the same transconductance is available at DC and at RF. However, a low frequency hysteresis of nanowire transistors transfer characteristics cannot be accepted for future circuit fabrication. There are a number of different dielectrics developed for high-k application in CMOS or III/V-MIS technology adopted for nanowire MISFET such as HfO_x, AlO_x/HfO_x, and MgO. Al_2O_3 deposited by atomic layer deposition gate dielectrics exhibiting almost no variation regardless of the drain bias selected [50].

14.4 Properties of GaAs Nanowire p–n Junction

14.4.1 Electrical Properties

Axial GaAs nanowires p–n junctions were grown in the VLS mode using DEZn and TESn as dopant precursors. For electrical measurements ohmic contacts have to be patterned on the p–n nanowire (inset Fig. 14.13). The use of multiple contacts

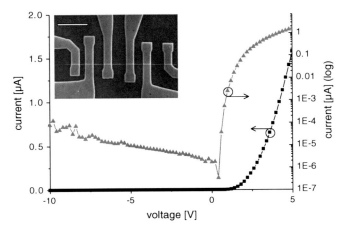

Fig. 14.13 Typical I–V characteristics of single pn-GaAs nanowire in the linear, as well as in the semi logarithmic scale. The inset shows a single pn-GaAs nanowire transferred to a carrier substrate and connected to six contacts via E-beam lithography (scale bar is five micron). [25]

enables both the characterisation of the whole p–n junction and also the investigation of the individually doped wire parts. While Pt/Ti/Pt/Au metallization was used for the p-doped part of the wire, typical Ge/Ni/Ge/Au annealed contacts were used for the n-GaAs wire.

Figure 14.13 shows the I–V characteristics of a single pn-GaAs nanowire structure both in the linear as well as in the semi logarithmic scale. The I–V characteristic clearly shows p–n diode behavior with excellent blocking in reverse direction in the low pA-regime and demonstrates current blocking up to at least 10 V. The forward current is about 6 orders of magnitude higher and reaches the µA range with an ideality factor of about 2 at low current levels. The diffusion voltage $V_D = 1.4$ V corresponds to the band gap of the GaAs material. Electroluminescence measurements show intense light emission at around 870 nm (refer to Sect. 14.4.2). Up to now no data on the abruptness of the p–n junction are available. However, a graded doping junction is expected due to the exchange of dopants in the Au-seed during growth. To our knowledge this is the first axial GaAs p–n diode realized in a single GaAs nanowire and in summary, diode I–V performance clearly exceeds previously published data of axial pn-InP nanowire diodes [16, 51].

14.4.2 Optoelectronic Properties

Micro photoluminescence was applied to investigate axial doping distribution and the sharpness of the fabricated junction in GaAs nanowire p–n diodes. Photoluminescence spectra (Fig. 14.14) were taken from 150 nm thick GaAs nanowire pn-diode, which was excited at different locations by a focused laser beam of 1 µm diameter

14 III/V Nanowires for Electronic and Optoelectronic Applications

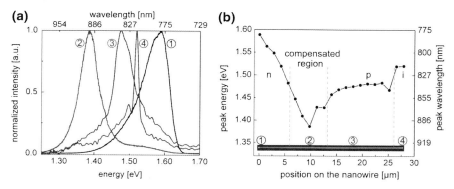

Fig. 14.14 a Micro photoluminescence spectra taken at different positions along the p–n junction of a GaAs nanowire. Positions on the nanowire are denoted by numbers in the (b). **b** Peak energy along the nanowire length. The SEM image of the investigated nanowire is shown in the inset [52]

and a wavelength of $\lambda = 532$ nm. The spectra were taken at the temperature of 9 K under optical excitation density of $8 \cdot 10^4$ W $\cdot$ cm^{-2}. The positions corresponding to the spectra are indicated by numbers on the inset in Fig. 14.14b. Four regions can be recognized on the plot of the peak energy along the nanowire length. At the non-intentional doped nanowire stump (Position 4) an exciton peak at 1.519 eV is observed (spectrum 4 in the Fig. 14.14a). As the laser spot moves further to the p-doped nanowire part the main peak shifts to the range 1.47–1.49 eV (Fig. 14.14b and spectrum 3 in the Fig. 14.14a). This peak is attributed to the recombination of electrons with holes via acceptor band, characteristic for the p-doped GaAs nanowires [52]. In the n-doped area close to the nanowire tip the position of the PL-peak was determined to be in the range between 1.48–1.58 eV. The position of the peak is explained by the Burstein-Moss shift observed in heavily doped n-GaAs nanowires. With a further increase of carrier concentration towards the nanowire tip, a band filling with electrons is taking place, which manifests itself in the Fermi-level shift and consequently in the shift of PL-peak to the higher energies.

In the area between n- and p-doped nanowire parts PL-peak position between 1.38 and 1.48 eV is observed (Fig. 14.14b as well as spectrum 2 in the Fig. 14.14a). We attribute these lines to the compensated region located in the area of the p–n junction. Due to the memory effect of the growth seed some zn is still present in gold after switching of the doping precursors from DEZn to TESn, yielding a region where both doping species are present. Such lines are illustrated in Fig. 14.15d. The emission is believed to come from the tunnelling-assisted transitions between spatially separated degenerate donor and acceptor states, so that emission lines with the energy much lower than the band gap may appear [53].

To investigate electroluminescent properties of single nanowire p–n diodes, the contacted nanowire samples (inset on the Fig. 14.13) were glued to the chip-carrier and wire bonded. The single nanowire p–n junction with a diameter of 200 nm and a diode like IV characteristic (Fig. 14.13) was excited by a constant current in forward

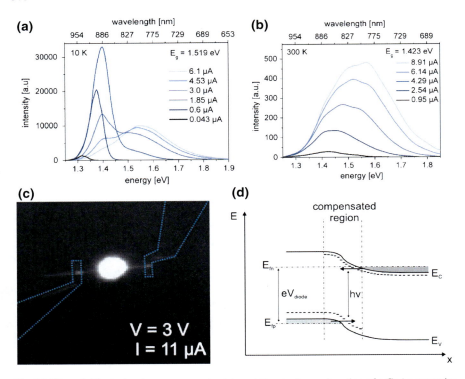

Fig. 14.15 a, b Electroluminescence spectra taken at the single p–n junction of a GaAs nanowire for different excitation levels at 10 and 300 K respectively. Band gap of GaAs at corresponding temperatures is indicated in the Figures. **c** Optical microscope image collected by CCD camera of a nanowire p–n diode at 300 K under forward bias of 3 V. Electrical contacts to the nanowire are plotted with dashed lines. **d** Model of a band structure for a diode with a compensated region biased in forward direction. Tunnelling assisted radiative transition in compensated region is indicated by an arrow. The figures **a, b, d** are adopted from [52] and **c** from [55]

direction while emissions spectra were measured. Figure 14.15a shows electroluminescent spectra from a single p–n junction nanowire taken at 10 K under different excitation currents. At low currents the emission peak has a maximum at 1.32 eV (Fig. 14.15a). For higher injection current the peak shifts to 1.4 eV and its intensity increases until an injection current of 1.85 μA. The observed emission peak energy corresponds to that observed in the compensated region via photoluminescence in pn-GaAs nanowires, which is described above. For that reason we attribute the peak to the tunnelling assisted transition between donor and acceptor band, taking place in the compensated region of the p–n junction (Fig. 14.15d).

This assumption of radiative tunnelling is supported by the shift of the emission peak to higher energies with an increasing excitation current. This shift is expected for tunnelling assisted transitions and is explained by shift of the quasi-Fermi levels with respect to each other [54]. The slope of the band structure at the junction flattens

14 III/V Nanowires for Electronic and Optoelectronic Applications

at higher bias voltages causing a reduction of tunnelling probability and a decrease of the tunnelling emission. For this reason tunnelling assisted emission peak diminishes for high injection levels and becomes dominated at $4.5\,\mu A$ by band-edge emission, appearing at $1.51\,eV$ for $10\,K$.

Scattering of free carriers by phonons increases at higher temperatures. This lowers the tunnelling probability and makes it more difficult to distinguish between two emission mechanisms. At room temperature, broad band–band emission dominates the whole spectrum even for low injection currents (Fig. 14.15b).

The population of states above the quasi-Fermi level increases with temperature and explains broadening of the emission peak at the high energy side while the low-energy tail stays saturated.

For the spatially resolved investigation of the electroluminescence optical microscope image of a forward biased nanowire p–n diode was made at $300\,K$ (Fig. 14.15c). The image was taken by CCD camera in the imaging mode collecting all light in a range of 350–$1,050\,nm$. To highlight the position of the contacted nanowire the sample was illuminated by scattered light from the side. In the Fig. 14.15c, strong electroluminescence in the middle of the contacted nanowire-diode at the expected position of the p–n junction is observed. This proves, that light emission originates from electroluminescence at the p–n junction and not from recombination at contacts.

Spatially resolved photocurrent spectroscopy was used to investigate the mechanism of carrier-photo generation in nanowire p–n diodes. I(V) characteristics of nanowire pn-diodes were measured, while nanowires were locally illuminated by focused CW laser ($\lambda = 532\,nm$) at different positions (Fig. 14.16b). The laser light was focused by a 50x objective lens yielding a spot of diameter $\sim 1\,\mu m$. Short circuit current and photocurrent are maximal when the diode is illuminated at the position of the p–n junction (position 5 in the Fig. 14.16a). Photocurrent was observed neither in the vicinity of contacts nor in the p- and n-diode parts.

Current voltage measurements under homogeneous AM 1.5 G illumination were done to determine the solar conversion efficiency of a fabricated photodiode (Fig. 14.17a). A short circuit current of $I_{SC} = 88\,pA$ and an open circuit voltage of $V_{OC} = 0.56\,V$ were obtained yielding a fill factor of $FF = 69\,\%$. The power conversion efficiency of the nanowire photovoltaic device under AM 1.5 G illumination was estimated with the formula:

$$\eta = \frac{V_{OC} \cdot I_{SC} \cdot FF}{P_{in}} = 0.09.$$

The input power P_{in} was calculated from the product of the illumination power density with the active projected absorption area of the nanowire device.

Power dependent photocurrent measurements made under monochromatic homogeneous laser illumination ($\lambda = 532\,nm$) demonstrate linear scaling of photocurrent with illumination intensity (Fig. 14.17b). This is in conformity with the relationship $I_{SC} = A \cdot q \cdot (L_e + L_h) \cdot G$, where A is an area of the p–n junction and G is a generation rate.

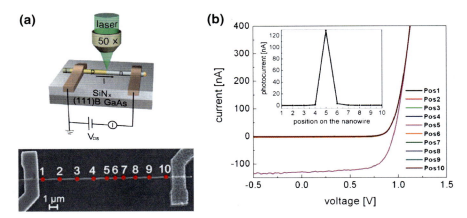

Fig. 14.16 **a** Schematic of a measurement setup for photocurrent microscopy. The lower inset shows a SEM micrograph of the investigated nanowire-diode. **b** I(V) characteristics of the GaAs nanowire pn-diode illuminated by focused laser spot at different positions. Upper inset shows photocurrent as a function of a laser spot position. The corresponding positions are denoted by numbers on the SEM image in Fig. 14.15a [55]

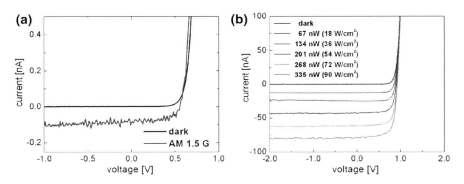

Fig. 14.17 **a** Dark and AM 1.5G illuminated I–V characteristics of nanowire pn-diode. **b** I–V characteristics of the nanowire pn-diode measured under monochromatic homogeneous laser illumination ($\lambda = 532$ nm) with various illumination powers [55]

To improve the sharpness of nanowire p–n junctions arrayed nanowire electroluminescent structures were fabricated: a p–n junction was formed between p-doped non-tapered GaAs nanowires grown on n-doped (111)B GaAs substrate as shown in the schematic in Fig. 14.18a. Contacted arrays contained approximately 50 nanowires. Durimide technology was adopted to form an isolator separating the substrate from the top-contact. Using oxygen plasma the durimide was etched down until the nanowire top became free. A SEM image of the fabricated structure is presented in the Fig. 14.18a. The fabricated pn-structure possessed a typical diode-like IV characteristic with quite high forward currents of about 2 mA at 2 V, and had an ideality factor of 1.5 in the low voltage range at room temperature (Fig. 14.18b).

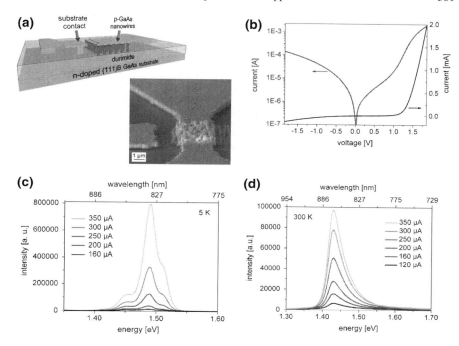

Fig. 14.18 a Schematics and an SEM micrograph of the fabricated arrayed nanowire LED. **b** I(V) characteristics of arrayed nanowire LED in semi-logarithmic (*left axis*) and linear (*right axis*) plot. **c** Electroluminescence spectra of the arrayed nanowire LED taken at 5 K. **d** Electroluminescence spectra of the arrayed nanowire LED taken at 300 K [52]

Figure 14.18c, d presents the electroluminescent spectra of the fabricated pn-diode. Arrayed nanowire electroluminescent structures demonstrate at 5 K strong electroluminescence peak at 1.488 eV, which is attributed to the electron-acceptor band recombination in the p-doped GaAs wires. Besides the main peak two shoulder peaks at 1.45 and 1.52 eV are present. We attribute the peak at 1.45 eV to tunnelling assisted photon emission, observed in abrupt GaAs p–n junctions at this energy. The peak at the 1.52 eV corresponds to the GaAs band gap at 5 K and originates from the band–band recombination at the junction. At room temperature one main peak was observed at 1.427 eV, only. This line energy corresponds to the band-gap of GaAs at room temperature and is therefore attributed to the band-edge emission due to thermal injection. Since the peak was very broad the other both lines could not be distinguished.

14.5 Conclusion

In the recent decade, III/V semiconductor nanowires have emerged to a degree of maturity making them suitable for both electronic and optoelectronic applications. Their flexibility in use of basically any substrate makes them especially useful for the combination with existing technologies like Silicon CMOS. Moreover, since high current density p–n junctions are now available after a careful elaboration of novel doping schemes, high performance optoelectronic application like light emitting devices on various substrates is perhaps the most promising candidate for the first commercialisation of nanowire devices. On a somewhat longer route the nanowires hold the promise to bring huge innovations to the photovoltaic business. This holds for the efficient collection of light but also for the high power/weight ratio for solar conversion. High current density axial p–n junctions for high yield conversion are demonstrated and coaxial structures with possibly waveguiding character are in vision for very high efficiency future solar cells.

Acknowledgments The authors are especially indebted to Einar Kruis and Thomas Weber for the aerosol preparation of nanoparticles, to Daniela Sudfeld, Zi-An Li, and Marina Spasova for excellent TEM analysis, to Matthias Offer, Stephan Lüttjohan, and Axel Lorke for high-resolution photoluminescence and photocurrent analysis, and to Benjamin Münstermann for high frequency measurements.

References

1. F. Glas, Critical dimensions for the plastic relaxation of strained axial heterostructures in freestanding nanowires. Phys. Rev. B **74**, 121302(R) (2006)
2. G.W. Sears, Growth of Hg-whiskers. Acta Metallurgica **1**, 457 (1953)
3. R.S. Wagner, W.C. Ellis, The vapour-liquid-solid mechanism of crystal growth and its application to silicon. Trans. Met. Soc. AIME **233**, 1053–1064 (1965)
4. E.I. Givargizov, Fundamental aspects of VLS growth. J. Cryst. Growth **31**, 20–30 (1975)
5. K. Hiruma, M. Yazawa, K. Haraguchi, K. Ogawa, T. Katsuyama, M. Koguchi, H. Kakibayashi, GaAs free-standing quantum-size wires. J. Appl. Phys. **74**(5), (1993)
6. M.S. Gudiksen, L.J. Lauhon, J. Wang, D.C. Smith, C.M. Lieber, Growth of nanowire superlattice structures for nanoscale photonics and electronics. Nature **415**, 617–620 (2002)
7. B.S. Sorensen, M. Aagesen, C.B. Sorensen, P.E. Lindelof, K.L. Martinez, J. Nygard, Ambipolar transistor behavior in p-doped InAs nanowires grown by molecular beam epitaxy. Appl. Phys. Lett. **92**(1), 012119 (2008)
8. E. Lind, M.P. Persson, Y.M. Niquet, L.E. Wernersson, Band structure effects on the scaling properties of [111] InAs nanowire MOSFETs. IEEE Trans. Electron Dev. **56**(2), 201–205 (2009)
9. V.G. Dubrovskii, G.E. Cirlin, I.P. Soshnikov, A.A. Tonkikh, N.V. Sibirev, YuB Samaonenko, V.M. Ustinov, Diffusion induced growth of GaAs nanowhiskers during molecular beam epitaxy: theory and experiment. Phys. Rev. B **71**, 205325 (2005)
10. M.T. Borgström, G. Immink, B. Ketelaars, R. Algra, E.P.A.M. Bakkers, Synergetic nanowire growth. Nat. Nanotechnol. **2**, 541–544 (2007)
11. M.T. Borgström, M.A. Verheijen, G. Immink, T. de Smet, E.P.A.M. Bakkers, Interface study on heterostructured GaP-GaAs nanowires. Nanotechnology **17**, 4010–4013 (2006)

14 III/V Nanowires for Electronic and Optoelectronic Applications

12. M.T. Björk, B. Ohlsson, T. Sass, A.I. Persson, C. Thelander, M.H. Magnusson, K. Deppert, L.A. Wallenberg, L. Samuelson, One-dimensional heterostructures in semiconductor nanowhiskers. Appl. Phys. Lett. **80**(6), 1058 (2002)
13. N. Sköld, L.S. Karlsson, M.W. Larsson, M.-E. Pistol, W. Seifert, J. Trägårdh, L. Samuelson, Growth and optical properties of strained GaAs-Ga$_x$In$_{1-x}$P core-shell nanowires. Nano Lett. **5**(10), 1943–1947 (2005)
14. K. Haraguchi, T. Katsuyama, K. Hiruma, K. Ogawa, GaAs p–n junction formed in quantum wire crystals. Appl. Phys. Lett. **60**(6), 745–747 (1992)
15. F. Qian, Y. Li, S. Gradačak, D. Wang, C.J. Barrelet, C.M. Lieber, Gallium nitride-based nanowire radial heterostructures for nanophotonics. Nano Lett. **4**, 1975–1979 (2004)
16. E.D. Minot, F. Kelkensberg, M. van Kouwen, J.A. van Dam, L.P. Kouwenhoven, V. Zwiller, M.T. Borgström, O. Wunnicke, M.A. Verheijen, E.P.A.M. Bakkers, Single quantum dot nanowire LEDs. Nano Lett. **7**(2), 367–371 (2007)
17. P. Velling, A comparative study of GaAs- and InP-based HBT growth by means of LP-MOVPE using conventional and non gaseous sources. Prog. Cryst. Growth Charact. Mater. **41**, 85 (2000)
18. O. Madelung, Grundlagen der Halbleiterphysik (Springer-Verlag, New York, 1970)
19. W. Prost, K. Blekker, Q.-T. Do, I. Regolin, F.-J. Tegude, S. Müller, D. Stichtenoth, K. Wegener, C. Ronning, Modeling the carrier mobility in nanowire channel FET. Mater. Res. Soc. Symp. MRS Proc. 1017-DD14-06 (2007). doi:10.1557/PROC-1017-DD14-06
20. Z.-A. Li, C. Möller, V. Migunov, M. Spasova, M. Farle, A. Lysov, C. Gutsche, I. Regolin, W. Prost, F.-J. Tegude, P. Ercius, Planar-defect characteristics and cross-sections of <001>, <111>, and <112> InAs nanowires. J. Appl. Phys. **109**, 114320 (2011)
21. K.A. Dick, K. Deppert, L. Samuelson, W. Seifert, Optimization of Au-assisted InAs nanowires grown by MOVPE. J. Cryst. Growth **297**, 326–333 (2006)
22. S.A. Dayeh, Electron transport in indium arsenide nanowires. Sem. Sci. Technol. **25** 024004 (2010)
23. P. Paiano, P. Prete, N. Lovergine, A.M. Mancini, Size and shape control of GaAs nanowires grown by metalorganic vapor phase epitaxy using tertiarybutylarsine. J. Appl. Phys. **100**, 094305 (2006)
24. C. Gutsche, I. Regolin, K. Blekker, A. Lysov, W. Prost, F.J. Tegude, Controllable p- type doping of GaAs nanowires during vapor-liquid-solid growth. J. Appl. Phys. **105**(2), 024305 (2009)
25. I. Regolin, C. Gutsche, A. Lysov, K. Blekker, Z.-A. Li, M. Spasova, W. Prost, F.-J. Tegude, Axial pn-Junctions formed by MOVPE using DEZn and TESn in vapour-liquid-solid grown GaAs nanowires. J. Cryst. Growth **315**, 143–147 (2011)
26. Q.-T. Do, K. Blekker, I. Regolin, W. Prost, F.J. Tegude, High transconductance FET with a single InAs nanowhisker channel". IEEE Electron Dev. Lett. **28**(8), 682 (2007)
27. C. Gutsche, A. Lysov, I. Regolin, K. Blekker, W. Prost, F.-J. Tegude, n-type doping of vapor-liquid-solid grown GaAs nanowires. Nano Res. Lett. **6**, 65 (2011)
28. H. Okamoto, T.B. Massalski, in *Phase Diagram of Binary Gold Alloys* (ASM International, Metals Park, OH, 1987), pp. 278–289H
29. S. Leu, H. Protzmann, F. Höhnsdorf, W. Stolz, J. Steinkirchner, E. Hufgard, Si-doping of MOVPE grown InP and GaAs by using the liquid Si source ditertiarybutyl silane. J. Cryst. Growth **195**, 91–97 (1998)
30. B. Lee, S.S. Bose, M.H. Kim, A.D. Reed, G.E. Stillman, W.I. Wang, L. Vina, P.C. Colter, Orientation dependent amphoteric behavior of group IV impurities in the molecular beam epitaxial and vapor phase epitaxial growth of GaAs. J. Cryst. Growth **96**, 27–39 (1989)
31. N. Ghaderi, M. Peressi, N. Binggeli, H. Akbarzadeh, Structural properties and energetics of intrinsic and Si-doped GaAs nanowires: First-principles pseudopotential calculations. Phys. Rev. B **81**, 155311 (2010)
32. C.-Y. Chai, J.-A. Huang, Y.-L. Lai, J.-W. Wu, C.-Y. Chang, Y.-J. Chan, H.-C. Cheng, Excellent Au/Ge/Pd Ohmic Contacts to n-type GaAs Using Mo/Ti as the Diffusion Barrier. Jpn. J. Appl. Phys. **35**, 2110–2111 (1996)
33. P.A. Smith, C.D. Nordquist, T.N. Jackson, T.S. Mayer, B.R. Martin, J. Mbindyo, T.E. Mallouk, Electric-field assisted assembly and alignment of metallic nanowires. Appl. Phys. Lett. **77**, 1399–1401 (2000)

34. K. Blekker, B. Münstermann, I. Regolin, A. Lysov, W. Prost, F.J. Tegude, in textitInAs Nanowire Transistors with GHz Capability Fabricated Using Electric Field Assisted Self-Assembly, 8th Topical Workshop on Heterostructure Microelectronics, Nagano, Japan, 26–28 Aug 2009
35. L.-E. Wernersson, C. Thelander, E. Lind, L. Samuelson, III-V nanowires-extending a narrowing road. Proc. IEEE **98**(12), 2047–2060 (2010)
36. W. Lu, P. Xie, C.M. Lieber, Nanowire transistor performance limits and applications. IEEE Trans. Electron Dev. **55**(1), 2859–2876 (2008)
37. S.E. Thompson, R.S. Chau, T. Ghani, K. Mistry, S. Tyagi, M.T. Bohr, In search of Forever continued transistor scaling one new material at a time. IEEE Trans. Electron Dev. **18**(1), 26–36 (2005)
38. C. Rehnstedt, T. Martensson, C. Thelander, L. Samuelson, L.E. Wernersson, Vertical InAs nanowire wrap gate transistors on Si substrates. IEEE Trans. Electron Dev. **55**(11), 3037–3041 (2008)
39. C. Thelander, C. Rehnstedt, L.E. Froberg, E. Lind, T. Martensson, P. Caroff, T. Lowgren, B.J. Ohlsson, L. Samuelson, L.E. Wernersson, Development of a vertical wrap-gated InAs FET. IEEE Trans. Electron Dev. **55**(11), 3030–3036 (2008)
40. F.-J. Tegude W. Prost, III/V semiconductor nanowire transistors, Advances in III/V Semiconductor Nanowires and Devices, chap. 7, ed. J. Li, D. Wang, and R. R. LaPierre, Bentham Science Publ. 2011
41. C. Soci, X.-Y. Bao D.P.R. Aplin, D. Wang, A systematic study on the growth of GaAs nanowires by metal–organic chemical vapor deposition. Nano Lett. **8**(12), 4275–4282 (2008)
42. S.A. Fortuna, X. L. Li, GaAs MESFET with a high-mobility self-assembled planar nanowire channel. IEEE Electron Dev. Lett. **30**(6), 593 (2009)
43. S. Vandenbrouck, K. Madjour, D. Théron, Y. Dong, Y. Li, C.M. Lieber, C. Gaquiere, 12 GHz F_{MAX} GaN/AlN/AlGaN Nanowire MISFET. IEEE Electron Dev. Lett. **30**(4), 322 (2009)
44. J. Noborisaka, T. Sato, J. Motohisa, S. Hara, K. Tomioka, T. Fukui, Electrical characterizations of InGaAs nanowire-top-gate field-effect transistors by selective-area metal organic vapor phase epitaxy Jpn. J. Appl. Phys. **46**(11), 7562–7568 (2007)
45. M. Choe, G. Jo, J. Maeng, W.K. Hong, M. Jo, G. Wang, W. Park, B.H. Lee, H. Hwang, T. Lee, Electrical properties of ZnO nanowire field effect transistors with varying high-k Al2O3 dielectric thickness. J Appl. Phys. **107**(3), 034504 (2010)
46. A. Wiersch, C. Heedt, S. Schneiders, R. Tilders, F. Buchali, W. Kuebart, W. Prost, F.J. Tegude, Room-temperature deposition of SiNx using ECR-PECVD for III/V semiconductor microelectronics in lift-off technique. J. Non-Cryst. Solids **187**, 334 (1995)
47. Q.T. Do, K. Blekker, I. Regolin, E. Schuster, R. Peters, W. Prost, F.-J. Tegude, *Magnesium oxide (MgO) as gate dielectric for n-doped single InAs nanowire field-effect transistor*, 7th Topical Workshop on Heterostructure Microelectronics (Japan, Aug, 2007)
48. J. Chaste, L. Lechner, P. Morfin, G. Fève, T. Kontos, J.M. Berroir, D.C. Glattli, H. Happy, P. Hakonen, B. Placais, Single carbon nanotube transistor at GHz frequency. Nano Lett. **8**(2), 525–528 (2008)
49. K. Blekker, B. Münstermann, A. Matiss, Q.T. Do, I. Regolin, W. Brockerhoff, W. Prost, F.J. Tegude, High frequency measurements on InAs nanowire field-effect transistors using coplanar waveguide contacts. IEEE Trans. Nanotechnol. **9**(4), 432–437 (2009)
50. Y. Otsuhata, T. Waho, K. Blekker, W. Prost, F.-J. Tegude, *On the temporal behavior of DC and rf characteristics, of InAs nanowire MISFET* (Int. Semiconductor Device Research Symposium, College Park, MD, USA, December, 2009), pp. 9–11
51. M.T. Borgström, E. Norberg, P. Wickert, H.A. Nilsson, J. Trägardh, K.A. Dick, G. Statkute, P. Ramvall, K. Deppert, L. Samuelson, Precursor evaluation for in situ InP nanowire doping. Nanotechnology **19**(44), 445602 (2008)
52. A. Lysov, M. Offer, C. Gutsche, I. Regolin, S. Topaloglu, M. Geller, W. Prost, F.-J. Tegude, Optical properties of heavily doped GaAs nanowires and electroluminescent nanowire structures. Nanotechnology **22**, 085702 (2011)
53. J.I. Pankove, *Optical Processes in Semiconductors* (Dover Publications, Inc., New York, 1971)

54. H.C. Casey, D.J. Silversmith, Radiative tunneling in GaAs abrupt asymmetrical junctions. J. Appl. Phys. **40**(1), 241–256 (1969)
55. A. Lysov, S. Vinaji, M. Offer, C. Gutsche, I. Regolin, W. Mertin, W. Prost, G. Bacher, F.-J. Tegude, Spatially resolved photoelectric performance of axial GaAs nanowire pn-Diodes. Nano Res. **4**(10), 987–995 (2011)

Chapter 15
Metal Oxide Thin-Film Transistors from Nanoparticles and Solutions

Claudia Busch, Simon Bubel, Ralf Theissmann and Roland Schmechel

Abstract This article compares several non-vacuum-based low-temperature deposition techniques of semiconducting oxides for thin-film transistor applications. After an introduction into basic thin-film transistor theory it summarizes in short the development in the field of semiconducting oxides. Three different deposition techniques are considered in more detail: (1) a direct deposition of semiconducting oxide nanoparticles from a carrier gas stream, on the example of SnO_x and In_2O_3, (2) a wet-deposition of nanodispersions of ZnO, and (3) a deposition of liquid precursors with subsequent transformation into the semiconducting oxide, on the example of ZnO. The advantages and disadvantages of the several methods are discussed critically also with respect to results from the literature.

15.1 Introduction

When in 1983 Ebisawa et al. [1] reported about a field effect device based on poly-acetylene, it was probably of pure academic interest. Few years later, in 1986 Tsumura et al. [2] presented a field effect transistor with polythiophene as semiconducting layer. Although the performance of the device was low, compared to its inorganic counterparts, there was a clear transistor characteristic. The fact that the active layer was made of a polymer, which in principle could be processed from solutions, opened the vision of fully printed transistors or even more—completely printed circuits as, for example, formulated by Gilleo in 1992 [3]. The first fully printed transistor was presented by Garnier et al. [4] in 1994. The advantages for printed electronics have been seen mainly in a low-cost market, where an acceptable performance has to be achieved at an extreme low prize level. RFID-tags [5] and smart packaging are quite

---

C. Busch · S. Bubel · R. Theissmann · R. Schmechel (✉)
Faculty of Engineering and CENIDE, University of Duisberg-Essen,
Bismarckstraße 81, 47057 Duisburg, Germany
e-mail: roland.schmechel@uni-due.de

A. Lorke et al. (eds.), *Nanoparticles from the Gas Phase*, NanoScience and Technology, 387
DOI: 10.1007/978-3-642-28546-2_15, © Springer-Verlag Berlin Heidelberg 2012

often considered as an example. And also large-area electronics, where simple electronic circuits have to be processed on large substrates, like for active-matrix flat panel displays, would benefit from such a printed electronic technology. Photovoltaics are another example, where simple electronic structures have to be processed on large areas. Despite of a large number of newly developed semiconducting polymers in the upcoming years, the effective mobility, as a key parameter remained below $1\,cm^2/Vs$ (see review by Facchetti [6] or Scherf [7]), which is too low for most electronic applications. In organic materials, an effective mobility above $1\,cm^2/Vs$ up to $60\,cm^2/Vs$ has been observed only in a few ultra-pure single crystals (reviewed by Gershenson et al. [8]), such as rubrene or pentacene, for example.

However, the idea of printable electronic devices, such as transistors, is still fascinating and motivated researches to reconsider, how far inorganic semiconductors could also be printed. Indeed, the charge carrier mobility in inorganic amorphous thin films reaches or even exceeds that in organic crystals, giving reason to the hope that higher mobilities could be obtained in printed thin films of inorganic semiconductors.

In fact, the attempt to replace the conventional silicon single crystal technology for transistors by thin-film technologies, in order to allow for a transistor processing on larger and cheaper substrates, started already in the middle of the 1970s. At that time, amorphous silicon thin-film transistors were produced by physical vapor deposition [9] or chemical vapor deposition [10] utilizing silane as precursor. Still, the active matrix flat panel display technology is mainly based on amorphous silicon thin-film transistors [11]. The charge carrier mobility in hydrogen-terminated amorphous silicon (a-Si:H) films reaches $1\,cm^2/Vs$ and can be increased by post annealing and recrystallization up to $100\,cm^2/Vs$, which becomes a kind of benchmark for all competing technologies. Despite some success, the amorphous silicon technology has essential drawbacks: (1) it is still a vacuum- / vapor-based process and (2) the process temperature is above $250\,°C$ up to $400\,°C$, which excludes low-cost polymer substrates. Therefore, there is still an increasing interest in printable inorganic semiconductors.

In order to make an inorganic material printable, it requires transferring it into a liquid form. There are mainly two strategies: (1) one produces nanoparticles of the desired material and transfers them into a nanodispersion with an adequate solvent or (2) one uses a liquid precursor and transforms the material subsequently to the deposition into the desired material. Both strategies will be considered here. However, first the basic working principle of a thin-film transistor will be summarized in Sect. 15.2 and 15.3.1 basic electronic properties of semiconducting metal oxides will be reviewed shortly, before thin-film transistors made from nanoparticles deposited directly from a carrier gas stream (in Sect. 15.3.2), and from a dispersion (Sect. 15.3.3) are discussed. Finally, the deposition from solutions or liquid precursors will be presented in Sect. 15.3.4.

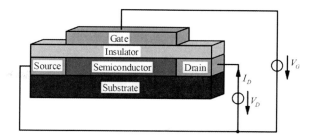

Fig. 15.1 Basic device structure of a thin-film transistor in top gate configuration

15.2 Operation Principle of Thin-Film Transistors

Thin-film transistors (TFTs) are field effect devices. The basic principle has already been patented in 1926 by Lilienfeld [12], but could not be realized due to technological deficiencies at that time. The first practical device was presented by Weimer in 1962 [13] utilizing CdS as semiconductor.

The basic device structure is sketched in Fig. 15.1. A semiconducting thin-film is connected by two metal electrodes, called source and drain, forming Ohmic contacts. A gate electrode is separated to the thin-film by a thin dielectric layer. A potential difference between gate electrode and semiconducting thin-film leads to charge carrier accumulation or depletion within the semiconducting layer, which allows for control over the sheet conductance of the semiconducting thin-film. Most TFTs are operating in an accumulation mode, which means the semiconducting thin-film has primarily a low conductivity, which can be increased by charge carrier accumulation. But in principle, also a depletion mode is possible, when the semiconducting thin-film already has a sufficient charge carrier concentration. In contrast to other transistor concepts, no p–n junctions are required, which makes TFTs quite simple. On the other hand, the charge transport takes place within the accumulation layer, which is usually only a few nanometers thick. This turns the morphology on the interface between gate dielectric and semiconducting thin-film into a crucial parameter. Further, the dielectric layer not only has to be thin in order to get a strong field effect, but it also has to be electrically leak-proofed.

The electric characteristics of a TFT are determined by the field effect and, therefore, can be described by the Shockley equation [14] in a first approach. The Shockley equation is obtained under the gradual channel approximation, which assumes that the electric field in the channel is dominated by the gate field. The expected output characteristic is presented in Fig. 15.2. As long as the drain-source voltage V_D is smaller than the gate-source voltage V_G ($V_D < V_G$) it holds:

$$I_D = \frac{WC}{L}\mu \left(V_D V_G - \frac{1}{2}V_D^2 \right), \qquad (15.1)$$

with I_D: drain current, W: channel width, L: channel length, C: gate capacitance per unit area, and μ: charge carrier mobility. For small V_D, the drain current is a linear

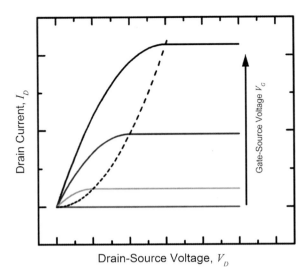

Fig. 15.2 Ideal output characteristic of a TFT according to the Shockley equations

function of V_D as well as V_G. For larger V_D the quadratic term becomes dominant leading to a nonlinear region. If V_D exceeds V_G ($V_D > V_G$) the drain current becomes independent from V_D. Then the TFT operates in the saturation region, where the saturation current is:

$$I_D = \frac{WC}{2L}\mu V_G^2. \tag{15.2}$$

Note, in principle the transistor would operate in an ambipolar mode in this region, where one type of charge carriers is injected from the source and the opposite type of charge carriers from the drain [15]. Under this condition, saturation is suppressed by the injection of charge carriers from the drain electrode. But usually, injection of the opposite type of charge carriers is hindered due to a high injection barrier on the drain. The occurrence of saturation is then a good indication that the channel is really pinched-off. If there are regions, which cannot be controlled by the field effect, for example, if the film is too thick, saturation will not occur.

Of course, the model expressed by Eqs. (15.1) and (15.2) is the most simple approach. There is a long list of reasons, why a real device will behave differently. Some important points will be discussed in the following. Equations (15.1) and (15.2) are based on the assumption that the concentration of mobile charge carriers in the channel is directly proportional to the applied gate voltage V_G. But $V_G > 0$ does not necessarily mean that there are mobile carriers available in the channel. There might be trap states that have to be filled up before mobile charge carriers are present in the channel. On the other hand, even for $V_G = 0$ there might be charge carriers present already, due to doping or band bending on the interface. All this effects can be taken into account roughly, when V_G in Eqs. (15.1) and (15.2) is replaced by a *corrected* or *effective* V'_G:

15 Metal Oxide Thin-Film Transistors from Nanoparticles and Solutions

$$V'_G = V_G - V_{th},$$
(15.3)

where V_{th} describes a threshold voltage. Utilizing this simple approach, the threshold voltage can be used in order to estimate the trap concentration or doping concentration in the transistor channel. If a positive threshold voltage has to be exceeded in order to open the transistor, the accumulated charge on the dielectric gate capacitor remains trapped as long as $V_G < V_{th}$: This yields an estimate for the trap density n_t

$$n_t = V_{th}\frac{C}{e_0},$$
(15.4)

with e_0: the elementary charge. In contrast, if the threshold voltage is negative, meaning a gate voltage is required in order to deplete the channel, Eq. (15.4) is an estimate for the concentration of free charge carriers, which would be a measure for the doping concentration. But a straightforward interpretation of the threshold voltage is dangerous. There might be other reasons for a non-zero threshold voltage. For example, if the source contact is not Ohmic but of Schottky-type, the threshold voltage reflects properties of the Schottky barrier and not of trap states.

The threshold voltage concept is useful for a first interpretation of experimental data, but it remains very rough. It ignores that the charge carriers have a thermal distribution in energy. If there are trap states, there remains a specific probability that charge carriers become excited into transport states, even if the gate voltage is below the threshold. Therefore, a real transistor will never switch on the threshold voltage as steep as predicted by Eqs. (15.1) and (15.2) at least at finite temperature. For that reason, the threshold voltage is obtained by an approximation of Eqs. (15.1) or (15.2) to the real measured I–V characteristics at voltages well above V_{th}. V_{th} is then derived from the fit parameter or a simple extrapolation of the model to small gate voltages. V_{th} is therefore rather a softly defined quantity than a hard measurable one. The range below the threshold voltage, which is not more described by Eqs. (15.1) or (15.2) is called subthreshold region. In this region, the drain current is determined by the thermal excitation of trapped charge carriers as a function of the gate voltage. In order to characterize this region, a subthreshold swing is defined:

$$S = \frac{d V_G}{d (\log I_D)},$$
(15.5)

which is a measure of how strong the gate voltage has to increase in order to increase the drain current by one decade. The subthreshold swing can be related to a trap density. Under the assumption that the trap density would be distributed homogenously in energy one obtains [16]:

$$n_t = \left[\frac{e_0 S \log e}{k_B T} - 1\right]\frac{C}{e_0}$$
(15.6)

The other basic assumption in Eqs. (15.1) or (15.2) is that the charge carrier mobility μ would be a constant of the material. But in reality, especially in defect containing thin films, the mobility will be a function (1) of the charge carrier concentration, due to partially trap filling, and (2) of the electric field strength, due to barrier lowering of localized shallow trap states. Therefore, Eqs. (15.1) and (15.2) can only be a rough estimate. In principle, one can use the I–V characteristics of a TFT in combination with Eqs. (15.1) or (15.2) in order to determine the charge carrier mobility. But again, the result depends on how the mobility was determined. The most common approaches will be discussed here.

In the linear region, the drain current can be described by:

$$I_D = \frac{WC}{L} \mu V_D (V_G - V_{th}).$$ (15.7)

If the drain voltage is low and the gate voltage high, the charge carriers are distributed almost homogenously within the channel. Therefore, the drain current will be mainly a drift current, while diffusion can be neglected. The charge carrier mobility can then be determined from the drain conductance:

$$g_D = \left. \frac{\partial I_D}{\partial V_D} \right|_{V_G = \text{const}} \quad \text{by}$$ (15.8)

$$\mu_{\text{eff}} = \frac{L}{WC(V_G - V_{th})} g_D$$ (15.9)

A mobility determined by this way is called *effective mobility*. It is indeed quite a good measure for the charge carrier mobility in the channel. But it will be a function of the gate voltage V_G, because the mobility depends on the charge carrier concentration. However, the model goes out of its range, if V_G is close to V_{th}, because one leaves the linear region and the model becomes quite inaccurate close to V_{th} as discussed before.

Alternatively, the mobility can be estimated from the transconductance:

$$g_m = \left. \frac{\partial I_D}{\partial V_G} \right|_{V_D = \text{const}} \quad \text{by}$$ (15.10)

$$\mu_{FE} = \frac{L}{WCV_D} g_m$$ (15.11)

A mobility determined in this way is called *field effect mobility*. But the method is critical, because one derives the mobility from the change of drain current by a change of gate voltage, while it is known that the mobility itself depends on the gate voltage. This quantity has more a technical meaning, because it is proportional to the transconductance, a key parameter for amplifying elements.

15 Metal Oxide Thin-Film Transistors from Nanoparticles and Solutions

A third method can be applied to the saturation region, where it holds:

$$I_D = \frac{WC}{2L}\mu\,(V_G - V_{th})^2 \tag{15.12}$$

From the slope of a plot of $(I_D)^{1/2}$ against $(V_G - V_{th})$ the *saturation mobility* μ_{sat} can be determined. Again this value will differ from the values obtained before. But the interpretation is difficult, because in the saturation regime the charge carrier concentration is very inhomogeneous within the channel leading to diffusion processes, which has been ignored here. Again the importance of this quantity has more technical reasons, because it is directly related to the saturation current.

Unfortunately, in the literature the terminology for the different mobilities is not as uniform as presented here. The term *field effect mobility* is often used in order to describe the fact that the mobility was derived from a field effect device, but not necessarily according to Eq. (15.11). Whatever transport parameter is obtained from the I–V characteristic of a TFT, one has to keep in mind that this parameter reflexes the special condition on the semiconductor-dielectric interface and might not be a good representative for bulk properties. For a more comprehensive discussion of TFT models it will be referred to a recent review by Kalb et al. [17].

15.3 TFTs with Semiconducting Metal Oxides

15.3.1 General Remarks

In the early state of TFT development there have been attempts to build up field effect transistors with ZnO [18], SnO_2, and In_2O_3 [19, 20] as active material. But the transistor performance expressed by the transconductance remained below the expectations, derived from the Hall mobility. While in the following decades semiconducting oxides became important for gas sensing (see review by Eranna et al. [21]), there was a long incubation period for semiconducting oxides in the TFT technology, which ended with the beginning of the 2000s.

Semiconducting oxides are usually n-type materials, where the bottom of the conduction band is formed by electronic states of the metal cations, while the valence band is built-up from the oxygen states. There are only few reports about p-type oxides, while the detailed mechanism of p-doping and p-conduction is still under controversial discussion [22]. In contrast, electron mobilities of several $100\,cm^2/Vs$ can be found in single crystals [23, 24], indicating interesting electronic properties. But many metal oxides have a high concentration of intrinsic defects, acting as electron donor. These are often oxygen vacancies, metal interstitials, or hydrogen impurities (a detailed study for ZnO is given by Janotti et al. [25]). In some oxides this becomes even more pronounced, if the stoichiometry between oxygen and the metal is not fixed, like in SnO_x. Therefore, the control over the doping concentration is indeed a challenge for many semiconducting oxides.

Nevertheless, there are several reasons, why a large potential is seen in oxide semiconductors [26]. (1) Many of them are widegap semiconductors, allowing for transparent electronics. This is of great importance not only for display technologies, but also for other optoelectronic applications. (2) Many metal oxides are inexpensive and abundantly available, which makes them attractive for low-cost electronics applications. (3) Many of them are nontoxic, which is a prerequisite for mass markets. (4) As an oxide, one would expect less sensitivity against ambient conditions, which should allow for simple processing. But this point turned out to be untrue. Humidity and reducing atmospheres have a negative effect on the durability [27]. There are special concepts required in order to stabilize oxides against ambient conditions. (5) Hosono et al. proposed a concept for transparent conducting or semiconducting amorphous materials in 1996 [28]. His argumentation is based on the fundamental electronic structure. The chemical bonds in conventional semiconductors, such as Si, GaN, or GaAs are based on highly anisotropic p-orbitals or sp^3-hybrid orbitals. This makes these materials very sensitive against structural defects and therefore requires the preparation of crystalline, or at least, polycrystalline films in order to keep the density of electronic defects sufficiently low. Hosono et al. instead proposed heavy metal oxides, where the metal cations have the electron configuration $(n-1)d^{10}ns^0$. The bottom of the conduction band is then formed by the isotropic s-orbitals of the metal cation. Therefore, the transport properties in the conduction band become insensitive against local disorder. Based on this concept, multi-cation oxides such as In–Ga–Zn–O were developed and the "age" of amorphous oxides semiconductors (AOS) was founded [29].

Most of the semiconducting oxides are formed by sputter processes, which leads to amorphous or polycrystalline layers. There are reports about ZnO TFTs [30–33], SnO_2 TFTs [34, 35], In_2O_3 TFTs [36], as well as TFTs with amorphous multi-cation oxides [37, 38], where effective mobilities up to $80\,cm^2/Vs$ have been obtained.

An interesting approach, in order to realize inorganic semiconducting thin films by low-temperature processing, is to utilize nanoparticles. The basic idea is to separate the material synthesis, which might require higher temperatures from the film-forming process.

15.3.2 TFTs with Nanoparticles from a Carrier Gas Stream

First, an attempt to deposit the nanoparticles directly from a carrier gas stream, without any solvents, will be considered. In order to deposit semiconducting particles directly from a gas stream, the deposition system shown in Fig. 15.3 is used. The target material is evaporated in a furnace where a nitrogen flow goes through. To exclude oxidation of the target material, it is important to start the nitrogen flow before the furnace is switched on. The vapor of the material is carried out of the furnace and condenses into nanoparticles. Oxygen is added to this flow to oxidize the aerosol particles in-flight and a second furnace is used for sintering to achieve spherical particles. Now, it is possible to deposit the particles of polydisperse size distribution

15 Metal Oxide Thin-Film Transistors from Nanoparticles and Solutions

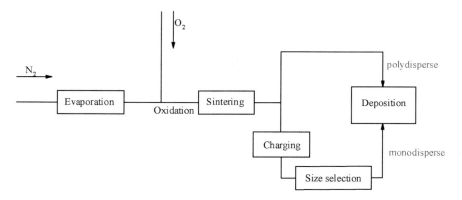

Fig. 15.3 Deposition system for nanoparticles from the gas phase

directly by leading the flow into an impactor, where the substrate is positioned (upper path in Fig. 15.3). A critical nozzle within the impactor causes an acceleration of the particle stream in the vacuum, before the particles impinge on the substrate.

In order to get a monodisperse size distribution of particles, anticipating a more homogeneous deposition, a differential mobility analyzer (DMA) is used. The size selection is realized by the following principle: charged particles getting accelerated in an electric field within the DMA. Perpendicular to the electric field a homogenous gas stream (sheath gas) is inserted, which deflects the particles from their original path depending on their size. This allows for a size selection of the particles and a monodisperse particle size distribution can be achieved. In order to charge the particles, a radioactive source (Kr^{85}) is used. After size selection, the particles are deposited on the substrate. A disadvantage of size selection is the high loss in particle concentration (more than 90% waste are possible). However, the removal of large agglomerates leads to a more homogeneous deposition and consequently a better TFT performance.

Nevertheless, the particle deposition remains a critical process. As can be seen in Fig. 15.4a, the deposition area (red circle) is inhomogeneous and difficult to align to the electrode structure. Reproducible deposition conditions are hard to realize. Therefore, the interpretation of device characteristics has to be done only qualitatively, keeping in mind that geometric quantities, such as film thickness or active channel width are not well-defined. Further, the particle layers are very porous as can be seen in Fig. 15.4b and c. Without size selection (Fig. 15.4b) large agglomerates are deposited, forming rough layers. With size selection (Fig. 15.4c) the layers are smoother, but still of fuzzy morphology.

For the bottom gate TFTs highly n-doped silicon substrates of 15 × 15 mm size and with thermally grown silicon dioxide of thickness 200 nm (±10 nm) are used. The highly doped silicon serves as the gate electrode. On top of the insulator 16 interdigitated gold electrodes of 30 nm thickness provide the source and drain contacts, 10 nm of ITO serves as adhesive layer between Au and SiO_2. The channel

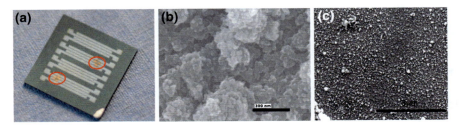

Fig. 15.4 a Silicon substrate with nanoparticles deposited directly from a carrier gas stream; *red circles* indicating the deposition area **b** In$_2$O$_3$ nanoparticles, deposited without size selection by a DMA; **c** In$_2$O$_3$ nanoparticle layer, deposited with size selection (mean particle size 16 nm)

width is 10 mm for all structures, while the channel length is available in the steps 2.5, 5, 10, and 20 μm. Beside these prestructured substrates there are also unstructured substrates available, in order to deposit source and drain electrode structures by physical vapor deposition through a shadow mask subsequently to the particle deposition (Fig. 15.4a). Prestructured as well as unstructured substrates were provided by Fraunhofer Institute Dresden.

The first devices are built with SnO$_x$ ($1 < x < 2$) as semiconducting layer. The deposition system for SnO$_x$ has been modified slightly as shown in Fig. 15.3. In contrast to Fig. 15.3 size selection is made before the sintering furnace, leading to a smaller particle size. About 10 vol.% of oxygen is injected at the middle of the second furnace. The evaporation temperature is in the range of 735 °C and the sintering temperature is 650 °C. The nitrogen flow is 2 L/min. The transistor is prepared in top contact configuration. Thus, subsequently to the particle deposition Ag source and drain electrodes are made by evaporating Ag through a shadow mask within a vacuum chamber. In Fig. 15.5a an output characteristic is shown. It is easy to see that the layer is very conductive and the device shows no field effect. Also, no accumulation of charge carriers can be seen. The reason for that is an under stoichiometric composition of the material. It has been shown by Ramamoorthy et al. that by this deposition method, in-flight oxidation, only SnO$_x$ ($1 < x < 2$) can be received [39] which is conductive because of its intrinsic n-doping by the excess Sn-atoms. However, it has been shown before that stoichiometric SnO$_2$ can be obtained, if the in-flight sintering step is replaced by a post-oxidation step of the deposited film at 300 °C [39]. Performing such a post-oxidation step, the intrinsic conductivity of the film is drastically reduced by the formation of SnO$_2$. In those layers the drain current can be controlled by the gate voltage [40] as presented in Fig. 15.5b. But the output characteristic shows no saturation and does not follow the basic TFT characteristic. The almost linear dependency within the semi-log plot indicates an exponential rising drain current. This might be an indication for tunnel processes within the current path. It is worth to mention that similar output characteristics have been observed before by other authors [20] on SnO$_2$ layers grown by vapor-phase reactions.

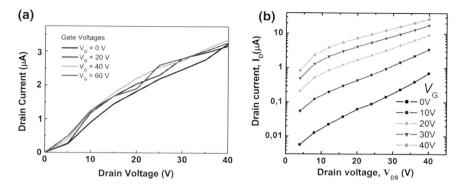

Fig. 15.5 Output characteristic obtained from a TFT with **a** SnO$_x$ (1<x<2) and **b** with SnO$_2$. Note, **b** is in semilogarithmic presentation [40]

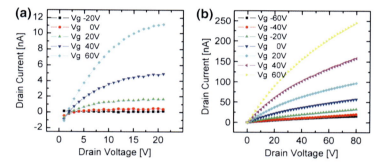

Fig. 15.6 Output characteristics of two identically produced TFTs with In$_2$O$_3$ nanoparticles, deposited directly from the carrier gas stream, indicting the low reproducibility due to uncertainties in the deposition process

With nanoparticulate indiumoxide films, better results are achieved because In$_2$O$_3$ can be easily generated by the oxidation of indium nanoparticles and, in contrast to SnO$_x$, it has a well-defined stoichiometry. In order to deposit In$_2$O$_3$ on the substrate the same flow rates for nitrogen and oxygen are used as for the SnO$_x$ deposition. The evaporation temperature is changed to 910 °C and the sintering temperature to 600 °C. In Fig. 15.6a, b two output characteristics of TFTs with polydisperse indiumoxide nanoparticles are shown. Both samples are prepared the same way and have the same transistor dimensions, but the output characteristics are significantly different. One can see that the currents are only in the range of nanoamperes, leading to a low on/off ratio of the devices. The sample in a) is more conductive; there is no off state even for $V_G = -60$ V and no saturation regime. In contrast, the characteristic given in b) shows saturation and the transistor almost exhibits an off state for $V_G = 0$V yielding a better control of the channel.

Thus, it has been shown that semiconducting nanoparticles can be deposited directly from a carrier gas stream onto a transistor structure and can be electrically analyzed subsequently. Clear field effect behavior could be proven. But it turns out that the layer morphology is hard to reproduce, leading to strong variations in the device parameters. Further, the very porous structure lowers the field effect, because only a few of the particles are in contact to the gate dielectric. Thus, the direct deposition of particles for transistor application has no technological advantages, while for a pure academic model system the deposition process is not reproducible enough.

15.3.3 TFTs with Nanoparticles from Dispersion

Nanoparticles can also be deposited from liquid dispersions. This is done by spin coating or inkjet printing, often followed by a post-treatment. The use of dispersions allows for a better control of the particle–particle interaction and in an optimal case, much denser films. Therefore, the deposition of nanoparticles for transistor applications from dispersions has been done by many other researches before. One of the first reports about a TFT made from nanodispersions was given from Ridley et al. in 1999 [41]. They used CdSe nanoparticles dispersed in pyridine and obtained mobilities around $1 \, cm^2/Vs$ if the films were post annealed at $350\,^{\circ}C$. ZnO nanoparticle-based TFTs were presented by Volkman et al. [42] obtaining effective mobilities around $0.1 \, cm^2/Vs$ after a post-annealing step. Higher effective mobilities without annealing of up to $0.9 \, cm^2/Vs$ have been obtained by Talapin et al. [43] with TFTs-based on PbSe nanoparticles. Like CdSe, PbSe might also not be the right material for a mass market due to toxicity. But this work clearly shows that the charge carrier mobility in nanoparticle films deposited at room temperature can exceed the mobility of charge carriers in aSi:H, processed at high temperatures. Further, in this work it was shown nicely, how the transport properties can be tuned from insulating over n-type semiconducting to p-type semiconducting, just by varying the additives filling the space between the particles. This emphasizes the crucial role of additives in dispersions.

In general, most dispersions need additives in order to prevent agglomeration of particles due to van-der-Waals interaction. Therefore, additives help to control the particle density in a film. For example, Fig. 15.7 compares two ZnO nanoparticle layers deposited by spin coating from a 2-methoxyethanol:ZnO dispersion, in case a) (upper figure) without further additives and in case b) (lower figure) with polyvinylpyrrolidone (PVP) as additive. One clearly sees that PVP allows for much denser and thinner films.

If no post-treatment is performed, the stabilizer remains in the film after deposition, affecting the electric properties. Usually, stabilizers are insulators and positive effects to the charge transport properties are not to expect [45]. But this is not necessarily the case, as will be shown here in the example of PVP.

Figure 15.8 summarizes several electric TFT parameters, measured in ambient air and in N_2 (glove box) for different PVP concentrations. For TFTs without PVP, there is a large difference between the behavior in ambient air and in glove box atmosphere. For example, the threshold voltage in ambient air (Fig. 15.8d) is about $-10 \, V$,

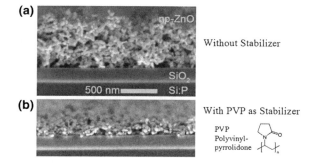

Fig. 15.7 ZnO nanoparticle film, deposited from a dispersion **a** without and **b** with PVP as stabilizer [44]

indicating a slight doping, but becomes about -275 V in glove box atmosphere, which indicates a large concentration of free charge carriers. Only a small amount of PVP is required in order to suppress this large threshold voltage and with increasing PVP content the difference in all these parameters between ambient air and N_2 becomes less pronounced. Obviously, the PVP makes the device performance independent from the surrounding atmosphere. Another interesting aspect is the increase of charge carrier mobility at low PVP content (Fig. 15.8a). Only for higher PVP content the mobility breaks down due to the isolating character of PVP. For device operation the threshold shift (Fig. 15.8b) is also a very important quantity. The threshold shift is the difference of threshold voltage between forward and backward sweep. Thus, it is a measure for the hysteresis in the device characteristics. Hysteresis is caused by carrier storage, often due to electrochemical reactions on surfaces. With higher PVP content the threshold shift vanishes, indicating well-reproducible characteristics without any storage phenomena.

This raises the question, why PVP has all this positive effects? An essential hint can be found by the work of Zhang et al. [46]. They used PVP in order to direct the crystal growth of ZnO from solutions. It turns out that PVP links selective to the ZnO surface, namely on the Zn-terminated (0001) surface and suppresses the crystal growth there. The adsorption process is described by a coordinated bond between the terminating metal cations of ZnO and O/N atoms from the pyrrolidone group. The unterminated (0001) surface is positively charged and acts most probably as a donor state, because the electron-accepting oxygen is missing due to the broken symmetry on the surface. These surface states are responsible for a high intrinsic conductivity, at least under inert N_2 atmosphere. This causes the large negative threshold voltage. PVP attaches on this surface and removes or compensates this donor state and reduces the threshold voltage to zero or shifts it even into the positive range. The increase of mobility for low PVP concentration is probably due to denser and smoother layer morphology. Since PVP attaches preferred on the (0001) surface, the other surfaces remain mainly uncovered. This allows for direct particle–particle contact, as long as the PVP content is not too high. A similar effect of PVP was reported by Königer et al. [47]. They found an improved conductivity of ITO nanoparticle films, if PVP was added up 40 vol.%. A second effect of PVP probably comes from its

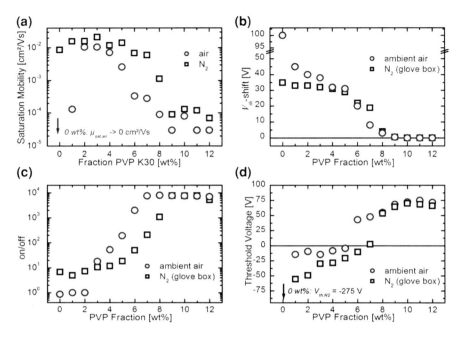

Fig. 15.8 Several ZnO-TFT parameters as function of the PVP content in the dispersion measured in ambient air and in N₂ (glove box); **a** saturation mobility, **b** threshold shift—the difference of threshold voltage between forward and backward sweep, **c** on/off ratio, **d** threshold voltage [44]

"water-bounding" property [48], acting as a getter and thus keeping water from the ZnO surface, making it more stable against humidity in ambient air.

Despite the positive effect of some additives, the particle–particle contact remains weak and hinders the charge transport. The obtained mobility in such systems usually does not exceed 1 cm²/Vs, if no post-annealing is performed. In order to reduce the number of particle boundaries along the transport path, anisotropic shapes, especially nanowires have been considered. This concept is, however, mainly restricted to ZnO, which grows easily anisotropic along its polar axis. While along a single ZnO nanowire high mobilities from 12 cm²/Vs up to 928 cm²/Vs have been reported [49–53], in printed or spin-coated films of nanowire conglomerates the mobility remains relatively low in the order of a few cm²/Vs [54, 55].

A fundamental problem in all TFTs-based on nano-object films is an inherent roughness between semiconducting film and gate dielectric. The charge carrier concentration in an accumulation channel decreases almost exponentially with distance to the dielectric interface. For a lateral transport along the interface, a roughness on the interface therefore corresponds to a strong modulation in the electric potential landscape. Okamura et al. [56] addressed this problem and have shown that small changes in the interface roughness have large effects on the effective charge carrier mobility.

15 Metal Oxide Thin-Film Transistors from Nanoparticles and Solutions 401

In order to overcome the fundamental problem of interface roughness an alternative transistor concept has been presented. Dasgupta et al. [57, 58] replaced the conventional dielectric by an electrolyte. The electrolyte acts as gate electrode, while an extreme thin electrochemical double layer of a few nanometer is self-assembled on the interface to the semiconductor. Since the electrolyte is able to fill out all accessible pores, the gate electrodes adapt automatically to the complex surface of the nanoparticle films. In this way, a huge field effect can be obtained and effective charge carrier mobilities of almost $25\,cm^2/Vs$ in In_2O_3:Sn (ITO) nanoparticle films have been reported. The principle of electrochemical gating is not new and has been used in a-Si:H TFTs [59], in polymer-based TFTs [60] or in order to control charge transport in carbon nanowires [61]. The main drawback of electrochemical gating is a relative low response time of the electrolyte, typically in the range of seconds to milliseconds, which limits the operation frequency of the transistor.

15.3.4 TFTs from Solutions (Liquid Precursors)

The intrinsic limitation of nanoparticle films is their granular structure. While the high amount of specific surface is important for gas-sensing applications [62], it becomes critical for TFT applications. First, because typically a TFT should mainly operate independent from its ambient conditions and second, the granular structure causes an interface roughness, which makes the coupling to an external electric field— necessary for a field effect device–more difficult. The concept of electrochemical gating can overcome this conflict; however, it is restricted to electrolytes, which usually have a low response time.

If one keeps conventional dielectrics, the semiconducting film has to be as smooth as possible, at least on the interface to the dielectric layer. Another approach to realize such smooth and printable semiconducting films is to start with a liquid precursor or a solution, containing all required ions and to transform it into the desired semiconductor. The challenge is not only to find adequate solutions or precursors for the different semiconductors, but also to allow for a transformation at acceptable temperatures. One example for such a system is an aqueous solution of zinc acetate ($[Zn(CH_3COO)_2 \cdot 2H_2O]$) which can be transformed into ZnO by annealing usually around $200\,^\circ C$ [63–66]. Also zinc nitride is known as a precursor, but requires higher temperatures for decomposition [67]. Additionally, similar systems are known for other metal oxides [68] and multi-cation amorphous oxides [69–72].

Meyers et al. [73] suggested an ammonia complex $Zn(OH)_x(NH_3)_y^{(2-x)+}$, which can be deposited by spin coating and gets transformed to ZnO by annealing. In order to get this complex they dissolve zinc nitrate in a caustic soda solution and zinc hydroxide (s) and sodium nitrate (aq) originates.

$$Zn(NO_3)_2 \cdot 6H_2O + 2NaOH \Rightarrow Zn(OH)_2 + 2Na(NO_3)$$

In the following, Na^+ and NO_3^- have to be removed in many cleaning steps. Subsequently the complex $Zn(OH)_x(NH_3)_y^{(2-x)+}$ is gained by adding aqueous ammonia.

Now, this solution can be processed by spin coating or printing and the water and ammonia will evaporate during annealing, forming a layer of ZnO. The disadvantage in this approach is the need for many cleaning steps to remove the ions, especially Na^+, which was introduced by adding NaOH. Na^+ is known to be extremely critical in transistor application, because it has a high diffusivity, and therefore incorporates easily in SiO_2 and other dielectric materials causing electric instabilities.

A more simple way to achieve the same complex can be gained by dissolving zinc oxide hydrate, $ZnO \cdot x\ H_2O$ (Sigma-Aldrich, 97 % purity) in aqueous ammonia (Sigma-Aldrich, ≥ 99.99 % purity, 28 % in H_2O) [74]. The idea is that the incorporated crystal water destabilizes the crystal and enhances its solubility. This yields the same aqueous ammonia complex without any intermediate steps leading to a simple one-step process. To produce layers of ZnO, in a first step the zinc oxide hydrate is dissolved under continuous stirring at 40 °C in an oil bath until a clear solution is obtained. For the devices shown, the molarity was in the range of 0.05–0.06 M/L. Then, test substrates Si/SiO_2 are impregnated with the solution by spin coating after filtering through a 700 nm glass filter. In order to transform the complex, annealing temperatures of only 125 °C are sufficient which offers the opportunity to work with flexible substrates. A cross-section of the layer was prepared utilizing a focused ion beam (FIB). Figure 15.9a presents a scanning transmission electron microscope (STEM) picture of this cross-section as well as an energy dispersive X-ray analysis (EDX) along the indicated line. The Zn signal in the EDX-spectra exhibits a strong localized Zn peak on the cross-over of the Si and Al signal, which is a proof for the presence of Zn, most probably from ZnO. The Al signal results from the top source/drain electrodes, while the Pt signal on the right side originates from Pt, which was deposited as a cover layer for the FIB preparation. From this measurement also the surprisingly low layer thickness of only a few nanometers can be seen. A more elaborated transmission electron microscope (TEM) analysis yields a mean layer thickness of approximately 7 nms(Fig. 15.9b). Also the homogeneity of the amorphous layer can be seen in this TE-micrograph, indicating a very homogeneous, smooth, and dense layer. There is no indication for a crystalline structure within this layer. An X-ray diffraction analysis (XRD) confirms the missing crystallinity. Such an amorphous ZnO layer deposited on a quartz substrate shows a transmission of almost 100 % in the visible range of light as can be seen in Fig. 15.9c and permits the fabrication of fully transparent devices. The absorption edge at around 400 nm corresponds to the fundamental absorption of ZnO. This is a clear proof for the presence of ZnO.

In order to analyze the electronic properties of the layer a test structure is used. Highly doped silicon serves as gate and thermally grown SiO_2 as insulator with a thickness of 200 nm. The transistors are built in bottom gate top contact layout and with the use of aluminum as top electrode material because of its work function of 4.28 eV fitting very well to the conduction band of the ZnO (4.3 eV) which makes us expect an Ohmic contact between intrinsic n-semiconducting layer and source and drain electrodes, obtained by thermal evaporation of aluminum through a shadow mask. All devices shown have the same channel length of $L = 100$ µm and channel width of $W = 7.4$ mm.

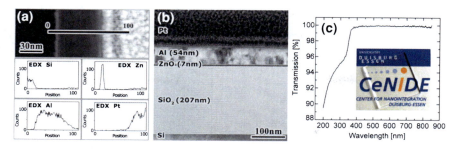

Fig. 15.9 **a** EDX linescan along the device cross-section to verify the composition of the layer **b** transmission electron microscope image of the device cross-section, **c** optical transmission spectrum of the ZnO layer on quartz [74]

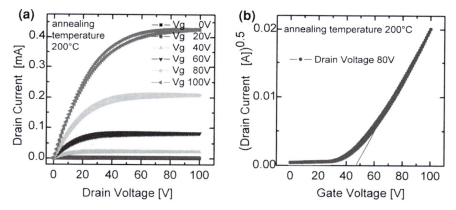

Fig. 15.10 Output and transfer characteristics, measured under nitrogen and *yellow* light. Field effect mobility of this device is 0.22 cm^2/Vs and the threshold voltage is 47 V

Figure 15.10 shows an output and transfer characteristic of a device prepared as described above. The annealing temperature used was 200 °C and the characteristics were measured under nitrogen atmosphere and yellow light, which are used as standard conditions for all electric measurements. One can see that there is an Ohmic behavior in the linear regime and clear field effect behavior. Almost no hysteresis indicates good layer properties. A clear saturation can also be seen and characteristic electrical properties specified by threshold voltage and mobility can be determined. In case of the device shown in Fig. 15.10 the threshold voltage is 47 V and the field effect mobility is 0.22 cm^2/Vs.

As mentioned above, 125 °C annealing temperature is sufficient to transform the ammonia complex into a layer of ZnO. To find the optimum temperature for a high mobility, transistor devices are prepared with different annealing temperatures varying between 125 and 500 °C. Depending on the annealing temperature one can see nearly no trend for the threshold voltage (Fig. 15.11a), but for the mobility. Figure 15.11b shows that there is an optimum for samples annealed at 300–350 °C.

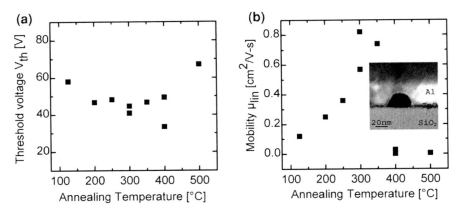

Fig. 15.11 Threshold voltage **a** and mobility **b** as function of the annealing temperature. Inset: ZnO crystal growth after annealing at 500 °C [74]

The decrease in mobility for higher temperatures can be caused by a decrease of semiconductor–insulator interface quality, caused by the beginning of a crystallization process in the layer, see inset in Fig. 15.11b, where a TEmicrograph is shown of a sample that was annealed at 500 °C. Compared to the micrograph shown in Fig. 15.9b the ZnO seems to be much more inhomogeneous and a pronounced crystallization has taken place.

So far, the samples were annealed under ambient atmosphere and owing to the change in moisture in the ambient atmosphere, a more defined annealing atmosphere is needed to increase the reproducibility of the TFTs. As defined atmospheres, nitrogen, synthetic air, and 5 % of forming gas (5 % H_2 in N_2) are used. In order to find the maximum of possible mobility in different annealing atmospheres, the annealing temperature for the samples is kept at 300 °C. The highest mobilities are achieved for samples that are annealed under forming gas atmosphere; this can be expected because hydrogen can be found, e.g., in antibonding configurations, acting as shallow donor [75].

Mobilities up to 1.2 cm^2/Vs are possible, but regarding all devices produced, there is a statistical variation between samples prepared the same way also. To stabilize the process further, ambient air exposure of the samples has to be avoided after electrode evaporation as can be seen from Fig. 15.12, where two output characteristics are shown. Both samples were annealed together for 45 min at 125 °C under ambient atmosphere. The sample in Fig. 15.12a was not exposed to air after electrode evaporation; the other sample was exposed to synthetic air for 30 minutes. It can be seen that the layer in Fig. 15.12a is very conductive; a resistance of 23.4 kΩ can be calculated from the *I–V* characteristics with zero volts at the gate. No saturation can be seen. This is probably due to a high concentration of free charge carriers in the channel preventing a pinch-off. In contrast to that, the sample that was exposed to synthetic air (Fig. 15.12b) shows clear transistor behavior with a visible off state for

15 Metal Oxide Thin-Film Transistors from Nanoparticles and Solutions

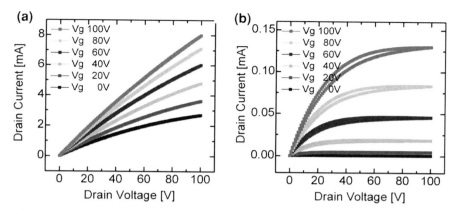

Fig. 15.12 Output characteristics of ZnO-TFTs **a** the device was kept in N_2 (*glove box*) atmosphere after electrode deposition; **b** the device was kept in synthetic air for 30 min after electrode deposition

$V_G = 0$ V and a clear linear and saturation regime. The characteristic parameters for this device are a threshold voltage of 14.2 V and a mobility of 0.05 cm^2/Vs.

The decrease in conductance from metallic to semiconducting behavior is assigned to adsorbates on the ZnO layer surface, saturating dangling bonds. These adsorbates are removed inside the vacuum of the evaporation chamber and brought back to the sample that has been exposed to synthetic air. In addition, the formation of O_2^- adsorbates on the ZnO surface is likely to form a surfacial depletion layer emphasizing the effect on the conductivity, e.g., as shown for ZnO nanowires [76, 77]. Caused by the small thickness of the solution deposited layers, the surface, and therewith adsorbates have a high influence on the channel performance. By changing the time the sample is exposed to synthetic air, it seems to be possible to optimize the layer properties and therewith the threshold voltage can be decreased.

15.4 Summary

Three different methods to build semiconducting layers for TFTs in a nanoscale range have been reviewed. (1) A direct deposition of gas carried nanoparticles has the advantage that no chemical additives are required. With respect to chemical purity, this method would have the highest potential. However, the crucial point is the morphology of the particle layer and the exact positioning of the deposition area. The last point might be overcome by more sophisticated deposition techniques, while the more porous or fuzzy morphology of the layers is an inherent problem, resulting from strong van-der-Waals interaction between the particles. Only particles with extremely high kinetic energy can overcome this problem.

(2) A deposition from dispersion allows for more control over the particle–particle interaction. More smooth and dense layers are possible in this way. Such dispersions might be spin coated or dip coated for large area deposition or printed for a more localized deposition. However, in order to control the particle–particle interaction suitable solvents and additives are usually required, but they are additional sources for chemical impurities in the thin-film. Additives improve the layer morphology, but might hinder the electronic transport, because most additives are electric insulators. There is a small process window, where some additives improve the morphology, compensate unwanted electronic states, and therefore enhance the overall electronic transport properties. This was presented for PVP here.

(3) The preparation of semiconducting oxide layers from solution or liquid precursors. This method overcomes the problem of the granular structure of the layer leading to an interface roughness, which is inherent if nanoparticles are used. However, some precursors or solutions are a source for additional impurities again. Small ions with high diffusion coefficient, such as Na^+, are extremely critical. These mobile ions are the origin for electric instabilities of the whole device. Using a low-temperature simple one-step process, an ammonia complex $Zn(OH)_x(NH_3)_y^{(2-x)+}$ could be prepared. It decomposes to ZnO at around $125\,°C$ or higher, while the chemical by-products are in gas phase and evaporate easily, which ensures high purity. This method allows for building up an amorphous, transparent, and surprisingly thin layer of ZnO. Those layers show a high performance in TFT devices with mobilities up to $1.2\,cm^2/Vs$ ($300\,°C$ annealing). This process also offers the opportunity to work on flexible substrates, because only an annealing temperature of $125\,°C$ is necessary to transform the complex into ZnO. Fully transparent devices can be built because of the excellent transparency of the films. If reproducibility can be upgraded this offers a great potential for electrical applications such as RFID tags, etc.

15.5 Conclusion

It has been shown that semiconducting oxide layers can be realized at temperatures well below the requirements for a common a-Si:H technology but keeping effective charge carrier mobilities at least in the same order of magnitude and maybe even better. In combination with the high transparency of such layers a clear advantage can be seen in metal oxide processes compared to conventional a-Si:H technology. The approach to use amorphous instead of polycrystalline layers seems to have further advantages, because critical surfaces are suppressed. Nevertheless, more elevated temperatures allow for more control over the chemical decomposition and leads to higher device performance. In order to combine the requirement of higher temperatures for the chemical decomposition and the low-temperature requirement for low-cost substrates, methods of local energy insertions are required. Laser annealing or rapid thermal annealing might be a way to overcome this contradiction.

15 Metal Oxide Thin-Film Transistors from Nanoparticles and Solutions

Acknowledgments This work was financially supported by the German Research Foundation (DFG) within the collaborative research program SFB 445. Further, part of the work was financially supported by the Dutch Polymer Institute (DPI) and the Center for NanoIntegration Duisburg-Essen (CENIDE).

References

1. F. Ebisawa, T, Kurokawa, S. Nara. J. Appl. Phys. **54**, 3255 (1983)
2. A. Tsumura, H. Koezuka, T. Ando, Appl. Phys. Lett. **49**, 1210 (1986)
3. K. Gilleo, National Electronic Packaging and Production Conference—Proceedings of the Technical Program (West and East) **3**, 1390–1401 (1992)
4. F. Garnier, R. Hajlaoui, A. Yassar, P. Srivastava, Science **265**, 1684 (1994)
5. W. Clemens, W. Fix, J. Ficker, A. Knobloch, A. Ullmann, J. Mater. Res. **19**, 1963 (2004)
6. A. Facchetti, Materialstoday. **10**, 29 (2007)
7. U. Scherf, Angew. Chem. Int. Ed. **47**, 2 (2008)
8. M.E. Gershenson, V. Podzorov, A.F. Morpurgo, Rev. Mod. Phys. **78**, 973 (2006)
9. G.W. Neudeck, A.K. Malhotra, Solid-State Electron. **19**, 721 (1976)
10. A. Deneuville, M.H. Brodsky, Thin Solid Films **55**, 137 (1978)
11. A. Sazonov, D. Striakhilev, Czang-Ho Lee, A. Nathan, Proc. IEEE, **93**, 1420 (2005)
12. J.E. Lilienfeld, Patent # CA272437 (1926)
13. P.K. Weimer, Proc IRE **50**, 1462 (1962)
14. W. Shockley, Proc. IRE **40**, 1365 (1952)
15. R. Schmechel, M. Ahles, H. von Seggern, J. Appl. Phys. **98**, 084511 (2005)
16. A. Rolland, J. Richard, J.-P. Kleider, D. Mencaraglia, J. Electrochem. Soc. **140**, 3679 (1993)
17. W.L. Kalb, B. Battlog, Phys. Rev. B **81**, 035327 (2010)
18. G.F. Boesen, J.E. Jacobs, Proc. IEEE **56**, 2094 (1968)
19. H.A. Klasens, H. Koelmans, Solid-State Electron. **7**, 701 (1964)
20. A. Aoki, H. Sakakura, Japan. J. Appl. Phys. **9**, 582 (1970)
21. G. Eranna, B.C. Joshi, D.P. Runthala, R.P. Gupta, Critical Reviews in Solid State and Materials Sciences **29**, 111 (2004)
22. A.N. Banerjee, K. K Chattopadhyay, Progress In Crystal Growth and Characterization of Materials **50**, 52 (2005)
23. K. Ellmer, J. Phys. D: Appl. Phys. **34**, 3097 (2001)
24. R.L. Weiher, J. Appl. Phys. **33**, 2834 (1962)
25. A. Janotti, C.G. Van de Walle, Reports On Progress In Physics **72**, 126501 (2009)
26. H. Ohta, H. Hosono, Materials Today **7**, 42 (2004)
27. S. Bubel, N. Mechau, H. Hahn, R. Schmechel, J. Appl. Phys. **108**, 124502 (2010)
28. H. Hosono, N. Kikuchi, N. Ueda, H. Kawazoe, J. non-cryst. solids **198–200**, 165 (1996)
29. T. Kamiya, K. Nomura, H. Hosono, Sci. Technol. Adv. Mater. **11**, 044305 (2010)
30. S. Masuda, K. Kitamura, Y. Okumura, S. Miyatake, H. Tabata, T. Kawai, J. Appl. Phys. **93**, 1624 (2003)
31. R.L. Hoffmann, B.J. Norris, J.F. Wager, Appl. Phys. Lett. **82**, 733 (2003)
32. E. Fortunato, A. Pimentel, L. Pereira, A. Gonçalves, G. Lavareda, H. Águas, I. Ferreira, C.N. Carvalho, R. Martins, J. Non-Cryst, Solids **338–340**, 806 (2004)
33. H. Frenzel, A. Lajn, M. Brandt, H. von Wenckstern, G. Biehne, H. Hochmuth, M. Lorenz, M. Grundmann, Appl. Phys. Lett **92**, 192108 (2008)
34. M.W.J. Prins, S.E. Zinnemers, J.F.M. Cillessen, J.B. Giesbers Appl, Phys. Lett. **70**, 458 (1997)
35. R.E. Presley, C.L. Munsee, C.-H. Park, D. Hong, J.F. Wager, D.A. Keszler, J. Phys. D: Appl. Phys **37**, 2810 (2004)
36. L. Wang, M.-H. Yoon, G. LU, Y. Yang, A. Facchetti, T. J. Marks. Nature Materials **5**, 893 (2006)

37. K. Nomura, H. Ohta, A. Takagi, T. Kamiya, M. Hirano, H. Hosono, Nature **432**, 488 (2004)
38. H. Yabuta, M. Sano, K. Abe, T. Aiba, T. Den, H. Kumomi, K. Nomura, T. Kamiya, H. Hosono, Appl. Phys. Lett. **89**, 112123 (2006)
39. R. Ramamoorthy, M.K. Kennedy, H. Nienhaus, A. Lorke, F.E. Kruis, H. Fissan, Sens. Actuators, B **88**, 281–285 (2003)
40. D.R. Chowdhury, A. Ivaturi, A. Nedic, F.E. Kruis, R. Schmechel, Physica E **42**, 2471 (2010)
41. B.A. Ridley, B. Nivi, J.M. Jacobson, Science **286**, 746 (1999)
42. S.K. Volkman, B.A. Mattis, S.E. Molesa, J.B. Lee, A.F. Vornbrock, T. Bakhishev, V. Subramanian, IEDM Tech. Dig. 1072 (2004)
43. D.V. Talapin, Ch.B. Murray, Science **310**, 86 (2005)
44. S. Bubel, N. Mechau, R. Schmechel, J Mater Sci **46**, 7776–7783 (2011)
45. N. Mechau, S. Bubel, D. Nikolova, H. Hahn. Phys. Status Solidi A, **207**(7), 1684–1688 (2010)
46. J. Zhang, H. Liu, Z. Wang, N. Ming, Z. Li, A.S. Biris, Adv. Funct. Mater **17**, 3897 (2007)
47. T. Königer, H. Münstedt, J. Mater. Sci. **44**, 2736 (2009)
48. M.J.A. de Dood, J. Kalkman, C. Strohhöfer, J. Michielsen, J. van der Elsken, J. Phys. Chem. B **107**, 5906 (2003)
49. H.T. Ng, J. Han, T. Yamada, P. Nguyen, Y.P. Chen, M. Meyyappan, Nano Letters **4**, 1247 (2004)
50. S.N. Cha, J.E. Jang, Y. Choi, G.A.J. Amaratunga, G.W. Ho, M.E. Welland, D.G. Hasko, D.-J. Ka, J.M. Kim, Appl. Phys. Lett. **89**, 263102 (2006)
51. S. Ju, J. Li, N. Pimparkar, M.A. Alam, R.P.H. Chang, D.B. Janes, IEEE Trans. Nanotechnol. **6**, 390 (2007)
52. S. Ju, A. Facchetti, Y. Xuan, J. Liu, F. Ishikawa, P. Ye, Ch. Zhou, T.J. Marks, D.B. Janes, Nature Nanotechnology **2**, 378 (2007)
53. H. Wu, D. Lin, R. Zhang, W. Pan, J. Am. Ceramic Soc. **91**, 656 (2008)
54. B. Sun, H. Sirringhaus, Nano Letters **5**, 2408 (2005)
55. B. Sun, H. Sirringhaus, J. Am. Chem. Soc. **128**, 16231 (2006)
56. K. Okamura, N. Mechau, D. Nikolova, H. Hahn, Appl. Phys. Lett. **93**, 083105 (2008)
57. S. Dasgupta, S. Gottschalk, R. Kruk, H. Hahn, Nanotechnology **19**, 435203 (2008)
58. S. Dasgupta, S. Dehms, R. Kruk, H. Hahn, Acta Physica Polonica A **115**, 473 (2009)
59. D. Gonçlves, D.M.F. Prazeres, V. Chu, J.P. Conde, IEEE Electron Device Lett. **29**, 1030 (2008)
60. J.D. Yuen, A.S. Dhoot, E.B. Namdas, N.E. Coates, M. Heeney, I. McCulloch, D. Moses, A.J. Heeger, J. Am. Chem. Soc. **129**, 14367 (2007)
61. T. Tang, A. Jagota, J. Comput. Theor. Nanosci. **5**, 1989 (2008)
62. O. Lupan, V.V. Ursaki, G. Chai, L. Chow, G.A. Emelchenko, I.M. Tiginyanu, A.N. Gruzintsev, A.N. Redkin, Sens. Actuators, B **144**, 56–66 (2010)
63. D. Levy, L. Irving, A. Childs, Digest of Technical Papers - SID International. Symposium **38**, 230 (2007)
64. J. Tellier, D. Kuščer, B. Malič, J. Cilenšek, M. Škarabot, J. Kovač, G. Gonçalves, I. Muševič, M. Kosec, Thin Solid Films **518**, 5134 (2010)
65. B.S. Ong, Ch. Li, Y. Li, Y. Wu, R. Loutfy, J. Am. Chem. Soc **129**, 2750 (2007)
66. H.-C. Cheng, C.-F. Chen, C.-Y. Tsay, Appl. Phys. Lett. **90**, 012113 (2007)
67. B.J. Norris, J. Anderson, J.F. Wager, D.A. Keszler, J. Phys. D: Appl. Phys. **36**, L105 (2003)
68. D.-H. Lee, Y.-J. Chang, G.S. Herman, C.-H. Chang, Adv. Mater. **19**, 843 (2007)
69. Y.W., X.W. Sun, G.K.L. Goh, H.V. Demir, H.Y. Yu. IEEE Trans. Electron Devices **58**, 480 (2011)
70. G.H. Kim, H.S. Shin, B.D. Ahn, K.H. Kim, W.J. Park, H.J. Kim, J. Electrochem. Soc. **156**, H7 (2009)
71. J.H. Lim, J.H. Shim, J.H. Choi, J. Joo, K. Park, H. Jeon, M. R. Moon, D. Jung, H. Kim, H. J. Lee, Appl. Phys. Lett. **95**, 012108, (2009)
72. Y. Wang, S.W. Liu, X.W. Sun, J.L. Zhao, G.K.L. Goh, Q.V. Vu, H.Y. Yu, J. Sol-Gel Sci, Technol. **55**, 322 (2010)
73. S.T. Meyers, J.T. Anderson, C.M. Hung, J. Thompson, J.F. Wager, D.A. Keszler, J. Am. Chem. Soc. **130**, 17603–17609 (2008)

74. R. Theissmann, S. Bubel, M. Sanlialp, C. Busch, G. Schierning, R. Schmechel, High performance low temperature solution-processed zinc oxide thin-film transistor, Thin Solid Films **519**(16), 5623–5628 (2011)
75. M.D. McCluskey, S.J. Jokela, K.K. Zhuravlev, P.J. Simpson, K.G. Lynn, Appl. Phys. Lett. **81**(20), 3807–3809, (2002)
76. Z. Fan, D. Wang, P.-C. Chang, W.-Y. Tseng, J.G. Lu, Appl. Phys. Lett. **85**, 5923 (2004)
77. Q.H. Li, Y.X. Liang, Q. Wan, T.H. Wang, Appl. Phys. Lett. **85**, 6389 (2004)

Index

A
Ab initio, 78, 90, 91, 95, 278
Admittance, 235
Aerosol, 330, 334, 335
Agglomerate, 63, 109, 163, 168
Agglomeration, 127, 139, 140, 146, 150–153, 155–157
Alloy nanoparticles, 111, 114
Alumina, 49
Amorphous oxides semiconductors, 394
Amorphous silicon, 388
Anatase, 53, 62, 141, 152–154, 156
Angular momentum conservation, 220
Anisotropy energy density, 297
Anisorophy of the hole effective mass, 322
Annealing temperature, 103, 107, 403
Antiferromagnet, 308
Arrhenius diagram, 13, 14
Asphericity, 171
Atomic layer deposition, 50
Atomic resonance absorption spectrometry, 8, 13
Au-Ge pair particles, 111
Auger electron spectroscopy, 36, 331
Averaging effect, 189, 193, 196
Axial GaAs nanowire p–n-junctions, 368
Axial heterojunctions, 360

B
Bain transformation, 81
Ballistic mean free path, 232
Band bending, 390
Bandgap, 199, 210, 214
Bandgap emission, 314
Bimetallic nanoclusters/nanoparticles, 110
Bimolecular reactions, 9, 11

Binary clusters, 85
Bipolar chargers, 102
Bipolar injection, 227
Blocking temperature, 274
Brick-layer-model, 233
Bright exciton, 216, 321
Brillouin scattering, 212
Brownian coagulation, 100
Bubbler, 60
Built in voltage, 201
Bulk diffusion, 165
Butane, 332, 341
Butterfly structure, 188, 189

C
Cantera, 26
Capacitance, 268
Carbon tetra bromide (CBr_4), 361
Carrier concentration, 365
Carrier confinement, 317
Carrier density, 233
Carrier injection, 212
Carriers, 357, 370
Catalysts, 5, 48, 355
CdSe, 304
Ceramics, 49
Ceramics, catalysts, fuel cells, photovoltaic devices, 49
CFD codes, 28
CFD simulation, 30
Characteristic length scales, 232
Charge carrier mobility, 388
Charge double-layer, 153, 154
Charge Transfer Multiplet Program (CTM4XAS), 285
Chemical composition, 67

C (cont.)

Chemical contrast, 290
Chemical kinetics mechanisms, 4
Chemical order, 283
Chemical potential, 164
Chemical vapor deposition, 49
Chemical vapor synthesis, 49
Chemistry modeling, 23
Cluster, 146, 150, 151, 155
Co, 355
Coagulation, 51
Coalesce, 61
Coalescence, 148, 163, 166, 168, 169, 174
Coalescence time, 165, 169
Coarsening, 180
Cobalt doped zinc oxide, 64
Co-discharging, 111
Coercive field, 284
Coercivity, 132
Coffee ring, 264
Cohesion, 162
Cohesive powders, 236
Coincidence site lattice index, 173, 174, 176, 177
Collective magnetic properties, 273
Collective properties, 274
Combustion synthesis
Common neighbor analysis, 83, 146, 147
Compaction, 236, 237
Compactification, 163, 180
Compensated region, 377, 378
Complex oxide material, 67
Composite oxides, 37
Composite particles, 37
Computer simulations, 167, 173
Concave, 169
Conductance, 235
Conductivity, 236, 266
Configurations, 167
Confinement energies, 216
Conjugate-gradient method, 147
Constant phase element, 257
Contact resistances, 197, 242, 244
Convex, 169
Cooling rate, 61
Coordination number at the surface, 297
CoPt, 124
Core conductivity, 235
Core resistance, 241
Core-shell, 360
Corona dischargers, 103
Co-sparking, 114
Creeping compaction, 268
Critical temperature, 144
Cross sensitivities, 330

Crystallinity, 53
Crystallographic orientation, 171, 173, 180
Crystalline quality of, 362
Crystal structure of pn-doped
    GaAs nanowires, 369
Cuboctahedron, 77, 81, 88, 95, 150, 155
Curie temperature, 145
Current assisted compaction, 248
Current assisted powder compaction, 246
Current density, 372
CVSSIN, 59, 61

D

Dangling bond, 405
Dark exciton, 216, 321
Dark excitons, 219
Debroglie wavelength, 232
Debye frequency, 167
Debye length, 232
Degussa P25, 5
Degree of agglomeration, 61–63
Denitriding, 132
Density functional theory, 140, 141, 152
Depletion length, 232
Deposition efficiency, 116, 118
Deposition techniques, 101
Diamagnetic shift, 211
Dielectric, 152
Dielectric response, 233
Dielectric spectroscopy, 235
Diethyl zinc (DEZn), 361
Differential mobility
    analyzer (DMA), 395
Dihedral angle, 172, 180
Dimer configuration, 292
Dipolar coupling, 297
Dipole interaction, 143
Direct band gap, 222
Discharging of particles, 107
Dislocations, 149
Dispersion, 388, 398
Dispersion number, 51
Dispersions, 238
Ditertiarybutyl silane (DitbBuSi), 361
Donor bound exciton emission, 307
Doping concentration, 391
Doping contrast
Doping gradient, 366
Doping of nanowires, 362
Doping transitions, 188, 197
Drain-source voltage, 389
Driving force, 164
Drude-type conductivity, 232
Dual conductance response, 350

Index

Dumbbell configuration, 172
Dynamic light scattering, 238
Dynamic properties, 273

E

Easy axis, 290
EDX, 278
Effect of particle size, 343
Effective carrier concentration, 195, 197
Effective mobility, 392
Electric modulus, 235
Electrical properties, 231
Electrochemical gating, 401
Electrode, 395
Electrode distance, 239, 329
Electroluminescence, 212, 224, 227, 379
Electroluminescent spectra, 381
Electron affinity, 188
Electron energy loss spectroscopy, 133
Electron-hole pair, 320
Electronic effect (EE), 352
Electrostatic deposition efficiency, 117
Electrostatic precipitation, 116
Electrostatic repulsion, 153
Elementary excitations, 95
Element-specific hysteresis loops, 273
Element-specific magnetic
    moments, 285
Emission, 376–379, 381
Energy barrier, 157
Energy-dispersive X-ray
    spectroscopy (EDX), 277
Equivalence ratio ϕ, 35, 41
Equivalent circuits, 234
Ethanol, 226, 254
Evaporation, 334, 337
Evaporation rate, 67
Extended X-ray absorption fine structure
    (EXAFS), 66, 70, 275, 278
Exchange field, 313, 321
Exchange interaction, 315, 317
Exciton, 211, 315
Exciton binding energy, 215
Exciton magnetic polaron, 313, 316
Exciton trapping, 211

F

Face-centered-cubic, 124
Faraday cups, 16
$Fe(CO)_5$, 9, 10, 19
Fe-Pt-Cu, 136
FePt, 124
Ferroelectric transition, 152, 153

Ferrofluids, 41, 48
Ferromagnetic alignment, 316
Ferromagnetic ordering, 85
Ferromagnetic resonance (FMR), 273, 294
Ferromagnetism, 304, 313
Field cooling, 308
Field effect mobility, 392
Field effect transistor, 387
Film formation, 101, 103, 105, 107, 109, 111,
    113, 115, 117, 119
Flame radicals, 29
Flame reactor, 4, 5, 13
Flame-based synthesis, 3, 5
Flame-synthesized particles, 5, 32
Flash evaporator, 52
Flat-flame reactor, 18, 19
Flexible substrates, 402
Focused ion beam, 402
Fractal dimension, 164
Free charge carriers, 262
Free energy, 95
Frustration, 293
FTIR, 66
Fuel cells, 49
Fumed silica, 49
Functional applications, 99
Functional nanoparticles, 5

G

GaAs nanowire pn junction, 375
GaAs nanowires, 362
GaAs/GaP heterostructure nanowire, 185
GaN, 304
Gas phase synthesis technique, 330
Gas sensors, 49, 245
Gate-source voltage, 389
Generalized gradient
    approximation (GGA), 78
Geometry factors, 245
G-factor, 297
Gilbert damping parameter, 297
Gold (Au) seeds, 359
Grain boundary, 171, 172, 178, 180, 235
Grain boundary diffusion, 163
Grain boundary energy, 175
Grain boundary pinning, 172, 179, 180
Green body, 162
Growth of III/V nanowires, 358
Growth rates, 191

H

Hard agglomerate, 162, 180
H-atom recombination cycle, 28

Index

H (*cont.*)
Hard direction, 290
Height above burne, 14, 35
Hematite, 285
Heterojunctions, 191
Heusler alloys, 81, 95
High-temperature plasma, 102
High-temperature superconductors, 37
Homogeneous nucleation, 104
Hot spot, 61
Hot-wall reactor, 60, 267
Humidity sensors, 265
Hybrid simulation technique, 173
Hydrogen plasma, 277
Hysteresis, 399
Hysteresis loop, 291

I
Icosahedral, 125
Icosahedron, 80, 81, 83–85, 88, 91, 95, 150
III/V nanowire, 359, 361, 363, 365, 367, 369, 371, 373, 375, 377, 379, 381, 383, 385
Impact ionization, 228
Impactor, 395
Impedance, 242, 243
Impedance spectroscopy, 235
In_2O_3, 397
InAs nanowire field effect transistors, 370, 373
InAs nanowires, 362
InAs/InP radial nanowire heterostructure, 185
Incipient ferroelectric, 152
Incorporation of the doping atoms, 364
Indirect bandgap, 222
Indium tin oxide, 178
Individual nanoparticles, 287
Individual properties, 274
Induction furnace, 60
In-flight annealing, 4, 125, 330, 331
Influence, 337, 339
Information technology, 64
Inkjet printing, 398
Ink-jet printing, 238, 264, 398
Interdigital structures, 264
Interdigitated, 395
Interfacial free energy, 172, 180
Interparticle transport, 234
Inter-particle collisions, 127
Intra-atomic dipole term, 282
Intraparticle transport, 234
Intrinsic magnetism, 273
Inverse spinel, 285
Internal resistance, 197

Ion implantation, 194
Iron, 150, 151
Iron clusters, 9, 11
Iron nitrides, 135
Iron oxide, 17, 19
Iron pentacarbonyl, 9, 45
Iron precursor, 39
Iron silicates, 39
Iron-oxide, 40
I-V characteristics, 404
I-V characteristics of a single pn-GaAs nanowire, 376

J
Jahn-Teller, 77, 81, 93

K
Kelvin probe force microscopy (KPFM), 186
Kinetic Monte Carlo simulation, 167
Kirchhoff's law, 242
Koch-Friedlander equation, 164, 168
Kurdjumow-Sachs transformation, 81

L
$L1_0$ phase, 124
Landauer-Büttiker approach, 232
Landau-Lifshitz-Gilbert equation, 294
Laser ablation, 102
Laser duty cycle, 56
Laser flash, 49
Laser flash Evaporation, 51, 64
Laser pulse repetition frequency, 54
Laser reactors, 6
Laser-induced fluorescence, 3, 17
Laser-induced incandescence, 3
Lateral, 192, 200
Lattice imperfections, 128
Lattice mismatch, 358
Lattice model, 247
Layer sensitivity, 348
Light emitting diode, 224
Light extinction, 9, 22
Light-to-current conversion, 358
Log-normal size distribution, 52
Long term stability, 329
Low pressure impaction (LPI), 116

M
Mackay transformation, 77, 81, 82, 93
Maghemite, 285

Index 415

Magic number clusters, 80, 82
Magnetic anisotropy, 283
Magnetic anisotropy constant, 86, 87, 91
Magnetic anisotropy energy, 123
Magnetic anisotropy energy density, 273
Magnetic data storage devices, 86
Magnetic dipolar coupling, 292
Magnetic dipole interactions, 283
Magnetic dopant, 315
Magnetic dopants, 65
Magnetic fluctuations, 293
Magnetic hysteresis loops, 290
Magnetic moments, 83, 85
Magnetic polaron, 321
Magnetic semiconductor, 64, 323
Magnetic semiconductors, 304
Magnetic storage, 123
Magnetic switching, 275
Magnetic switching, 275
Magnetisation curve, 293
Magnetisation dynamics, 294
Magnetism, 79
Magnetite, 285
Magnetite maghemite, 41
Magnetization, 308, 317, 322
Magnetocrystalline anisotropy, 77, 87–90, 93, 283
magnetocrystalline anisotropy energy, 274
Magsilica®, 41
Mars van Krevelen mechanism, 351
Mass spectroscopy, 79
Mass transport, 164
Massively parallel computing, 78
Material, 357, 358
Maximum stable gain (MSG), 374
Maxwell–Boltzmann statistics, 219
Melting temperature, 180
Metallic particles, 108
Metal-organic vapour-phase epitaxy, 358
Metastable, 180
Micromagnetic simulations, 291
Microscopy, 362
Microsensor, 333
Mixed oxides, 37
Mixing energy, 85
Mobility, 165, 172
Mobility enhancer, 129
Mobility-based diameter, 103
Molecular dynamics, 214
Molecular resonance absorption spectroscopy, 9
Molecular-beam sampling, 15
Molecular-dynamics simulations, 140, 142, 146, 150
Monodisperse, 105, 276

Monodisperse distribution function, 24
Monodisperse nanoparticles, 101, 103, 105, 107, 109, 111, 113, 115, 117, 119
Monodisperse particles, 102
Monosized SnO_x, SnO_x:Pd, SnO_x:Ag or pure Pd, 350
Morphology, 150–152
MOVPE, 358
Multiferroic nanocomposites, 140
Multilayers, 240
Multiple domains, 109
Multiple twinning, 107
Multiply twinned, 4, 87, 125

N

Nanocomposites, 37
Nanocrystalline oxides, 49
Nanocrystalline solid, 181
Nanoparticle ensembles, 294
Nanoparticle pair, 113
Nanopowder, 161
Nanostructured thin films, 329
Nanowire, 400
Nanowire FET, 372
NASA database, 26
NBE emission, 311
Near bandgap emission, 305
Nearest neighbor bonds, 85
Neck growth, 162, 175
Néel relaxation time, 87
Néel surface anisotropy model, 91
Nickel, 146, 149, 167
Nishiyama-Wassermann transformation, 81
Nitriding, 132
Noncontact mode, 188
Non-radiative energy transfer, 313
Non radiative recombination, 218
n-type response, 341
Nyquist diagram, 235, 257
Nyquist plot, 251, 253

O

OH imaging, 19
Omega shaped gate MISFET, 371
Optical band gap, 210
Optical communication, 225
Optical diagnostics, 22
Optical spectroscopy, 304
Optically induced magnetization, 319, 323
Orbital magnetic moment, 283, 284
Order/disorder transformation, 125
Orientation time, 175
Orientational mismatch, 173

O *(cont.)*
Oscillator strength, 221
Output characteristics, 372
Oxide-free nanoparticles, 277
Oxidic nanocomposite materials, 152
Oxidic semiconductors, 394
Oxidizing radicals, 35

P

Pair distribution function, 71, 83
Pair particle formation, 112
Paramagnetic behaviour, 308
Particle configurations, 274
Particle growth, 50
Particle mass, 15, 44
Particle mass spectrometer, 15
Particle reorientation, 173, 175
Particle size, 50
Particle size distribution, 105
Particle-particle contact, 400
Particulate networks, 241
P-doping level, 194
Percolation, 245
Percolation threshold, 245
Phase coherence length, 232
Phase diagram, 359
Phenomenological theory, 167
Phonon confinement, 211
Phonon replica, 312
Phonon-less transitions, 217
Phonons, 316
Phosphor, 224
Photocatalysis, 59
Photo-generated charge carriers, 316
Photolithography, 226
Photoluminescence, 210, 305, 376
Photon correlation spectroscopy, 53
Photovoltaic devices, 49
Photovoltaics, 59
Plasma reactors, 6
Plasma cleaned, 297
p–n junction nanowire, 200
Poisoning, 341
Polarons, 317
Polycrystalline, 127
Polydisperse, 100
Population balance equations, 24
Population balance model, 30
Pore coordination number, 178
Pore shrinkage, 162
Porosity, 236, 267
Porous silicon, 225
Post-oxidation, 396
Post-preparation annealing, 136

Powder porosity, 243
Precursor, 361
Precursor delivery system (bubbler), 60
Primary particles, 59, 104
Printed electronics, 387
Printing, 237, 402
Production rate, 105
Propane, 332, 341
p-type response, 350
Pulsed precursor delivery, 50
Pure Pd nanoparticle, 352
PVP, 398

Q
Quantum confinement, 211, 313
Quantum efficiency, 212, 220, 227
Quantum yield, 305

R
Radial, 360
Radial growth, 192
Radiative recombination, 218, 227
Radiative tunnelling, 378
Radius of gyration, 169
Raman spectroscopy, 211
Random orientation, 176, 178
Rate coefficients, 8
Ratio of orbital-to-spin magnetic
    moment, 290
Reaction-coagulation-sintering model, 59
Reactive flows, 23
Reactive sputtering, 125
Recombination decay time, 315
Recombination lifetime, 306, 322
Recombination rate, 217, 219
Recrystallization, 172
Reduced magnetisation, 283
Reevaporation of particles, 334
Relative permittivity, 235
Relaxation processes, 294
Relaxation time, 164, 170
Remanence magnetisation, 291, 292
Reorientation, 173, 175
Residence time, 61
Residual vector minimization, 79
Resistance, 235
Response and recovery times
Reverse monte carlo method, 66
Rietveld method, 52
Rigid body dynamics, 173
RMC modeling, 70
Room temperature, 304
Rutile, 53, 141, 152, 153, 157

Index

S

Saturation magnetization, 41, 144
Saturation mobility, 393
Scaling variable, 144
Scanning mobility particle sizer (SMPS), 331
Scanning transmission electron microscope (STEM), 276
Schottky barrier, 391
Schottky contacts, 235
Second harmonic generation, 212
Seed splitting, 364
Segregation, 85, 95, 129
Selection rule, 216
Self-preserving, 54
Semiconducting nanoparticles, 108
Semiconducting oxides, 393
Semiconductor, 185, 217, 231, 302, 329, 357
Sensing mechanism, 254
Sensitivity, 257
Sensitivity of a gas sensor, 341
Sensor to sensor reproducibility, 349
Shallow donor, 261
Sheath gas, 102
Shellwise Mackay transformed cluster, 150
Shockley equation, 389
Shock-tube, 8
Shrinkage, 162, 179
Side wall diffusion, 364
Side wall incorporation, 198
Silica, 32, 40, 41, 49
Silicon, 210, 267
Simulation, atomic, 167, 173, 181
Simulation, particle-based, 173, 180
Single oxides, 34
Sinks, 165, 169
Sinter neck, 148–150, 154, 156
Sintering, 59, 107, 140, 146, 147, 149, 150, 161–163, 165, 167, 169, 171, 173, 175, 177, 179, 181, 183, 394
Sintering, current assisted, 180
SiO_2/SiO, 210
Site-selective magnetic hystereses, 287
Size distribution, 50, 54, 100, 110, 172, 180, 217
Size effects, 232
Size-selected, 109
Size-selected nanoparticles, 273
SnO, 330, 396
SnO_x, 329
SnO_x: M (with M=metallic: Pd, Ag), 330
Soft agglomerate, 162, 180
Solar conversion efficiency, 379

Solid state lighting, 224
Sooting flame, 3
Sources, 165, 169
Spark discharge, 103
Spark frequency, 105
Spark plasma sintering, 233, 246
Spatially resolved hysteresis loop, 289
Spatially resolved photocurrent spectroscopy, 379
Spectrometer, 15
Spectro-microscopy, 289
Spherical particles, 107
Spin alignment, 317
Spin canting, 282
Spin coating, 398, 402
Spin magnetic moments, 282
Spinel, 309
Spin-flip, 219
Spin–orbit coupling, 294
Spin-polarised relativistic Korringa–Kohn–Rostoker method, 282
Spintronic, 304, 324
Sputtering, 123
SQUID, 296, 308
Stacking fault, 146, 147, 149
Static properties, 273
Stokes shift, 211, 315
Stokes-Cunningham model, 337
Stoner-Wohlfarth model, 283, 287
Stray fields, 290
Streak camera, 314
Structural phase transition, 140, 150
Structural transformation, 79, 80, 150, 151
Sub-oxides, 35
Suface diffusion, 129
Superconducting oxides, 35
Superconducting quantum interference device (SQUID), 296, 308
Superparamagnetic, 37, 273
Superparamagnetism, 123
Superparamagnetic limit, 87, 89, 274
Supersaturation, 359
Supersonic free jet, 15
Surface activity, 267
Surface area, 164
Surface band bending, 188
Surface diffusion, 129, 162
Surface free energy, 165
Surface growth, 6
Surface melting, 246
Surface plasmon resonance, 110
Surface plasmon, 226
Surface potential, 185

418

Index

Surface tension, 165
Surface to volume ratio, 232
Surfactant, 179, 180, 276
Synchroton radiation, 273
Synthesis, 101, 103, 105, 107, 109, 111, 113, 115, 117, 119

T

Tandem DMA measurement, 114
Tapering effect, 362
Temperature field, 19
Temperature imaging, 17
Tertiary butyl arsine (TBAs), 361
Tetra ethyl tin (TESn), 361
TFT, 389
Thermal evaporation, 102
Thermoelectric applications, 233
Thermometry, 17
Thermophoresis, 60
Thermophoretic deposition, 116
Thermophoretic forces, 8
Thermophoretic losses, 103
Thermophoretic particle collector, 52
Thermophoretic sampling, 16
Thin-film transistors, 389
Threshold, 245
Threshold voltage, 391
Tight-binding method, 215
Time-temperature profile, 59, 61
Tin dioxide, 250
Tin monoxide, 36
Tin oxide, 259
Titania, 49
Titanium-tetraisopropoxide, 60
Tomography, 276
Transconductance, 372, 392
Transistors, 387, 389, 391, 393, 395, 397, 399, 401, 403, 405, 407, 409
Transition metal clusters, 77, 78, 89, 94
Transmission electron microscopy, 52
Transport mechanisms, 165
Trap concentration, 391
Trap states, 390
Trial wavefunctions, 79
Trimer, 293
Trimethyl gallium (TMGa), 361
Trimethyl indium (TMIn), 361
Tubular flow reactor, 54
Tungsten oxide, 254
Tunnel resistance, 241
Twin boundary, 5, 127, 174, 178
Twinned, 132

U

Ultrahard nanomagnets, 78
Ultra-high mobility, 374
UV photoionizers, 103

V

Vacancies, 125
Van-Hove singularities, 213
Vapor–liquid–solid (VLS) technique, 186
Vapor–liquid–solid (VLS) growth, 358, 359
Vegard's law, 38
Vertical all-around-gate MISFET, 371
Vienna ab initio simulation package (VASP), 78, 79, 95
Voltage, 104

W

Wavelet transform, 275, 278, 279
Widegap semiconductors, 394
Wire kinking, 364
Work function, 186
Wurtzite, 66, 69, 314

X

XAFS, 66, 69
XANES, 69, 277, 282, 285, 286, 309
XAS, 66
XMCD, 281, 282, 285, 286, 289
XPEEM, 275, 287, 288
X-ray absorption near-edge structure (XANES), 69, 277, 282, 285, 286, 289
X-ray absorption, 66
X-ray absorption spectroscopy, 133
X-ray diffraction, 52
X-ray magnetic circular dichroism (XMCD), 281, 282, 285, 286, 309
X-ray photoemission electron microscopy (XPEEM), 275, 287, 288
X-ray photoelectron spectroscopy, 135

Y

Young's equation, 172

Z

Zeeman energy splitting, 308
Zero field cooling, 308
Zinc oxide, 259
Zirconia, 49
ZnO, 64, 304, 398

Printed by Publishers' Graphics LLC
SO20120731